ADVANCED PLASMA SPRAY APPLICATIONS

Edited by **Hamidreza Salimi Jazi**

Advanced Plasma Spray Applications
Edited by Hamidreza Salimi Jazi

Published by InTech
Janeza Trdine 9, 51000 Rijeka, Croatia

Publishing Process Manager Silvia Vlase

Technical Editor Teodora Smiljanic

Cover Designer InTech Design Team

Additional hard copies can be obtained from orders@intechopen.com

Advanced Plasma Spray Applications, Edited by Hamidreza Salimi Jazi
p. cm.
ISBN 978-953-51-0349-3

For EU product safety concerns:
IN TECH d.o.o., Prolaz Marije Krucifikse Kozulić 3, 51000 Rijeka, Croatia,
info@intechopen.com or visit our website at intechopen.com.

Meet the editor

Dr Hamidreza Salimi Jazi graduated from Isfahan University of Technology (IUT) with B.Sc. (1994) and M.Sc. (1999) in Materials Engineering Department. He received his PhD degree from the Department of Mechanical and Industrial Engineering at the University of Toronto in 2005. Before joining Isfahan University of Technology, he held postdoctoral and instructor positions at the Centre for Advanced Coating Technologies, University of Toronto and collaborated in some research projects with Pratt & Whitney Canada, GE Global research US, PyroGenesis Canada. He joined the Department of Materials Engineering at the Isfahan University of Technology in 2010. His main research interests are the study of thermal spray coatings, near-net shape forming of alloys, composites, and sandwich structures, thermal spray coatings for Solid Oxide Fuel Cells, and nano structure materials.

Contents

Preface

I have been familiar with plasma spraying in one of the leading thermal spray research laboratories in North America; Centre for Advanced Coating Technologies at the University of Toronto since I started my PhD thesis under supervision of Professors Javad Mostaghimi and Thomas W. Coyle in 2001. Having access to various methods of thermal spraying, diagnostic tools, and processes simulation, I gained experiences in process, deposition, characterization, and simulation of various types of thermal spray coatings.

One member of the family of thermal spray processes is plasma spraying which has a direct current (dc) arc or radio frequency (RF) discharge as a heat source. The plasma spray torch, generally, consists of a cathode and anode. The anode is usually high purity oxygen free copper and the cathode is made from tungsten. A gas, such as argon or nitrogen or a mixture of these with hydrogen or helium, flows around the cathode and through the anode which serve as a constricting nozzle. A direct current (dc) arc, initiated with a high frequency discharge is maintained between the anode and the cathode. The power of the gun varies between 5 to 250 kW depending on the type of torch and the operation parameters. The plasma gas generated by the arc consists of free electrons, ionized atoms, and some neutral atoms and undissociated diatomic molecules if hydrogen or nitrogen is used. The temperature of the core of the plasma jet may exceed up to 30,000 K. Gas velocity in the plasma spray torch can be varied from subsonic to supersonic using converging-diverging nozzles (between 600 ms-1 with pure Ar and 2300 ms-1 with Ar-H2).

In an ideal plasma spray process, powders are fed into the high velocity, high temperature gas jet, generated by a plasma gun, using a feeder system with a specific feeding rate (107–108 particles per second). Feeding powders have a relatively wide size distribution and the melting behavior and momentum of powders will be different. Depending on the powder size distribution and plasma spraying process parameters, the powders are completely or partially melted and accelerated (depending on their masses and trajectories) toward the substrate or mandrel mold. Therefore, the molten or semi-molten powders have various temperature and axial velocity distributions at the deposition point. Finally, the droplets solidify after impacting the substrate with a very high cooling rate (~104–108 ºC/sec) forming lamellae (splats) and build up deposits by laying of splats.

The temperature in the plasma jet core is high enough to melt any material, given sufficient time. Heat transfer in the plasma jet is primarily the result of the recombination of the ions and re-association of atoms in diatomic gases on the powder surfaces and absorption of radiation. Due to the nature of the plasma plume and plasma deposition, plasma spray has received a large amount of attentions from highly trained researchers and engineers to be used in various applications in industries such as, polymer, biomedical, electronic, automobiles, and etc.

The current book gives the state-of-the-art researches of new and advanced applications in the fields of plasma spraying. Some new applications of coatings deposited by plasma spraying are comprehensively discussed in Chapter 1. In Chapter 2, a few application of plasma spraying in biomaterials is addressed. Nano powder production and deposition of nanostructure coatings by plasma spraying are investigated in Chapter 13. In Chapter 4, the application of plasma plume in polymer industry for surface modification and treatment is addressed.

Dr. Hamidreza Salimi Jazi
Isfahan University of Technology,
Isfahan,
Iran

Part 1

Plasma Spray for Corrosion and Wear Resistance

1

Thermal Sprayed Coatings Used Against Corrosion and Corrosive Wear

P. Fauchais and A. Vardelle
SPCTS, UMR 7315, University of Limoges, France

1. Introduction

Coatings have historically been developed to provide protection against corrosion and erosion that is to protect the material from chemical and physical interaction with its environment. Corrosion and wear problems are still of great relevance in a wide range of industrial applications and products as they result in the degradation and eventual failure of components and systems both in the processing and manufacturing industries and in the service life of many components. Various technologies can be used to deposit the appropriate surface protection that can resist under specific conditions. They are usually distinguished by coating thickness: deposition of thin films (below 10 to 20 μm according to authors) and deposition of thick films. The latter, mostly produced at atmospheric pressure have a thickness over 30 μm, up to several millimeters and are used when the functional performance and life of component depend on the protective layer thickness. Both coating technology can also be divided into two distinct categories: "wet" and " dry " coating methods, the crucial difference being the medium in which the deposited material is processed. The former group mainly involves electroplating, electroless plating and hot-dip galvanizing while the second includes, among others methods, vapor deposition, thermal spray techniques, brazing, or weld overlays. This chapter deals with coatings deposited by thermal spraying. It is defined by Hermanek (2001) as follows , "Thermal spraying comprises a group of coating processes in which finely divided metallic or non-metallic materials are deposited in a molten or semi-molten condition to form a coating". The processes comprise: direct current (d.c.) arcs or radio frequency (r.f.) discharges-generated plasmas, plasma transferred arcs (PTA), wire arcs, flames, high velocity oxy-fuel flames (HVOF), high velocity air-fuel flames (HVAF), detonation guns (D-gun). Another spray technology has emerged recently ; it is called cold gas-dynamic spray technology, or Cold Spray (CS). It is not really a thermal spray technology as the high energy gas flow is produced by a compressed relatively cold gas (T < 800°C) expanding in a nozzle and will not be included in this presentation.

Most processes are used at atmospheric pressure in air, except r.f. plasma spraying, necessarily operated in soft vacuum. Also, d.c. plasma spraying can be carried out in inert atmosphere or vacuum and Cold Spray is generally performed at atmospheric pressure but in a controlled atmosphere chamber to collect and recycle the spray gas (nitrogen or helium) because of the huge gas flow rates used (up to 5 $m^3.min^{-1}$). In the following only processes operated in air at atmospheric pressure will be considered, except when the coating material is very expensive, such as platinum that must be sprayed in a chamber to recover the over-spray.

The coating material may be in the form of powder, ceramic rod, wire or molten materials. The central part of the system is a torch converting the supplied energy (chemical energy for combustion or electrical energy for plasma- and arc-based processes), into a stream of hot gases. The coating material is heated, eventually melted, and accelerated by this high-temperature, high-velocity gas stream towards a substrate. It impacts on the substrate in the form of a stream of droplets that are generated by the melting of powders or of the tips of wires or rods in the high-energy gas stream. The droplets flatten or deform on the substrate and generate lamellae called "splats". The piling up of multiple layered splats forms the coating.

Thermal spray processes are now widely used to spray coatings against, wear and corrosion but also against heat (thermal barrier coating) and for functional purposes. The choice of the deposition process depends strongly on the expected coating properties for the application and coating deposition cost. Coating properties are determined by the coating material, the form in which it is provided, and by the set of parameters used to operate the deposition process. Thermal spray coatings are generally characterized by a lamellar structure and the real contact between the splats and the substrate or the previously deposited layers determine to a large extent the coating properties, such as thermal conductivity, Young's modulus, etc. The real contact area ranges generally between 20 to 60 % of the coating surface parallel to the substrate. It increases with impact velocities of particles provided that the latter are not either too much superheated or below their melting temperature. That is why roughly the density of coatings increases from flame, wire arc, plasma, HVOF or HVAF and finally D-gun spraying and self-fluxing alloys flame sprayed and then re-fused.

Also thermal spray coatings contain some defects as pores, often globular, formed during their generation, un-molten or partially melted particles that create the worst defects, exploded particles, and cracks formed during residual stress relaxation. The cracks appear as micro-cracks within splats and macro-cracks running through layered splats especially at their interfaces and tending to initiate inter-connected porosities. Moreover, when the spraying process is operated in air, oxidation of hot or fully melted particles can occur in flight as well as that of splats and successive passes during coating formation. Thus, depending on the spray conditions and materials sprayed, the coatings are more or less porous and for certain applications must be sealed by appropriate means.

This chapter will present successively:

- the following thermal spray processes: flame, High Velocity Oxy-Fuel (HVOF), D-gun, plasma, wire arc and Plasma Transferred Arc (PTA)). The possibility to use them to manufacture coatings on site will also be mentioned. The coating structures (lamellar or granular) with their void content and the inter-connected porosities and crack networks will be linked to their corrosion resistance. The different sealing processes will also be discussed according to service temperature of coatings.
- A short introduction to the main modes of wear (abrasion, erosion and adhesion) linked to corrosion.
- Coatings used against atmospheric or marine corrosion (sacrificial coatings).

- Coatings used against high-temperature corrosion: carburization, nitriding, sulfidation, molten salt, and molten glass.
- Coatings used against high-temperature oxidation.
- Coatings used against corrosive wear at different temperatures
- Examples of industrial applications to illustrate the interest of thermal sprayed coatings.

2. Thermal spray

In the following, we will only present the processes that are used in air at atmospheric pressure. Figure 1 shows the general concept of thermal spray, *Fauchais et al (2012)*. The coating material can be fed in the hot gas stream as powder or wire or rod. Coatings are built by the flattening and solidification of droplets impacting onto the part to be covered. These droplets can be partially or totally melted when they are issued from powders or totally melted when they result from the atomization of melting wires or rods. The microstructure of the coating formed by the piling up of these particles depends on (i) particle impact parameters (particle temperature, molten state, velocity and size), (ii) substrate conditions (shape, roughness, surface chemistry…), (iii) the temperature control of substrate and coating before (preheating) during and after (cooling) spraying and (iv) the spray pattern.

Some general remarks can be expressed:

- Different materials require different deposit conditions,
- Specific coating properties (high density or desired porosity) may require specific particle velocity/temperature characteristics,
- The heat fluxes to the substrate depend on the coating method and for some substrate materials they have to be minimized,
- Substrate preheating and temperature control during spraying strongly affect coating properties and in particular residual stresses,
- And frequently a trade-off exists between coating quality and process economics.

For instance, if the plasma spray process can offer a high-temperature and high-velocity environment for the injected powders, it will bring about a strong heating of the substrate, and the powder material may undergo chemical change during the deposition due to excessive heating, e.g. WC-Co powders may decompose.

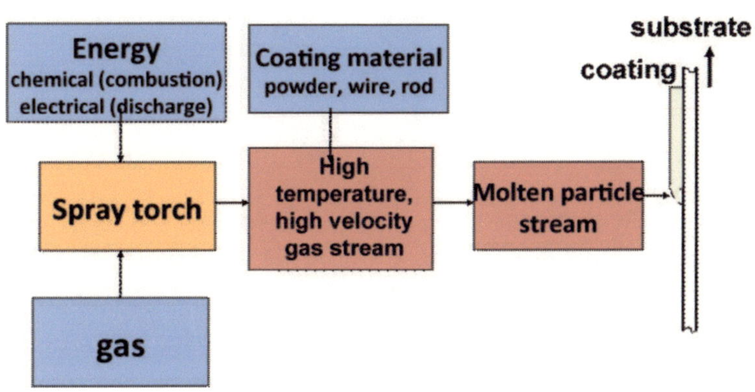

Fig. 1. Schematic of the thermal spray concept, *Fauchais et al (2012)*

2.1 Thermal spray processes

2.1.1 Plasma-based processes

They comprise d.c. plasma spraying, plasma transferred arc and wire arc spraying.

d.c. plasma torches: they generate a plasma jet from a continuously flowing gas heated by conversion of electrical energy into thermal energy thanks to an electric arc striking between a thoriated cathode and a concentric anode that plays also the role of nozzle. The cathode is mostly a stick with a conical extremity for arc power levels below 60-80 kW and some times a button for arc power levels up to 250 kW. The plasma torches work with Ar, Ar-H_2, Ar-He, Ar-He-H_2, N_2 and N_2-H_2 mixtures resulting in temperatures above 8000 K and up to 14000K, and velocities between 500 and 2800 m/s at the nozzle exit. Most of applications use solid feedstock in the form of powders but recently some of them use liquid feedstock in the form of suspensions or solutions. Figure 2 from *Gärtner et al (2006)* illustrates mean particle temperatures and velocities achieved with the different spray processes.

Most of the plasma torches have one cathode; their electrical power level ranges from 30 to 90 kW. For electrical powers in the range 40-50 kW, the powder feeding rate is between 3 and 6 kg.h^{-1} and the deposition efficiency around 50%. With high-power plasma torches (250 kW) powder flow rates can reach 15-20 kg.h^{-1}. Tri-cathode torches have appeared more recently on the market with electrical power varying from 60 to 100 kW.

Generally, plasma-sprayed coating porosities vary from 3 to 8 %, the oxygen content of metal or alloy coatings is between 1 and 5 % and their adhesion is good (>40-50 MPa). Plasma spray processes are mainly used to spray oxide ceramics.

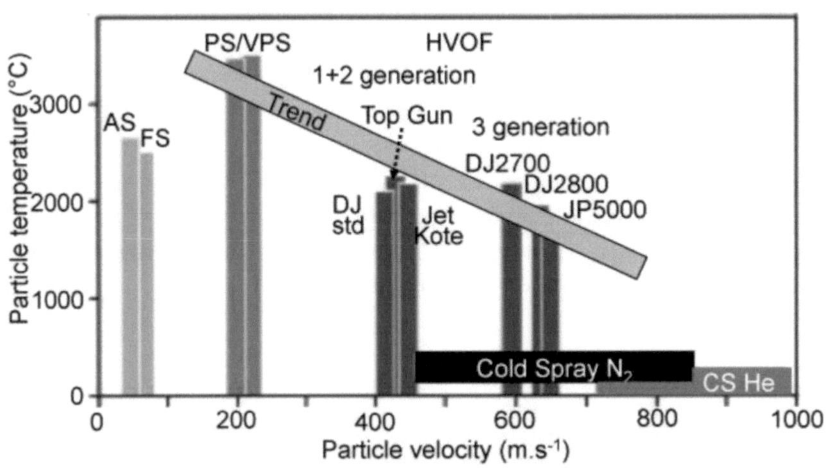

Fig. 2. Particle temperatures and velocities obtained in different thermal spray processes, as measured for high-density materials. The bar indicates the observed trend of recent developments (AS: Powder flame spraying, FS: Wire flame spraying, PS: Air plasma spraying, VPS: Vacuum plasma spraying, C.S.: Cold Spray) Gärtner et al (2006).

Wire arc spraying: instead of using solid electrodes, the arc strikes between two continuously advancing consumable conductive wires, one being the cathode and the other the anode. The melted tips of the wires are fragmented into tiny droplets, a few tens of µm in diameter, by the atomizing gas blown between both wires. The latter is generally compressed air but non oxidizing gas as nitrogen or argon is also sometimes used. The droplet high temperatures at impact produce metallurgical interactions or diffusion zones or both between the impinging droplets and the substrate or previously deposited layers. The wires are necessarily made of ductile material, or of a ductile envelope filled with a non-ductile or ceramic material. They are, then, called cored wires. All particles issuing from the ductile wires are fully melted at impact and their impact velocity can be as high as 120 $m.s^{-1}$.

The maximum arc current that can vary from 200 up to 1500 A characterizes the different arc spray guns. This process has higher spray rates (5 to 30 $kg.h^{-1}$) than other spray processes and its deposition efficiency is about 80 %; it is the most economical for materials that can be processed in form of wires or cored wires. The oxide content of coatings depends strongly on the sprayed materials and the atomization gas; it is generally rather high, e.g. over 25% for Aluminum coatings. Using nitrogen instead of air as atomizing gas can reduce it, but according to the high gas flow rates (around $1m^3.mn^{-1}$ or more), the process becomes expensive. Coating porosity is usually over 10% and its adhesion is medium in the 40-MPa range. An advantage of this process is the little heating of the substrate while the divergence of the spray pattern is a disadvantage. As other spray processes it is noisy and dusty. Coatings can be used against abrasion and adhesion (friction) under low load but their main uses are protection against atmospheric or marine corrosion and electrical applications.

Plasma transferred arc (PTA) deposition process: PTA is different from the other thermal spray processes because the substrate serves as one electrode, usually the anode, for the arc that heats the process gases and the sprayed material. Therefore, it requires electrically-conductive substrates. The transferred arc induces a local melting of the substrate and the particles, heated in the plasma column, stick to the molten pool and are melted by the transferred arc. The process is a combination of welding and thermal spraying processes. This feature allows the attainment of excellent bonding between the substrate and the coating, and very high coating densities. The feedstock is either powders with particle sizes in the one hundred micrometer range or wires. Metals, alloys and cermets can be sprayed with this technique. The PTA guns differ in maximum arc current varying from 200 to 600A. The coatings are thicker than those manufactured by other thermal spray processes, their thickness can reach centimeters, and they are metallurgically bonded to the substrate that must be kept horizontal or close to it during spraying. Coatings do not show porosity and the process deposition efficiency is over 90 %. Powder flow rate can be 18 $kg.h^{-1}$ or even more with high-power PTA devices. The coating resistance to wear is excellent as well as that to corrosion at high temperature.

2.1.2 Combustion-based processes

They mainly comprise flame spraying, High Velocity Oxy-fuel Flame (HVOF) or High Velocity Air-fuel Flame (HVAF) and Detonation gun (D-gun).

Flame Spray torches: they work at atmospheric pressure using mostly oxy-acetylene mixtures achieving combustion temperatures up to about 3000 K. Sprayed materials are introduced axially into the spray torch either as powders or wires, rods or cords. Flame velocities at torch exit are below 100 $m.s^{-1}$ and particle velocities at impact are around 50 m.s^{-1}, as shown in Figure 2. Sprayed materials are mainly metals or polymers and many materials are commercially available, which are very easy to spray. Flame-sprayed coatings generally present a high porosity (> 10%), low adhesion (< 30MPa) and oxide content between 6 and 12%. The deposition efficiency is around 50%. The particle flow rate is between 3 and 7 kg.h^{1}. Substrate and coating must be cooled during spraying.

Flame spraying can be used to deposit self-fluxing alloys ; the coating process is then followed by a fusing process. Self-fluxing alloys contain Si and B (e.g. CrBFeSiCNi) which act as deoxidizers. Rather dense coatings can be observed when spraying self-fluxing alloys, containing Si and B with a reaction of the type:

$$(FeCr)_xO_{x+y} + 2\,B + 2\,Si \rightarrow x\,Fe + x\,Cr + B_2O_x.SiO_y$$

Where $B_2O_x.SiO_y$ is a borosilicate. This process requires a thermal post treatment of the coating. This "fusing" step is commonly carried out with oxy-acetylene torches very well suited to reheat the coating over the minimum of 1040°C. During reheating oxide diffuses towards the coating surface where it is mechanically removed (turning, milling…). The process results in rather dense and hard coatings *Lin and Han (1998)*. These self-fluxing alloys can be reinforced with hard ceramic particles such as WC, *Harsha et al (2007)*.

Post-treated coatings have almost no porosity and exhibit an excellent adhesion thanks to diffusion at the substrate-coating interface. Self-fluxing alloys can also be deposited by plasma and HVOF spraying. These materials of brazing type are very easy to deposit. However, these coatings are limited to substrates that can tolerate the fusing temperature and possible induced distortion. So, this process can be used for steel substrates but not for Aluminum alloys! Such coatings are used against abrasion (friction, erosion) and corrosion (cold or hot).

When the coating material is in the form of wire (metallic and ductile) or cored wires (with ceramics or non-ductile materials), rod or cord (for ceramics), a compressed gas jet is used to atomize the melted tip and the noise level of the process is not negligible. However, compared to powders the variety of sprayed materials is larger. Wires of self-fluxing alloys can be sprayed. Oxides are of course generated during the spray process but less than with powders and the coating oxide content is generally about 4 to 8%; the deposition efficiency is also better than that achieved with powders; it is around 70%. Compared to powders the material flow rate varies from 5 to 15 kg.h^{-1}. With wires the coating adhesion is slightly better than that obtained with powders. The porosity is similar to that obtained with powders except for ceramic materials that are more porous but present an excellent wear resistance. Generally these coatings are used against abrasion or adhesion under low load and against atmospheric corrosion.

High Velocity Oxy-fuel Flame (HVOF) and high velocity air-fuel flame (HVAF). These processes use significantly higher upstream pressures than flame spray processes and a de Laval nozzle; they are characterized by supersonic speeds of gas flow. The combustion of a hydrocarbon molecule (C_xH_y) either as gas or liquid (kerosene) is achieved with an oxidizer, either oxygen or air, in a chamber at pressures between 0.24 and 0.82 MPa or slightly more for high-power guns. A convergent-divergent de Laval nozzle follows the combustion chamber achieving very high gas velocities (up to 2000 m/s). The last trend is to inject nitrogen (up to 2000 slm) in the combustion chamber to increase the gas velocity and decrease its temperature. Mostly powders are used, which are injected either axially or radially or both, depending on the gun design. Few guns have been designed to use wires or cored wires. Also, recently liquid feedstock injection (suspensions or solutions) has been developed, mainly for axial injection. The particle velocities and temperatures achieved with different guns (Top gun, Jet Kote,DJ standard, DJ 2700, DJ 2800 and JP 5000 ones) are presented in Figure 2. Substrate and coating must be cooled during spraying.

Power levels for HVOF guns working with gases is about 100-120 kW, while they can reach 300 kW for guns working with liquid. Globally this process, working mainly with metals, alloys and cermets (one of the most successful applications) has deposition efficiencies of about 70% for powder flow rates up to 7.2 $kg.h^{-1}$ for gas-fuel guns and up to 12 $kg.h^{-1}$ for liquid-fuel guns. Resulting coating porosities are a few %, with a good adhesion to substrate (roughly 60 to 80 MPa) and low oxygen content (between 0.5 and a few %). The process is rather noisy, dusty with large quantities of explosive gases. As for detonation-gun (see below), the main applications of coatings are protection against abrasion and adhesion (friction) under low load as well as protection against corrosion.

Detonation gun (D-gun). The detonation is mainly generated in acetylene- or hydrogen-oxygen mixtures (with some nitrogen to modify the detonation parameters) contained in a tube closed at one of its ends. The shock wave created by the combustion in the highly compressed explosive medium results in a high pressure wave (about 2 MPa) pushing particles heated by the combustion gases. Gas velocities of more than 2000 m/s are achieved. Contrary to the flame and HVOF devices where combustible gases and powders are continuously fed within the gun, combustible gases and powder are fed in cycles repeated at a frequency of 3 to 100 Hz.

The resulting deposits are dense and tightly bonded to the substrate. The process is the nosiest of all the thermal spray processes (more than 150 dBA). Coating porosity is low (below 1 %) and its oxygen content is between 0.1-0.5%. The deposition efficiency is about 90%, for powder flow rates of 1 to 2 $kg.h^{-1}$. The sprayed materials are mainly powders of metals, alloys and cermets; some oxides can be sprayed but with particle sizes in the 20-μm range or below. Substrate and coating must be cooled during spraying. The main applications are coatings against abrasion and adhesion (friction) under low load as well as coatings against corrosion.

On-site spraying: Many applications, especially those on big parts, e.g. a bridge, require that spraying is performed on site. This is feasible with wire arc, flame and in certain cases HVOF.

2.2 Coating formation

Compared to many other material processes, coatings produced by thermal spraying generally contain many defects. (i) The real contact area between the splat and the substrate or the previously deposited layers determines the coating properties. Inter-lamellar pores exist between layered splats, or first splats and substrate; their thickness is between a few hundredths to a few tenths of micrometers. The real contact between splats increases from about 20 to 60 % with particle impact velocities, provided that particles are not either too much superheated or below their melting temperature. (ii) Splashing of the melted particles during flattening upon impact can significantly affect the coating properties *Gawne (1995)*. Splats deposited on splashed material exhibit lower adhesion and this effect is more significant when spraying metals because the splashed material is oxidized rather fast due to the small droplet sizes. (iii) Substrate geometry may affect the flow of impacting and splashing particles *Racek (2010)*. (iv) Pores, often called globular, are formed during coating generation because of shadowing effect, narrow holes in valleys between splats not completely filled, un-molten or partially melted particles and exploded particles. These globular pores are distributed more or less homogeneously through the coating and their potential to deteriorate coating properties is proportional to their size, *Ctibor et al (2006)*. Therefore, to reduce the coating porosity, the incorporation in the coating of large un-melted particles, which are sufficiently heated to stick on the substrate, or of partially melted particles, must be avoided. This can be done by choosing carefully the particle size distribution and optimizing the particle injection. However, sometimes a compromise has to be found between having most particles fully melted but with rather high oxide content, and more un-melted particles with lower oxide content. (v) Coatings also contain cracks formed during residual stress relaxation and that often contribute to the open porosity of coatings. The micro-cracks within splats result from quenching stress relaxation and are observed in ceramic materials. The macro-cracks are often due to the relaxation of an expansion mismatch stress; they run through layered splats especially at their interfaces and, so, tend to initiate inter-connected porosities. Other stress relaxations can occur but they can be avoided by optimizing spray or service conditions. (vi) At last, another source of defects is the particle impact angle that reduces the normal impact velocity, resulting in elongated splats. Spraying with an angle above a certain value, depending on particle and substrate materials, will promote splashing even on substrates preheated above the so called "transition temperature". Above this temperature, the splats exhibit a regular disk shape on a smooth substrate while below they have an irregular shape. Typically, depending on the spray process used and the material sprayed total porosity varies between 0.5 and 15 %. It does not mean necessarily that porosities are inter-connected, but it can happen.

Figure 3 illustrates schematically the structure of a thermal-sprayed coating. Figure 4a presents the cross section of a stainless steel coating (304L) deposited by air plasma spraying on a low carbon (1040) steel substrate and Figure 4b that of an yttria partially stabilized zirconia (Y-PSZ). Figure 4a and Figure 4b show all the coating characteristics presented in the schematic cross section in Figure 3, except that the Y-PSZ coating (Figure 4b) does not present, of course, oxidation and its lamellar structure is more pronounced.

It should be noticed that microstructural homogeneity, process reproducibility and precise dimensional tolerances are indispensable requirements in manufacturing of coatings. Net shape and high precision dimensional deposition of coatings are essential for cost efficiency, especially in the case of hard coatings that require machining and finishing. The movement and velocity of the robot-guided torch has a crucial importance, especially for net-shaped coatings on complex 3D geometries, on the final properties of coating: thickness, roughness, adhesion, porosity, thermal stress distribution, etc. It is thus mandatory to develop software toolkits for the off-line generation, simulation and implementation of movement (position and orientation) of the spray torches *Candel and Gadow (2006)*.

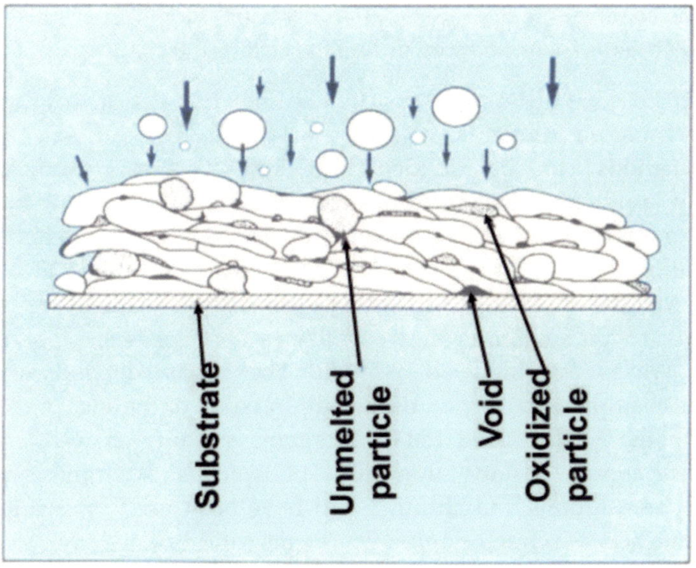

Fig. 3. Schematic cross section of a thermally-sprayed coating

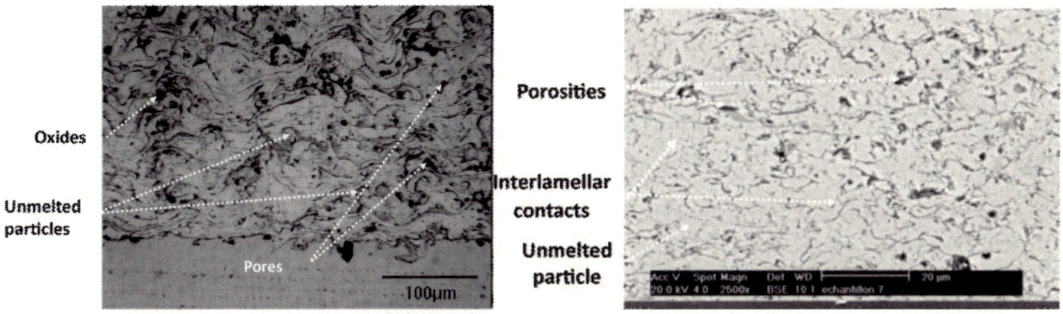

Fig. 4. (a) Stainless steel coating (304L) deposited by air plasma spraying on a low carbon (1040) steel substrate, (b) Y-PSZ (8wt%) coating deposited by air plasma spraying on a super alloy

2.3 Coating sealing or post-treatment

When thermally-sprayed coatings are porous, especially with connected porosities, they are not very suitable for corrosion protection. This situation can be drastically improved by curing their inter-connected porosities through sealing or by post-treatments that are generally thermal. The sealing or post-treatment process depends on the coating material and its service conditions and, also, on the cost that the customer can accept. The advantages, *Davis (2004)*, are the following: prevention of the penetration within coating of the corrosive substance (liquids or gases) and the attack of the coating/substrate interface, limitation of the lodging of wear debris in the coating, enhancement of inter-splat cohesion, granting of special surface properties such as non-stick surfaces, extending the life of aluminum or zinc coatings on steel to prevent corrosion. However it must be kept in mind that such post-treatments increase more or less the cost of the coating.

Knuuttila et al (1999) gave a description of the sealing process presenting various sealant types and impregnation methods, as well as the factors influencing impregnation. Impregnation methods can be divided into four categories: atmospheric pressure impregnation, low-pressure impregnation, overpressure impregnation, and a combination of the previous methods. The choice depends on the size of the coated component, the required penetration depth, and the sealant. The choice of the latter also depends on the coating material and the application. The often-used organic sealants are based on epoxies, phenolics, furans, polymethacrylates, silicones, polyesters, polyurethanes, and polyvinylesters. Waxes can be used as well. The choice depends upon the service temperature. For example wax can be used only in cold conditions, phenolic resins up to 150-260°C and methacrylate up to 150°C. Inorganic sealants are mostly used for high-temperature applications. Besides aluminum phosphates, sodium and ethyl silicates, various sol-gel type solutions and chromic acid have been used for sealing purposes. For example phosphate acts as refractory glue that forms solid bridges over pores and cracks; it has also a beneficial effect in transforming the residual stresses of the coating into compressive stresses. Sol-gel processes refer to the formation of a stable sol, the hydrolyzing of the sol to a gel, and the calcination of the gel at elevated temperature to oxide. The sol includes a variety of metal alkoxides, nitrates, or hydroxides. Molten metals are also used for sealing and strengthening purposes. The metal-coating wetting is, then, a key issue. For example zirconia is completely wetted by liquid pure manganese. Electroplating or glazing (enamel deposition) can also be used.

Other solutions can be used to improve coating density and two examples are presented below.

- Coating properties can be improved by annealing. By definition annealing corresponds to a heat treatment that alters the microstructure of a material causing change in properties such as strength, hardness, ductility, etc. For thermally-sprayed coatings, annealing or heat-treating is performed *Davis (2004)* at high temperatures, but well below the coating melting temperature, T_m. It takes place at atmospheric pressure in air or more generally in a controlled atmosphere or under vacuum and may induce changes in coating microstructure and thermo-mechanical properties. Annealing may result in the formation of amorphous phase, the relaxation of residual stress (temperature must be over that of recovery $\approx 0.4xT_m$), diffusion between substrate and coating improving the bond strength, the coating densification by increasing inter-splat cohesion. It also may lead to recrystallization and grain growth occurring through sintering, reduction of coating oxide content when heated in hydrogen atmosphere or complete oxidation when heated in oxygen atmosphere, carbide precipitation from solid solutions with cobalt, thus enhancing the coating toughness.
- Another means to promote diffusion at the substrate and coating interface and to collapse the internal pores is to use some austempering process that consists of quenching the part directly into a liquid salt bath . For example *Lenling et al (1991)* used this process for plasma-sprayed composite coatings consisting of WC-Co and Ni-base, about 250-µm thick deposited onto AISI 5150 steel substrates. The samples were

immersed in a 870°C neutral salt for 30 min, quenched in a neutral salt bath at 315°C for 30 min and finally cooled and rinsed with room temperature water. After this treatment, the samples showed an increase in the percentage of WC phase and less desirable carbide phases formed during plasma spraying were eliminated. The bonding to substrate was improved, the coating hardness increased and residual compressive stresses appeared.

- The advantage of laser heat treating and glazing is that the amount of thermal energy applied to the surface and the location on the surface to be treated are well controlled *Davis (2004)*. For treated surfaces, the intensity of the hemispherical reflection coefficient, R of the coating, must be considered, the energy or power absorbed by the surface being proportional to (1-R). Using a wide range of energy densities makes it possible to get accurate temperature profiles at precise locations at different depths in the coating. The treatment enables the heat treating or melting of the surface (laser glazing). It must be kept in mind that coatings with low thermal conductivities (below 30-40 $W.m^{-1}.K^{-1}$) may exhibit high temperature gradients that induce important stresses. The latter are generally relaxed by macro-cracks orthogonal to the coating.

3. Coatings and corrosion

The different types of corrosive attack, especially for coatings can be classified as (i) general corrosion, corresponding to about 30% of failure, where the average rate of corrosion on the surface is uniform and as (ii) localized corrosion, corresponding to about 70% of failures. The latter comprises: (i) galvanic corrosion occurring when two dissimilar metals are in contact with each other in a conductive solution (electrolyte), the more anodic metal being corroded, while the more cathodic one is unaffected. The electrolyte plays a key role, as well as the relative surface contact area; the smaller the anodic to cathodic area ratio is, the more severe is the anodic metal corrosion. For example the protectection of low-carbon iron part from atmospheric corrosion by a coating may use either an anodic coating (nickel) or cathodic one (aluminum or zinc). In the first case no discontinuity in the coating can be tolerated, while it has no importance with cathodic coating, as illustrated in Figure 5, (ii) inter-granular corrosion, occurring when a chemical element is depleted during the coating or bulk material manufacturing, e.g. during heat treatment, (iii) pitting, which is a localized corrosion characterized by depression or pit formation on the surface. It occurs for example when stainless steel is corroded by chloride-containing solutions, (iv) transgranular corrosion is mainly due to high static tensile stress in the presence of a corrosive environment. It can be intergranular but also transgranular when cracking occurs. The coating material and its microstructure play an important role in this type of corrosion.

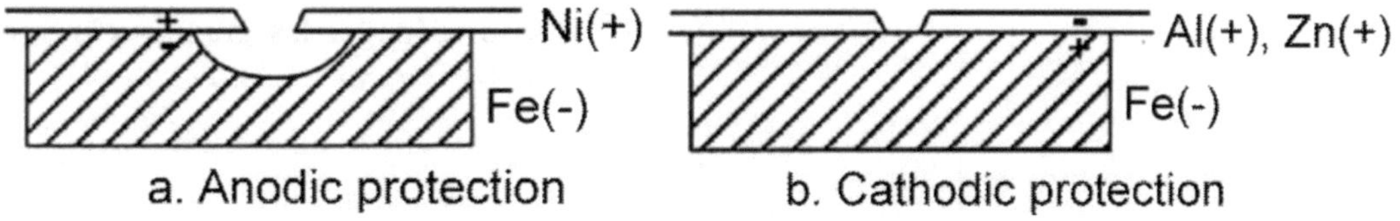

Fig. 5. Examples of protective coatings: a). Anodic (no discontinuity possible in the coating) , b). Cathodic (discontinuity possible in the coating, resulting in no corrosion of iron).

Thus coatings can be used against corrosion:(i) As sacrificial coatings (cathodic behavior relatively to ions, for example Zn or Al on steel): the thicker they will be the longer will be the protection (typical thickness varies between 50 and 500 µm, the most frequent one being around 230 µm),

(ii) As dense as possible (even sealed) if they have an anodic behavior and used against either atmospheric or marine corrosion, and high temperature corrosion: oxidation, carburization, nitriding, sulfidation, molten salt, molten glass…

Corrosive wear occurs when the effects of corrosion and wear are combined, resulting in a more rapid degradation of the material surface. A surface that is corroded or oxidized may be mechanically weakened and more likely wear out at an increased rate. Furthermore, corrosion products, including oxide particles, which are dislodged from the material surface can subsequently act as abrasive particles. Stress corrosion failure results from the combined effects of stress and corrosion. At high temperatures reactions with oxygen, carbon, nitrogen, sulfur or flux result in the formation of oxidized, carburized, nitrided, sulfidized, or slag layer on the surface. Temperature and time are, then, the key factors controlling the rate and severity of high temperature corrosive attack *Chattopadhyay (2001).*

4. Applications of thermal coatings against corrosion

4.1 Land-based and marine applications

4.1.1 Sacrificial coatings

The corrosion protection of large steel structures such as bridges, pipelines, oil tanks, towers, radio and television masts, overhead walkways and large manufacturing facilities in the metallurgical, chemical, energy, and other industries is a key issue. The protection of structures exposed to moist atmospheres and seawater such as ships, offshore platforms, seaports, is even more difficult *Evdokimenko (2001)*. In most cases the surface to be protected is thousands and even tens of thousands of square meters, requiring that coating costs are competitive with those of traditional painting methods. The coating rate must be at least 10 m^2/h and coating must be, if possible, deposited in one unique pass; the equipment must be mobile and autonomous for operation in field conditions and can work under manual control, automation being generally difficult for large scale operations, and at last the spray gun can be up to 30 m away from other elements of the equipment *Evdokimenko (2001).* Flame and wire-arc spraying meet such requirements. These equipments are widely used in industry because the investment is rather low and also coating adherence is generally good (over 20 MPa), with almost no heating of the substrate. However coatings obtained by these methods are relatively porous (up to 20 %). This porosity can be reduced by shot peening just after spraying. For example the initial porosities of 4 to 14 % (depending on spray conditions) of aluminum coatings deposited by wire- arc was reduced to 0.16 to 0.83 % after being shot peened with SiC glass beads of 0.21 to 0.3 mm in diameter *Pacheo da Silva (1991).*

Therefore, the main use of thermal sprayed coatings is as sacrificial coating with typical thicknesses between 50 and 500 µm. Referring to section 3, such coatings must have a cathodic behavior relatively to the ions of the metal to be protected, in almost all cases steels. As illustrated in Figure 5b, the cathodic protection can be porous without any corrosion of the underneath metal. Metals used are then zinc, aluminum and zinc-aluminum. Zinc performs better than aluminum in alkaline conditions, while aluminum is better in acidic conditions. If resistance to wear must be improved, aluminum coatings can be sprayed with alumina particles, e.g. by using cored-wires. For the protection of steel reinforcement in concrete, zinc is generally used, but titanium has also been used. In that case the coating is applied directly on the concrete substrate, *Davis (2004)*. However with wire-arc spraying the atomization of the wire tips must be performed with nitrogen to limit as much as possible the formation of γ TiO_2 *Holcomb (1997)*. Aluminum must be avoided where thermite sparkling may occur. That is due to the reaction of rusted steel and aluminum smears when this combustible mix is ignited by an impact *Davis (2004)*. Thus they must be avoided whenever there is a thermite-sparking hazard. Another interest of such coatings is their anti-fouling properties. Marine bio fouling is the undesirable accumulation of marine organisms on artificial surfaces that are immersed in the sea. When marine bio fouling occurs on ship hulls, it leads to an increase in the weight of the ship and friction to sail. *Murakami and Shimada (2009)* have studied the corrosion and marine fouling behaviors of various flame - sprayed coatings. They used the following powders as coating material: aluminum-copper alloy powders, aluminum-copper blend powders, aluminum-zinc blend powders and a zinc powder. After immersion in the sea, the aluminum-copper coatings showed poor anti-corrosion and anti-fouling properties. The aluminum-zinc coating with high zinc content and the zinc coating possessed the best anti-corrosion and anti-fouling properties. For example sluice gates and canal lock gates of the St. Denis Canal in France that have been zinc coated in the early 1930s have remained in perfect condition with virtually no maintenance for decades. According to Davis (2004) the lifetime of a 255-µm thick zinc or zinc aluminum coating is about 25 years and it can be extended by 15 years by sealing it with vinyl paint. Besides painting, impregnation with special compositions (epoxy resin, silicon resin…) is also intensively used as sealer. As soon as sealing is considered, the porosity of these coatings becomes an advantage for the adhesion of the sealer. Figure 6 represents an iron bridge arc-sprayed with zinc and then painted.

Fig. 6. Iron bridge arc-sprayed with zinc and then painted *Ducos (2006)*.

Sealers can also present anti-fouling properties. *Chun-long et al (2009)* have arc-sprayed aluminum on steel panels and then sealed coatings with nano-composite epoxies especially developed for this application. Test panels were tested during three years in the East China

Sea. They were disposed in the marine atmosphere zone, seawater splash zone, tidal zone and full-immersion zone. The tests included marine atmospheric outdoor exposure test, seawater exposure corrosion test and coating adhesion test. It was found that the appearance of coated panels was as fine as original but with a little sea species adhering to panels when they were immersed in tidal zone and full-immersion zone. Basically no change in the morphology, bond strength and no visible coating crack, blister, rust and break off was observed.

4.1.2 Non-sacrificial coatings

Austenitic stainless steels, aluminum bronze, nickel-base alloys, super-alloys MCrAlY where M is Ni, Co or Ni-Co, cermets (metal matrix re-enforced with WC, Cr_2C_3, and matrices containing chromium or nickel or both are used against corrosion, often when it is associated with wear. However such coatings, presenting no galvanic protection, will never protect the substrate if connected porosities and oxide networks exist, which is the case in most of thermal-sprayed coatings. Therefore, the substrate protection requires using a protective bond coat or producing dense coatings, or sealing them. The latter operation is not always possible if the service temperature is over a few hundreds of Celsius degrees. A few examples will be presented below. According to *Moskowitz (1993)*, the use of vacuum chambers or post-treatments can eliminate most defects, but these methods are costly and impractical on a large scale. Thus, he proposed using modified HVOF process with unique inert gas shrouding to achieve highly-dense, low-oxide coatings of metallic alloys. Coatings of corrosion-resistant alloys for severe petroleum industry corrosion applications, such as type 316L stainless steel and Hastelloy C-276, were shown to act as true corrosion barriers. The oxide content also plays a role. For example 316L stainless steel coatings formed by HVAF and HVOF where applied on carbon steel panels and their resistance to salt spray was tested as sprayed or sealed *Zeng et al (2008)*. When coatings were sealed, corrosion was less on the HVAF coatings than that on the sealed HVOF coating, and almost no corrosion was observed on the sealed coating sprayed with powder of the largest particles (highest porosity) after even 500 h of salt spray testing. While the amount of through pores dominates the corrosion resistance of as-sprayed coatings, the degree of oxidation of the coatings (much less with HVAF) determines the corrosion resistance when sealing is applied.

For applications in oil and gas industry on parts subjected to severe wear, cermets are used, with however some problems for offshore installations. According to *Meng (2010)* the choice between the wide varieties of tungsten carbides with different alloying binders is not simple. The corrosion resistance of coatings can be improved by the proper choice of binder. For example, *Souza and Neville (2003)* have shown that WC-CrNi exhibits passive behavior, as stainless steel, and would be compatible for use as a coating/substrate system when exposed to seawater, which is not the case for WC-CrC-CoCr. As previously mentioned the coating porosity and its oxide content must be as low as possible. It is why HVOF guns are largely used as they result in higher impact velocities (and so low coating porosity), and lower particle temperatures (and so less particle oxidation) as shown by *Ishikawa et al (2005)*. They used a commercial HVOF gun with a gas shroud attachment (GS-HVOF) to prepare WC-CrC-Ni coatings. Results of corrosion test indicate that through porosity was eliminated at velocities above 770 m/s with a lower degree of WC Degradation, wear resistance and hardness of coatings prepared by GS-HVOF were superior to those prepared by the conventional HVOF. *Fedrizzi et al (2007)* showed that Cr_3C_2–NiCr coatings, in sodium chloride solution under sliding wear, presented good barrier properties and substrate corrosion was never observed. Moreover, when chromium was added to the metal matrix of WC–Co based systems, tribo-corrosion behavior was enhanced and the lower tribo-corrosion rates were measured. Plasma sprayed Cr_2O_3-8 wt.%TiO_2 coatings were used on hydraulic cylinder piston rods and rolls but the substrate was rapidly corroded by the diffusion of the corrosive solution through pores: the bond coating was destroyed by the aggressive solution and ceramic coating flakes dropped off. *Zhang et al (2011)* have sealed coatings with epoxy and silicone resins. The sealing treatment improved significantly the corrosion resistance of coatings by blocking the open pores and cracks of the coating. Sealed by silicone resin, the coating gained remarkable anti-corrosion properties and after 1200-h salt spray test, no rust was observed on the silicone resin sealed coating.

The selection of bulk materials and coatings of valve components is an important factor for the economic success of oil and gas production activities in the petrochemical field. Particle erosion and surface wear are associated to corrosion by hydrogen sulphide during oil and gas flow. For such applications, *Scrivani et al (2001)* have characterized the following HVOF - coatings : NiAl and composite material WC/intermetallic compounds containing Ni, Cr, Co and Mo. WC–CoCr carbide coatings showed high erosion resistance due to their elevated micro hardness. Also WC/Mo compound, because of its carbide content, showed fairly good behavior in an erosive environment while WC/Mo compound showed an higher erosion resistance than Inconel 625 and NiAl.

Al alloys find increasingly large industrial use, in accordance to their high strength-to-density ratio and elastic modulus-to-density ratio. These alloys depending on their composition, besides their poor tribological behavior, can be prone to localized corrosion (e.g., pitting) in some particularly aggressive environments (e.g. seacoast atmosphere). To prevent these drawbacks, protective coatings or surface treatments are often employed. *Barletta et al (2010)* have studied the resistance of WC-CoCr coatings HVOF-sprayed onto AA6082T6 substrates. Coatings with thickness ranging between 50 and 150 μm were produced by stepwise increase of the number of torch scans. An increase in coating thickness came along with an increase in coating density because of peening effect and modifications of the splat formation mechanism. Thanks to coating densification, the hardness, wear and impact resistance, and corrosion protectiveness of the layers increased with the number of torch scans. The largest improvement occurred from 2 to 3 torch scans. When compared to anodized films these coatings had superior wear and impact resistance but offered less corrosion protection.

4.3 Low or moderate temperature coatings

4.3.1 Polymers

Thermal spraying of polymers (see for example *Petricova et al (2002), Leivo et al (2004), Zhang T. et al (1997), Chen et al (1999), Zhang C. et al (2009)* and *Henne and Schitter (1999)*) is one-coat process that acts as both the primer and the sealer, with no additional cure process. Polymer thermal spraying is ideally suited for large structures that otherwise could not be dipped in a polymer suspension. Moreover it seems that functionalized polyethylene polymers such as ethylene methacrylic acid copolymer (EMAA) and ethylene acrylic acid (EAA) can be applied in high humidity. Of course the use of polymer coatings depends strongly on their service conditions. Especially it must be kept in mind that melting temperatures vary from 40-60 °C for ethylene methacrylic acid copolymer (EMAA) to 300 °C for polyimide.

Compared to thermal spraying of conventional materials (metals, cermets and ceramics) the first necessity with polymers is to adapt the spray conditions and the polymer particle sizes to melt them partially with no overheating. It requires first to eliminate the smallest particles that are easily over heated, as well as the bigger ones that are not enough melted. Then, the residence time of particles must be adapted in order to avoid their over heating. However, polymer spraying has a certain number of limitations, particularly on coating quality, e.g. high porosity, low interfacial adherence. For that reason a thermal post-processing step is often necessary. For example *Zhang et al (2009) and Soveja et al (2010)* have studied the effect of laser (Nd:YAG, CO_2 or diode lasers) heat treatment on the morphological structure (compactness) and mechanical properties (adherence and tribological properties) of flame-sprayed PEEK and PTFE coatings. Whatever the laser wavelengths used, the laser treatment resulted in an improvement of the compactness and adherence to the substrate of both polymer coating.

Polymers may be deposited onto metals, ceramics, cermets, and composites substrates … As they present a high chemical resistance, a rather high impact and abrasion resistance at low temperatures, they are used in many industries, especially in food industry. Figure 7 represents flame-sprayed polyamide coating, 3-mm thick, deposited on a cylinder used in food industry.

Fig. 7. Flame-coating of a cylinder used in food industry: 3-mm thick polyamide coating *Ducos (2006).*

In food industry, polymer coatings replace paints on the wall because they have a much better resistance to the chemical products used for cleaning (about one week for paint against about four for the polymer coating). They are even used on the floors where polymer coatings doped with alumina particles provide an excellent anti-slip lining, the alumina particles rippling out when people walk on it.

In petroleum industry, the protection of external steel structures, pipes, tanks... is achieved with Al, Zn or Zn-Al wire arc sprayed *Davis (2004)*, as well as polymers sprayed by using flame, HVOF or plasma according to the polymer melting temperature *Petrovica et al (2002)*.

4.3.2 Pulp and paper

Machine producing paper and cardboard comprise many parts that can have large dimensions (e.g. rolls over one meter in diameter and ten meters in length) and are subjected to high wear and corrosion problems. Coatings are used for several types of rolls and cylinders, including for instance, center press rolls, dryer cylinders, calender rolls, traction rolls, and Yankee cylinders *Vuoristo and Nylen, (2007)*. Coating materials used are iron- and nickel-base alloys, nickel-chromium self-fluxing alloy, carbides, oxide ceramics, and various multi-layers depending on the application. Figure 8 shows a roll coated by flame-spraying of NiCrBSi and Figure 9 another roll coated by HVOF spraying of WC-Co, which size is underlined by the person next to it.

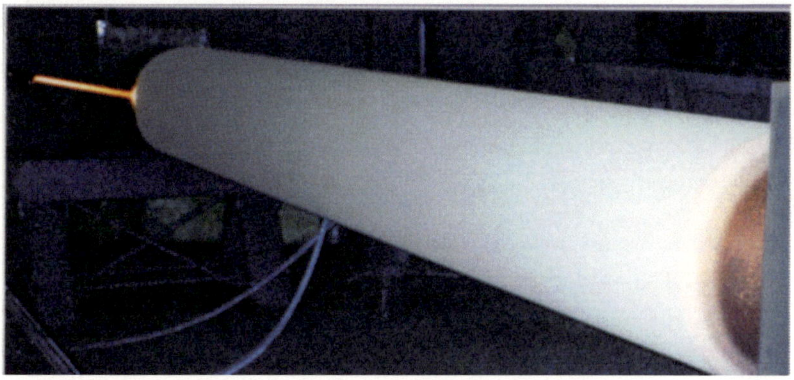

Fig. 8. Paper machine roll coated by NiCrBSi (self fluxing alloy) *Vuoristo and Nylen, (2007)*.

When oxide coatings are used to protect such rolls, one of the problems is the expansion mismatch between the ceramic coating and the metal (for example 5-6 $10^{-6}K^{-1}$ for iron and zirconia). Functionally graded (FG) coatings help to increase the compatibility between ceramic coatings and metallic substrates as coating properties such as coefficient of thermal expansion (CTE) and elastic modulus gradually change in order to reduce the thermal stresses within the coating *Hannula et al (2009)*. Post-treatments, such as the deposition of coating with fluoro-polymers or the sealing of coatings to protect them against corrosive environment, are also used. Few examples are presented below:

- Press rolls are used to remove water from the sheet using mechanical forces. First they were made of granite rock and were very expensive with limited rotation velocity. Now cast iron or steel roll bodies with coatings are used. The coatings are application-specific and tailored to perform optimally under production conditions of various types of paper. Factors such as wear and corrosion resistances, and the functionality of the roll surface in paper manufacturing process, are key properties in these applications *Vuoristo and Nylen, (2007)*. In most cases alumina-titania and chromia coatings are used with HVOF-deposited bond coat *Davis (2004)*. According to the size of the coated cylinders, high power plasma torches (250 kW), such as Plazjet, are generally used for coating deposition.

- Dryer cylinders of paper machines and large Yankee drying cylinders of tissue paper machines are protected by HVOF-sprayed coatings of cermets containing various carbides. According to their weight and size, these components are commonly coated on-site and all the coating process stages (surface preparation, coating, surface finish) are carried out in the paper factory *Vuoristo and Nylen, (2007)*. Figure 9 represents HVOF WC-Co coating of a paper machine after finishing.

Fig. 9. Paper roll HVOF coated with WC-Co (87-13 wt %) *Ducos (2006)*.

The coatings are generally top-coated with a fluoro-polymer layer to improve the release properties of the roll surfaces *Vuoristo and Nylen, (2007)*. Steels with high molybdenum contents (better corrosion resistance and thermal conductivity) HVOF sprayed are also used *Davis (2004)*.

Carbon Fiber-Reinforced Plastics rolls (CFRP roll) exhibiting excellent characteristics including lightweight, high-stiffness, and low-flexure have been increasingly employed in manufacturing industrial fields. Compared to conventional metal rolls, CFRP rolls are lighter and stiffer, and exhibit lower inertia moment *Nagai et al (2009)*. Unfortunately carbide-cermet coatings failed because of their poor thermal shock resistance on CFPR rolls. Ni-base composite porous coating including ceramics particles developed by *Nagai et al (2009)* showed high thermal-shock resistance. Five coated rolls were installed in the actual papermaking line. They achieved successful results with 10% improvement of the line speed, whereby outstanding performance and maintenance-free has been confirmed even after 4 years of use. *Yoshiya et al (2009)* have developed a 5.4-m length thermal sprayed carbon roller, installed in a paper slitter/winder, that is moving stably at ultrahigh speed, 2300 $m.min^{-1}$, with no abrasive wear. The roller has a three-layer structure: a tungsten carbide cermet layer on grooved metal sleeve, which covers the CFRP substrate roller shell.

The thermal spay process is the finishing process as no grinding is performed after cermet coating deposition.

4.3.3 Printing industry

According to *Döering et al (2008)* who made an extensive review of thermal spray processes in printing: "the printing processes need all an ink transferring unit, a print form and an impression cylinder". The requirements for an appropriate coating in terms of the function but also the requirements for the grinding and finishing steps after the coating process are rather high, as illustrated with few examples:

- The ink transporting systems in offset printing machines consist of several different rollers, providing a homogenous film of ink for the print form. For example the ductor-roller elevates the ink from the ink box to the ink conditioning system. Its surface roughness Rz must be better than 2 µm and tolerances must be better than 2/1000 mm in concentricity. Such conditions are achieved with chromia rich coating consisting of a metal bond coat (Ni, NiCr, NiAl) and a ceramic topcoat *Döering et al (2008)*.
- In contrast to common offset inking units, the anilox unit consists of a laser-engraved roller, taking the ink from a chamber doctor blade system *Döering et al (2008)*.The gravure procedure is performed after finishing the roller surface. Producing pure chromia coatings implies using carefully adapted plasma spray conditions and adapted powders, for more details see the paper of *Pawlowski (1996)*.
- Many other rolls are used in printing machines *Döering et al (2008)* with plasma sprayed oxide coatings (chromia, alumina-titania (3 or 13 wt %)). Recently *Lima and Marple (2005)* have proposed to replace the alumina-titania coatings that are conventionally plasma sprayed, by HVOF-sprayed nanostructure titania coating exhibiting a very dense (nearly pore free) and uniform isotropic microstructure with an excellent wear and corrosion resistance.
- Other rolls such as blanket cylinders are coated with Hastelloy C by either APS or HVOF and at last draw rolls are coated with nickel-chromium *Davis (2004)*.

4.4 High temperature coatings

Hot corrosion degradation of metals and alloys is a serious problem for many high temperature applications in aggressive environment, such as boilers, internal combustion engines, gas turbines, fluidized bed combustion, industrial waste incinerator. The depletion of high-grade fuels and use of residual fuel or oil in energy generation systems contribute to corrosive degradation. Residual fuel oil contains sodium, vanadium, and sulfur as impurities. The latter react together to form low melting point compounds, known as ash, which settle on the surface of materials and induce accelerated oxidation (hot corrosion). Corrosion occurs when these molten compounds dissolve the protective oxide layers that naturally form on materials during boiler/gas turbine operation. In the following a few examples will be presented according to industries. However it must be underlined that the same coating can be used for example against corrosion in gas turbines, internal combustion engines, and boilers and industrial waste incinerators. Solutions were found in the use of sophisticated materials, that are resistant to oxidation and corrosion at high and low temperatures and exhibit high-strength, e.g. nickel- base superalloy Inconel 625. However coatings of these materials on cheaper metals would be cheaper than bulk super-alloy.

Tuominen et al (2000) have shown that HVOF-sprayed Inconel 625 coatings presented mechanical and corrosion properties typically inferior to wrought materials due to the chemical and structural inhomogeneity of the thermal-sprayed coating material. Laser re-melting, with high-power continuous wave Nd:YAG laser equipped with large beam optics, resulted in the homogenization of the sprayed structure and strongly improved the performance of the laser-remelted coatings in adhesion, wet corrosion, and high-temperature oxidation testing.

4.4.1 Metal processing industries

Many parts in metal working industries are submitted to severe wear and corrosion and thermal sprayed coatings can help to maintain the parts. The main parts that can be protected by thermal spraying are *Davis (2004)*: components of electric arc furnace (EAF) and basic oxygen furnace (BOF), molds, casting dies, casting salvage, molten metal containment and delivery, steel mill rolls working in both wet and dry mill environment: entrance and exit rolls of steel processing line, rolls for galvanized and aluminized steel sheets.

Components of furnaces: A wide variety of components associated with electric arc furnace (EAF) and Basic Oxygen Furnace (BOF) are subjected to severe attack from heat, particulate and acidic gases. Water-cooled components, in the off-gas duct systems such as pans, roofs, boxes and panels, are subjected to high-velocity combustion gases that contain a number of corrosives chemicals that condense and attack the heat transfer surfaces. Coatings used are those developed for high-temperature wear and corrosion resistance (see the articles of *Wang and Verstak (1999), Higuera Hidalgo et al (2001), Sidhu et al (2005),(2006), Kaushal et al (2011))*.

Molds: In continuous casting the cast shell in the lower half of the mold abrades and wears the bottom of the mold. The diffusion of the copper substrate from the mold into the surface of the cast product leads to a quality defect called "star cracking". Chrome- and nickel-based coatings protect copper molds from wear, and also enhance caster product quality by greatly reducing cast product contamination and star cracking problems. Thermal barrier coatings are also used to control the heat flow and retard rapid chilling *Sanz (2001), Davis (2004)*. For very corrosive melts, pure yttria is used instead of zirconia partially stabilized with yttria. They are deposited onto a metal bond coat. Multi-layer coatings can also be used to achieve a good compliance between the expansion coefficients of mold and topcoat. *Sanz (2001)* has studied different coatings to protect the mold wall. *Gibbons and Hansell (2006)* have shown that Cr_2C_3-25(Ni-20Cr) and WC-10Co-4Cr coatings deposited using JP5000 HVOF hardware, offer properties that could enable low-cost, low-volume production aluminum injection mold tooling to be upgraded to higher volume production tooling.

Casting salvage is also achieved by filling the voids of porosity or wear zones, after grinding, with plasma or wire-arc sprayed coatings that are then re-machined *Davis (2004)*.

Die casting: Hot dipping rolls: MoB/CoCr, a novel cermet material for thermal spraying, with high durability in molten alloys has been developed for aluminum die-casting parts, and for hot continuous dipping rolls in Zn and Al-Zn plating lines, *Mizuno and Kitamura (2007)*. The tests revealed that the MoB/CoCr coating has a high durability without dissolution in the molten Al-45wt.%Zn alloy. Using undercoat is effective to reduce the effect of large difference in thermal expansion between the MoB/CoCr topcoat and substrate of stainless steel of AISI 316L, widely used for the hot continuous dipping *Mizuno and Kitamura (2007)*. MoB-based cermet powders (MoB/NiCr and MoB/CoCr) were deposited on SKD61 (AISI H-13) substrates used as a preferred die (mold) material *Khan et al (2011)*. The durability of these coatings on cylindrical specimens against soldering has also been investigated by immersing them in molten aluminum alloy (ADC-12) for 25 h at 670 °C. The comparison with the durability of NiCr and CoMoCr coatings under the same condition showed that both types of MoB-based cermet coatings have high soldering resistance as negligible intermetallic formation occurred during the immersion test.

Weiss et al (1994) have used arc-sprayed steel-faced tooling to create matched die sets for injection molding applications.

Entrance and exit rolls of steel processing line: when tungsten carbide coatings are used on bridle and accumulator rolls in entrance and exit ends of a steel processing line damage of the roll surface is eliminated, proper grip provided and slippage prevented. The surface coating is properly textured to provide the required characteristics or profile on the strip surface. Multi-component white cast iron is a new alloy that belongs to system Fe-C-Cr-W-Mo-V; it seems promising for rolls when deposited by HVOF spraying *Maranho et al (2009)*.

Galvanized and aluminized steel sheets: They require very high surface quality, particularly in exposed panels. In continuous galvanizing and aluminizing, the steel strip is dipped in the molten bath through a series of rolls, which control the speed and tension of the strip and guide the steel strip through the molten metal bath. The rolls operating in the molten Zn-Al alloy are subjected to severe corrosive environment and require frequent change and repair. *Seonga et al (2001)* have shown that WC-Co coatings are not very good with molten Zn-Al. By coating the sink rolls and stabilizer rolls with molybdenum boride, tungsten carbide and other materials, the rolls remain smoother and produce an improved strip surface.

Sheet metal forming dies: Conventionally mold and dies are manufactured by machining from bulk metallic materials. Tooling by using arc spray process to spray metal directly onto a 3D master pattern is an alternative method to manufacture mold and dies for plastic injection molding and other applications, *Seong (2009)*.

4.4.2 Chemical industry

Coatings in chemical industry are used for pressure and storage vessels with Hastelloy B or C, Inconel 600. In heat-affected zones, the solutions used in gas turbines (see section 4.4.6) are often employed. In some chemical reactors dealing with strong acids in combination with organic solvents, glass lining are used and can be repaired by APS sprayed tantalum coatings, bonding well to the glass with an overlay of chromium oxide *Davis (2004)*. For oil, gas, and petrochemical industries the following components are protected with thermal sprayed coatings : mud drill rotors, pump impellers, plunger, turbine, rotor shaft of centrifugal compressor/pump, pump shafts, boiler tube, mixing screw, mandrels, actuator shafts and housings, housings and valves, valve gates and seats, ball valve with large diameter, progressive cavity mud motor rotors, rock drill bits, riser tensioner rods, impeller /blade drilling and production risers, sub-sea piping, wellhead connectors, fasteners, compressor rods, mechanical seals, pump impellers, tank linings, external pipe coatings, structural steel coating….

4.4.3 Electrical utilities

Coatings against corrosion and wear (C-W) are used in fluidized-bed combustor (FBC) and conventional coal-fired boilers:

Fluidized-bed combustor boilers: the problem is linked to the finely divided mixtures of coal and limestone particles eroding and corroding steam pipes and boiler walls, as well as the high sulfur content of coals or low grade combustibles resulting in corrosion at high-temperature. Different coatings are used:

- Cr_2O_3 (20 wt %)-Al_2O_3 on NiCrAlY bond coat that protects against corrosion through the porous ceramic coating, *Davis (2004)*. The addition of approximately 20 wt % chromia results in the formation of one solid solution of $(Al\text{-}Cr)_2O_3$ in the α-phase modification; working temperatures can reach 1000 °C and the transformation of γ phase starts around 900 °C.
- HVOF sprayed Cr_3C_2-NiCr coatings with high compactness and fine grain size, *Wang (1996)*. The wear resistance is due to the particles of hard carbide homogeneously distributed within the coating, the ductile matrix being corrosive-resistant.

Coal-fired boilers: Plasma-sprayed stellite-6 coating has been found to be effective in increasing the erosion-corrosion resistance of boiler steels in the coal-fired boiler environment. A less porous structure obtained after laser re-melting was found to be effective for increasing erosion-corrosion resistance, *Sidhu and Prakash (2006)*. Inconel systems or high chromium alloys or chromium-nickel alloys coatings, presenting a good resistance to sulfur have also been used, sprayed with plasma, wire-arc or HVOF *Davis (2004)*. *Notomi and Sakakibara (2009)* have proposed low cost plasma-sprayed coatings with high hardness, developed by adding carbon and hardening elements to high chromium cast iron (C-Si-Mn-Cr-Mo-V-other-Fe) used for wear resistant material. Nitrogen gas atomization was applied to manufacture the powder in order to prevent the oxidation of particles. These coatings showed the same or better erosion resistance than Cr_3C_2-NiCr cermet coating, and had higher reliability for long period operation and higher practicality.

4.4.4 Ceramic and glass manufacturing

For the production of a variety of glasses, platinum is currently used to withstand the abrasive action of molten glass, because of its high melting point, strength and resistance to corrosion. Rhodium is often alloyed with platinum to increase the strength of the alloy and extend the life of the equipment. According to the prices of these metals, instead of using them as plates or sheets, they are often plasma-sprayed in inert atmosphere chambers where over-sprayed powder can be collected. A ceramic coating that is re-applied regularly according to its wear *Davis (2004)* can protect these precious metal coatings.

The electric glass melting for homogenizing, feeding and shaping was developed with molybdenum electrode materials. Since stirrers, mixing paddles, and some mold surfaces are plasma coated with molybdenum and its alloys *Davis (2004)*.

Mold glass material must have sufficient strength, hardness and accuracy (no deforming process) at high temperature and pressure. Also, its oxidation resistance must be good, its thermal expansion low and its thermal conductivity high. Therefore, the mold material choice depends critically on the transition temperature of the glass material. For low temperature-transition-glasses, steel molds with a nickel alloy coating can be used, e.g. self-fluxing alloy NiCrBSi flame-sprayed and refused. For higher transition temperatures NiCr-Cr_2C_3 or TiC cermets are used.

4.4.5 Aerospace

As illustrated in Figure 10, many parts (hundreds of components) are covered by thermal sprayed coatings in aircraft engine. If, at the beginning coatings were APS and VPS sprayed, HVOF spraying has been rapidly adapted to needs and twin-arc spray process is now explored. Coatings are used against fretting wear, for friction, reduction, for clearance control and for high temperature protection (thermal barrier coatings, TBC). They are also used as seals and to replace hard chromium in landing gear *Davis (2004)*.

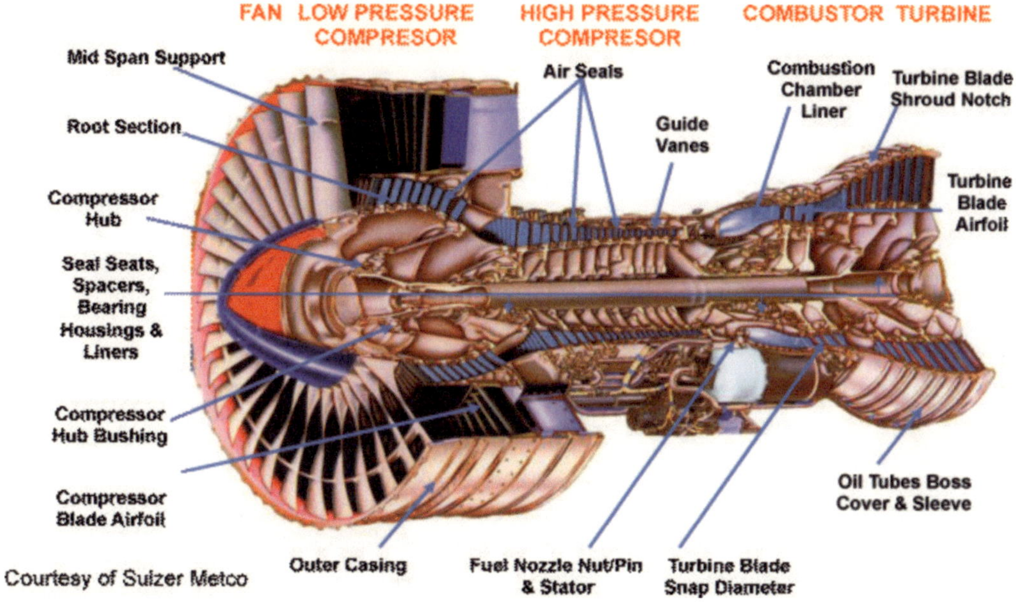

Fig. 10. Plasma-sprayed coatings on aircraft turbine engine parts (*Courtesy of Sultzer-Metco*)

TBCs were first successfully tested in the turbine section of a research gas turbine engine in the mid-1970s. By the early 1980s they had entered revenue service on the vane platforms of aircraft gas turbine engines, *Miller (1997), Bose and de Masi-Marsin (1997)*, and today they are flying in revenue service on vane and blade surfaces, *Golosnoy et al (2009)*. Thermal insulation benefits provided by TBCs and the resulting impact on component creep and thermo-mechanical fatigue life have made them enablers of high-thrust gas turbine engines. As underlined by Pratt and Whiteny, of particular importance are the TBCs elaborated by EB-PVD (Electron beam physical vapor deposition) that present an excellent compliance upon thermal cycles and can improve blade life by a factor of three *Bose and de Masi-Marsin (1997)*. The aging of TBC's topcoat depends strongly upon the spray conditions and powder morphologies used to spray or deposit it, conditions acting on its sintering *Golosnoy et al (2009), Cipitria et al (2009), Markocsan et al (2009)*. The second problem is the oxidation of bond coat with the formation of Thermally Grown Oxide (TGO) *Feuerstein et al (2008)* as well as the bond coat corrosion with oxides such as CMAS (calcium-magnesium-aluminosilicate) *Li et al (2010), Vassen, Giessen et al (2009)* or vanadium oxide *Chen et al (2009). Feuerstein et al (2008)* have shown that the most advanced thermal barrier coating (TBC) systems for the hot section components (combustors, blades and vanes) of aircraft engines and power generation systems, consist of EBPVD-applied (Electron Beam Physical Vapor Deposition) yttria-stabilized zirconia coating and platinum modified diffusion aluminide bond coating. Thermally-sprayed ceramic and MCrAlY bond coatings, however, are still used extensively for combustors and power generation blades and vanes. *Feuerstein et al (2008)* have reviewed and compared the state of the art of processes for the deposition of TBC systems: shrouded plasma and HVOF for MCrAlY bond coat, plasma for low density YSZ and dense vertically cracked Zircoat, platinum aluminide diffusion coatings, EBPVD TBC. They outlined and compared the key features and cost of coatings actually used in industry. *Vassen, Stuke et al (2009), Vassen et al (2010)* have presented the last developments using advanced processing (e.g. methods making it possible to obtain highly segmented TBCs) and the important recent directions of development for TBC systems, including improved processing routes and advanced TBC materials. Many works are devoted to the bond coat resistance, *Pint et al (2010)* and deal with the ways to improve it, *Toscano et al (2006), Schulz et al (2008),* by bond coat treatment or choice of the spray process *Richer et al (2010).* For the TBC topcoat, promising innovative technologies such as TF-LPPS (Thin Film-Low Pressure Plasma Spraying) *Hospach et al (2011)* or plasma spraying of liquid feedstock (solution or suspension) or of nanometer sized agglomerated particles will probably play an increasing role in future applications *Lima and Marple (2007), Fauchais et al (2011).*

In the lower and higher-temperature sections of engines, carbide cermets are the common materials used against fretting. Below 540 °C tungsten carbides in cobalt matrices (6 to 12 wt %) with chromium addition (4 to 12 wt %) are used and at higher temperatures chromium carbides in NiCr matrices are used *Davis (2004).* HVOF spraying is often used to limit the decomposition of carbides.

In compressors, gas turbines and turbochargers, dimensional changes take place between the rotor and stator components because of thermal and mechanical effects during operation. These dimensional changes affect sealing, so, clearance control systems are used. They consist of a sacrificial element and a cutting component. Thermal spray coatings, called abradables, and honeycomb seals form effective sacrificial systems. Abradable coatings are machined in-situ and consist of a soft metal with polymer particles in cold sections and Ni-graphite of MCrAlY with polyester or BN particles in hot areas *Davis (2004), Ma and Matthews (2009), Johnston (2011).* Additives provide the necessary friability, as well as aid in dry lubrication. Other thermal sprayed coatings can also be used on the 'cutting' side of the clearance control system, and also when the dynamic member of the system is too soft to cut without a coating.

4.4.6 Land-based turbine

Compared to aero engines, land-based turbines work under different conditions: the external environment might range from cold (- 40 °C) to high temperature (55-60 °C) and corrosive and erosive contaminants due to the environment and fuel are present. Coatings are applied onto bearing journals, bearing seals, sub shaft journal, labyrinth seals, blades, tip seals, inlet and exhausts and housing *Davis (2004).* A detailed description of the different land-based turbines can be found in the paper of *Lebedev and Kostennikov (2008)*. *Pomeroy (2005)* gives a detailed description of coatings for gas turbine materials and long-term stability issues. Due to the demand to increase turbine inlet temperatures and thus cycle efficiencies, ceramic insulating coatings can be applied to decrease the temperature of the hottest parts of the turbine components by up to 170 °C. Turbines are exposed to excessive amounts of moisture and chlorides. In addition, two types of hot corrosion and oxidation occur:

- Type II hot corrosion occurring at temperatures in the range 500–800 °C and involving the formation of base metal (nickel or cobalt) sulphates which require a certain partial pressure of sulphur trioxide for their stabilization,
- Type I hot corrosion, observed in the range 750–950 °C, involving the transport of sulphur from a sulphatic deposit (generally Na_2SO_4) across a preformed oxide into the metallic material with the formation of the most stable sulphides. Once stable sulphide formers (e.g. Cr) are fully reacted with the sulphur moving across the scale, then base metal sulphides can form with catastrophic consequences as they are molten at the temperatures at which Type I hot corrosion occurs.

At last oxidation starts at temperatures over 1000 °C.

Sprayed materials with good corrosion and oxidation resistance are nickel- and/or cobalt-based alloyed coatings *Davis (2004)*, e.g. NiCrMo (Hastelloy or Nistelle), CoCrSiMo (Triballoy), and MCrAlYs, that are either plasma or HVOF sprayed. Oxidation- and corrosion-resistant coatings are applied on air-inlets, combustor liners, injectors, turbine tip shoes, nozzles and exhausts. MCrAlY coatings are used on blades and vanes as bond coats and for corrosion-oxidation protection.

For thermal barrier coatings, mainly ZrO_2-Y_2O_3 (6 to 8 wt %), ceria or dysprosia stabilized zirconia are used *Curry et al (2011)* and also ceria and yttria stabilized zirconia or, for certain applications, calcium titanate *Davis (2004)*. It is worth underlying that for abradable and seals in the low-temperature areas, where moisture is important porous aluminum-base coatings containing polyester, polyimide or BN as well as nickel-graphite coatings are used. For higher temperatures (over 450 °C) abradables are made of MCrAlY with BN or polyester *Davis (2004), Wilson et al (2008)*. Ceramic abradables have also been introduced but the expansion mismatch with the metal substrate must be accounted for with the cooling of the super alloy.

4.4.7 Other industries

As previously described, thermal spray coatings provide superior wear (abrasive, erosive, fretting…) resistance and corrosion protection with coatings having low porosity, high hardness, good toughness for cermets, and great flexibility of composition. Thus they are also used in:

Nuclear industry: the use of cobalt-based alloys (stellites or cermets matrix) is limited, thus coatings against wear and corrosion rely on nickel alloy based coatings (NiCr-WC or NiCr - Cr_3C_2). Cermets with hafnium-carbide that present a large neutron cross-section have been developed. However, thermal sprayed coatings are essentially used in pumps, turbines, heat exchangers, vanes… For example Figure 11 presents a spherical vane that was HVOF coated.

In numerous applications developed at CEA-DEN, French Atomic Agency, Atomic Energy Department, particularly those encountered in the processing of nuclear wastes, metallic components are subjected to extreme environments in service, in terms, for example, of ageing at moderated temperature (several months at about 300 °C) coupled to thermal shocks (numerous cycles up to 850 °C for a few seconds and a few ones up to 1500 °C) under a reactive environment made of a complex mixture of acid vapors in the presence of an electric field of a few hundred volts and a radioactive activity *Berard et al (2008)*. The authors have tested alumina plasma-sprayed coatings manufactured with feedstock of different particle size distributions, graded alumina-titania coatings, and phosphate-sealed alumina coatings to improve the properties of metallic substrates operating in such extreme environments. The effects of particle size distribution, phosphate sealant, and graded titania additions on the dielectric strength of the as-sprayed, thermally cycled and thermally aged coatings were investigated. Thermal ageing test was realized in furnace at 350 °C for 400 h and thermal shocks tests resulted from cycling the coating between 850 and 150 °C using oxyacetylene flame and compressed air-cooling. Aluminum phosphate impregnation appeared to be an efficient post-treatment to fill the connected porosity of coatings. Alumina as-sprayed coatings manufactured with +22 -45 μm and +5 -20 μm particle size distributions exhibited good dielectric strengths after thermal solicitations compared to coatings manufactured with bigger size distributions or to graded titania coatings.

Fig. 11. HVOF coated spherical vane for nuclear industry Ducos (2006).

- *Cement industry*: Again the main role of sprayed coatings is wear and corrosion protection of mechanical seal, sleeve, burner tip, boiler tube, thermo-well, kiln support roll, pinion shaft, coating for cement preheat tower, impeller blade, calender roll, cone crusher and hydraulic rams with acid resistant coatings…
- Drawing Machine: guide roller and ceramic disc, rod breakdown drawing machine, fine drawing machine and other accessories
- *Waste treatment*: The oxidation of steel tubes causes an important problem in Municipal Solid Waste Incinerator (MSWI) plants due to burned wastes containing high concentrations of chemically active compounds of alkali, sulfur, phosphorus and

chlorine. Ni-based HVOF coatings are a promising alternative to the MSWI conventional protection against chlorine environments, *Wang (1996)*, *Guilemany et al (2007)*. To protect substrates from hot corrosion *Shidu et al (2007)* studied WC-NiCrFeSiB coatings HVOF sprayed to provide nickel and iron-based super alloys with the necessary resistance against oxidation and hot corrosion under the given environmental conditions at 800 °C. The oxides of active elements of the coatings, formed in the surface scale as well as at the boundaries of nickel and tungsten rich splats, contributed for the oxidation and hot corrosion resistance of WC-NiCrFeSiB coatings, as these oxides act as barriers for the diffusion/penetration of the corrosive species through the coatings. According to the study of *Lee et al (2007)* there are 88 waste-to-energy (WTE) plants in the U.S. and over 600 worldwide. In total, they combust close to 143 million metric tons of municipal solid wastes (MSW) and generate about 45 billion kW.h of electricity and an equal amount of thermal energy for district heating and industrial use. The presence of various impurities, especially HCl and chloride salts, in the combustion gases results in much higher corrosion rates of boiler tubes and has led to the development of special alloys and also metal protection techniques, including HVOF-sprayed coatings. The application of corrosion resistant materials by means of the HVOF or plasma spray processes has shown to be the best combination of erosion-corrosion resistance among all other thermal spray processes *Uusitalo et al (2002)*. In practical applications, NiCrSiB alloy HVOF coatings and Inconel 625 plasma sprayed coatings have been used successfully on water-wall tubes while TiO_2-Al_2O_3/625 cement HVOF coatings applied on super-heater tubes show long-term durability of more than 3 years *Fukuda et al (2000)*. *Shidu, Prakash and Agrawal (2006)* have characterized the HVOF sprayed Cr_3C_2-NiCr coating on Ni-based superalloys and evaluated their performance in an aggressive environment of Na_2SO_4-60%V_2O_5 salt mixture at 900 °C under cyclic conditions. The coating was resistant to hot corrosion in the given molten salt environment at 900 °C, which has been attributed to the formation of oxides of nickel and chromium, and spinels of nickel-chromium.

Coal-Gasification: Shidu et al (2007) have studied the resistance to corrosion (exposure to molten salt at 900 °C under cyclic conditions) of Cr_3C_2-NiCr, NiCr, WC-Co and Stellite-6 alloy coatings sprayed on ASTM SA213-T11 steel using the HVOF process. Liquid petroleum gas was used as the fuel gas. Hot corrosion studies were conducted on the uncoated as well as HVOF sprayed specimens All these overlay coatings showed a better resistance to hot corrosion as compared to that of uncoated steel. NiCr coating was found to be most protective followed by the Cr_3C_2-NiCr coating. WC-Co coating was least effective to protect the substrate steel. It was concluded that the formation of Cr_2O_3, NiO, $NiCr_2O_4$, and CoO in coatings might contribute to better hot-corrosion resistance. The uncoated steel suffered corrosion in the form of intense spalling and peeling of the scale, which may be due to the formation of un-protective Fe_2O_3 oxide scale.

4.5 Coatings against corrosive wear

Corrosive wear occurs when the effects of corrosion and wear are combined, resulting in a more rapid degradation of the material's surface. A surface that is corroded or oxidized may be mechanically weakened and more likely to wear at an increased rate. Furthermore, corrosion products including oxide particles that are dislodged from the material's surface can subsequently act as abrasive particles. Stress corrosion failure results from the combined effects of stress and corrosion. At high temperatures reactions with oxygen, carbon, nitrogen, sulfur or flux result in the formation of oxidized, carburized, nitrided, sulfidized, or slag layer on the surface. Temperature and time are the key factors controlling the rate and severity of high temperature corrosive attack *Chattopadhyay (2001).*

A few examples are presented below. An important concern in the oil and gas production industry is the behavior of materials in an aggressive environment with the presence of suspended sand particles, which contribute to corrosion, erosion and overall wear of the surface. *Al-Fadhli et al (2006)* have HVOF-sprayed Inconel-625 onto stainless steel components used in oil/gas industry. Coatings were applied on three different metallic surfaces: (a) plain stainless steel (SS), (b) spot-welded stainless steel (SW-SS), and (c) a composite surface of stainless steel and carbon steel welded together (C-SS-CS). These coated surfaces were tested in a jet impingement rig under two fluid conditions: (i) free from added solids, (ii) containing 1% silica sand. The coating was found to be highly sensitive to the presence of sand particles in the impinging fluid. As the period of coating exposure to the flow of slurry fluid increased, weight loss increased significantly. This increment was dependent on the type of substrate material.

WC-Co HVOF-sprayed coatings present poor resistance to corrosive wear: the tungsten carbide in HVOF coatings dissolves as well as the cobalt-chromium matrix, leading to cobalt in solution. WC and Co go through an oxidation process before dissolution, the oxidation of WC to WO_3 makes the pH drop, accelerating the dissolution of cobalt and corrosion of hard phase leading to its removal. So, there are serious implications when coatings are used in corrosive-erosive environments *Souza and Neuville (2006)*. These authors have tested WC-Co-Cr HVOF-sprayed coatings. They have shown that the corrosion of these coatings is very complex and corrosion rate increases with temperature. However, chromium forms an oxide layer, which protects from dissolution and retards the corrosion. *Toma et al (2001)* found similar results, concluding that due to its low erosion-corrosion rate the HVOF sprayed Cr_3C_2-NiCr coating can be considered to be an excellent replacement for the thermal sprayed Cr_2O_3 coatings. *Espallargas et al (2008)* found that both WC–Ni and Cr_3C_2–NiCr coatings are promising alternatives to hard chromium from the point of view of erosion–corrosion resistance.

Plasma-sprayed aluminum oxide and chromium oxide coatings are widely used to improve the resistance of metallic components against various types of wear and corrosion. However their corrosion resistance depends strongly on their porosity, especially open pores. *Leivo et al (1997)* used aluminum phosphates to seal the structures of Al_2O_3 and Cr_2O_3 coatings. The abrasive wear resistance of sealed coatings did not decrease after immersion tests of 30 days in liquids of pH 0 to 10, except for the Al_20_3 coating, which corroded in pH 0 and pH 14 solutions. No corrosion was found with aluminum phosphates in very acidic solutions. Aluminum phosphate is a good candidate to seal oxide coatings that are exposed in corrosive environments, excluding high basic environments of approximately pH 14.

In diesel engines, sulfur contain in the fuel induces corrosive attack possibility. *Uusitalo et al (2005)* have tested the newly developed ferrous powder (Fe–C–Ni–Cr–Cu–V–B alloy) plasma sprayed with the Rota-Plasma® of Sultzer-Metco on Al-13Si cylinder wall. It presented excellent corrosion and wear resistances, compared with currently used bulk casting materials such as Fe – C – Si – B alloy and Fe – C – Si – Mo – B alloy for cylinder liners.

Basak et al (2006) have tested the corrosion and corrosion–wear behavior of thermal sprayed nanostructured FeCu/WC–Co coating in Hank's solution and compared the results with that of stainless steel AISI 304 and nanostructured WC–Co coatings. The multiphase structure of the FeCu/WC–Co coating induces a complex corrosion behavior. Under corrosion–wear conditions, the nanostructured FeCu/WC–Co coating exhibited a depassivation/repassivation behavior comparable to that of stainless steel AISI 304 and nanostructured WC–Co coatings.

Generally speaking, thermal sprayed coatings behave better when their density is improved. For example *Liu et al (2007)* have investigated the effects of laser surface treatment on the corrosion and wear performance of Inconel 625, and Inconel 625-based WC HVOF-sprayed metal matrix composite coatings. Significant improvement of corrosion and wear resistance were achieved after laser treatment as a result of the elimination of discrete splat-structure, micro-crevice and porosity, and also the reduction of micro-galvanic driving force between the WC and the metal matrix. In addition, the formation of faceted dendritic structure of the WC phase was considered to be beneficial for the wear performance.

5. Conclusions

Thermal sprayed thick (from 50 to 3000 μm) coatings, including cold spray coatings are more and more used in industry for the following reasons: (i) They provide specific properties onto substrates which properties are very different from those of the sprayed coating; (ii) They can be applied with rather low or no heat input to substrates (allowing for example spraying ceramics onto polymer substrates); (iii) Virtually any material that melts without decomposing or vaporizing can be sprayed including cermets or very complex metal or ceramic mixtures, allowing tailoring coatings to the wished service property; (iv) Sprayed coatings can be strip off and the worn or damaged coatings recoated without changing part properties and dimensions; (v) Some spray processes can be moved on site, allowing spraying rapidly big parts, which displacement would otherwise be rather long and expensive.

The main thermal-sprayed coatings drawbacks are the following: (i) They are a line-of-sight technology, e.g. making it impossible to coat small and deep cavities; (ii) Most coatings have lamellar structures with contacts between layered splats that represent between 15 and 60 % of the splat surfaces depending on spray conditions; (iii) They have pores, cracks… that can be connected, depending on the spray process and spray conditions, and that must be sealed for certain applications.

Most bulk materials used in corrosion conditions can be sprayed, however splat boundaries and cracks (for ceramics coatings) often dominate the corrosion properties of coatings. Sacrificial coatings (cathodic behavior relatively to ions, for example Zn or Al on steel) are extensively used for the protection of large steel structures such as bridges, pipelines, oil tanks, towers, radio and television masts, overhead walkways and large manufacturing facilities, as well as for structures exposed to moist atmospheres and seawater such as ships, offshore platforms and seaports. Their porosity does not affect the anodic material protection, except when the whole cathodic coating structure is completely corroded. Painting as sealing or densification by shot peening are often used to extend their lifetime. No-sacrificial coatings, against corrosion or corrosive wear are extensively used in many industries: aerospace, land-based turbines, automotive, ceramic and glass manufacturing, printing industry, pulp and paper, metal processing, chemical, nuclear, cement, waste treatment… However, in almost all cases these coatings must be retreated to get rid of their porosity. This is achieved by using self-fluxing alloys that are fused after spraying, heat treating or annealing, laser glazing, austempering, sealing with organic, inorganic, metals… peening densification, diffusion… Such post-treatments increase the cost of coatings. However, in many cases the cost of retreated coatings is lower than the use of bulk materials and this is especially the case for the repair of parts.

6. References

Al-Fadhli H.Y., J. Stokes, M.S.J. Hashmi, B.S. Yilbas, (2006) The erosion–corrosion behavior of high velocity oxy-fuel (HVOF) thermally sprayed inconel-625 coatings on different metallic surfaces, Surface & Coatings Technology 200, 5782–5788

Barletta M., G. Bolelli, B. Bonferroni, and L. Lusvarghi, (2010) Wear and Corrosion Behavior of HVOF-Sprayed WC-CoCr Coatings on Al Alloys, Journal of Thermal Spray Technology 19(1-2) 358-367

Basak A.K., P. Matteazzi, M. Vardavoulias, J.-P. Celis, (2006) Corrosion–wear behavior of thermal sprayed nanostructured FeCu/WC–Co coatings, Wear 261, 1042–1050

Berard G., P. Brun, J. Lacombe, G. Montavon, A. Denoirjean, and G. Antou, (2008)®Influence of a Sealing Treatmenton the Behavior of Plasma-Sprayed Alumina Coatings Operatingin Extreme Environments, Journal of Thermal Spray Technology17(3)410-419

Bose S. and J. de Masi-Marcin, (1997) Thermal Barrier Coating Experience in Gas Turbine Engines at Pratt & Whitney, Journal of Thermal Spray Technology 6(1) 99-104

Candel A.and R. Gadow, (2006) Optimized multiaxis robot kinematic for HVOF spray coatings on complex shaped substrates, Surface & Coatings Technology 201, 2065-2071

Chattopadhyay R., (2001) Surface Wear: Analysis, Treatment, and Prevention (pub.) ASM Int. Materials Park, OH, USA, 307 p.

Chen Z., J. Mabon, J.-G. Wen, R. Trice, (2009) Degradation of plasma-sprayed yttria-stabilized zirconia coatings via ingress of vanadium oxide, Journal of the European Ceramic Society 291647–1656

Chen H., H. Zhao, J. Qu and H. Shao, (1999) Erosion-Corrosion of Thermal-Sprayed Nylon Coatings, Wear 233-235 431-435

Chun-long Y., A. Yun-qi, S. Ya-tan, (2009) Three Years Corrosion Tests of Nanocomposite Epoxy Sealer for Metalized Coatings on the East China Sea, in Thermal Spray 2009: Proc. of the International Thermal Spray Conference (eds.) B.R. Marple, M.M. Hyland, Y.-C. Lau, C.-J. Li, R.S. Lima, G. Montavon (pub. ASM Int, Materials Park, OH, USA 1090-1093

Cipitria A., I.O. Golosnoy, T.W. Clyne, (2009) A sintering model for plasma-sprayed zirconia TBCs. Part I: Free-standing coatings, Acta Materialia 57 980–992

Ctibor P., R. Lechnerová, V. Beneš, (2006) Quantitative analysis of pores of two types in a plasma-sprayed coating, Materials Characterization 56 297–304

Curry N., N. Markocsan, X.-H. Li, A. Tricoire, and M. Dorfman, (2011) Next Generation Thermal Barrier Coatings for the Gas Turbine Industry, Journal of Thermal Spray Technology 20(1-2) 108-115

Davis J. R. (ed.) (2004) Handbook of Thermal Spray Technology (pub.) ASM Int. Materials Park OH, USA.

Döring J.-E., F. Hoebener and G. Langer, (2008) Review of applications of thermal spraying in the printing industry in respect to OEMs, in Thermal Spray Conference: Crossing the Border (ed.) E. Lugsheider (pub.) DVS, Düsseldorf, Germany e-Proc.

Ducos M. (2006) Costs evaluation in thermal spray, Continuing education course, ALIDERTE, Limoges, France

Espallargas N., J. Berget, J.M. Guilemany, A.V. Benedetti, P.H. Suegama, (2008) Cr_3C_2–NiCr and WC–Ni thermal spray coatings as alternatives to hard chromium for erosion–corrosion resistance, Surface & Coatings Technology 202 1405–1417

Evdokimenko Yu. I., V. M. Kisel', V. Kh. Kadyrov, A. A. Korol', and O. I. Get'man, (2001) High-Velocity Flame Spraying of Powder Aluminum Protective Coatings, Powder Metallurgy and Metal Ceramics 40(3-4) 121-126

Fauchais P., G Montavon , R S Lima and B R Marple, (2011) Engineering a new class of thermal spray nano-based microstructures from agglomerated nanostructured particles, suspensions and solutions: an invited review, J. Phys. D: Appl. Phys. 44 093001

Fauchais P., J. Heberlein and M. Boulos, (2012) Thermal spraying, (pub.) Elsevier Amsterdam, NL, to be published

Fedrizzi L., L. Valentinelli, S. Rossi, S. Segna, (2007) Tribocorrosion behaviour of HVOF cermet coatings, Corrosion Science 49 2781–2799

Feuerstein A., J. Knapp, T. Taylor, A. Ashary, A. Bolcavage, and N. Hitchman, (2008) Technical and Economical Aspects of Current Thermal Barrier Coating Systems for Gas Turbine Engines by Thermal Spray and EBPVD: A Review, Journal of Thermal Spray Technology 17(2) 199-213

Fukuda Y., K. Kawahara, and T. Hosoda, (2000) Application of High Velocity Flame Sprayings for Superheater Tubes in Waste Incinerators, Corrosion 2000, 00264.1-00264.14

Gärtner F.,T. Stoltenhoff, T. Schmidt, and H. Kreye, (2006) The Cold Spray Process and Its Potential for Industrial Applications, Journal of Thermal Spray Technology,15(2) 223-232

Gawne D.T., B.J. Griffiths, and G. Dong, (1995) Splat Morphology and Adhesion of Thermally Sprayed Coatings, in Thermal Spraying: Current Status and Future Trends Kobe, Japan (1995) (Ed.) A. Ohmori,), (Pub.) High Temperature Society of Japan,pp 779-784

Golosnoy I.O., A. Cipitria, and T.W. Clyne, (2009) Heat Transfer Through Plasma-Sprayed Thermal Barrier Coatings in Gas Turbines: A Review of Recent Work, Journal of Thermal Spray Technology 18(5-6) 809-821

Guilemany J.M., M. Torrell, J.R. Miguel, (2007) Properties of HVOF Coating of Ni Based Alloy for MSWI Boilers Protection, in Thermal Spray 2007: Global Coating Solutions (eds.) B.R. Marple, M.M. Hyland, Y.-C. Lau, C.-J. Li, R.S. Lima, and G. Montavon(pub.) ASM Int., Materials Park, OH, USA, (2007) e-proc.

Hannula S.-P., E. Turunen, J. Koskinen, O. Söderberg, (2009) Processing of hybrid materials for components with improved life-time, Current Applied Physics 9, S160–S166

Gibbons G. J. and R. G. Hansell, (2006) Down-Selection and Optimization of Thermal-Sprayed Coatings for Aluminum Mould Tool Protection and Upgrade, Journal of Thermal Spray Technology 15(3) 340-347

Harsha S., D.K. Dwivedi , A. Agrawal, (2007) Influence of WC addition in Co–Cr–W–Ni–C flame sprayed coatings on microstructure, microhardness and wear behavior, Surface & Coatings Technology 201 5766–5775

Henne R. H. and C. Schitter, (1995) Plasma spraying of high performance thermoplastics, in Thermal Spray Science and Technology (eds.) C. C. Berndt and S. Sampath (pub.) ASM Int., Materials Park, OH, USA, 527-532

Hermanek, F.J., (2001) Thermal Spray Terminology and Company Origins (pub.) ASM International, Materials Park, Ohio, USA.

Higuera Hidalgo V., F.J. Belzunce Varela, A. Carriles Menéndez, S. Poveda Martinez, (2001)A comparative study of high-temperature erosion wear of plasma-sprayed NiCrBSiFe and WC–NiCrBSiFe coatings under simulated coal-fired boiler conditions, Tribology International 34, 161–169

Holcomb G.R., S.D. Cramer, S.J. Bullard, B.S. Covino, Jr, W.K. Collins, R.D. Govier, G.E. McGill, (1997) Characterization of thermal-sprayed titanium anodes for cathodic protection, in Thermal Spray: A United Forum for Scientific and Technological Advances (ed.) C.C. Berndt (pub.) ASM Int., OH, USA 141-150

Hospach A., G. Mauer, R. Vaßen, and D. Stöver, (2011) Columnar-Structured Thermal Barrier Coatings (TBCs) by Thin Film Low-Pressure Plasma Spraying (LPPS-TF), Journal of Thermal Spray Technology 20 (1-2) 116-120

Ishikawa Y., J. Kawakita, S. Osawa, T. Itsukaichi, Y. Sakamoto, M. Takaya, and S. Kuroda, (2005) Evaluation of Corrosion and Wear Resistance of Hard Cermet Coatings Sprayed by Using an Improved HVOF Process, Journal of Thermal Spray Technology 14(3) 384-390

Johnston R.E., (2011) Mechanical characterisation of AlSi-hBN, NiCrAl-Bentonite, and NiCrAl-Bentonite-hBN freestanding abradable coatings, Surface & Coatings Technology 205, 3268–3273

Kaushal G., H. Singh, and S. Prakash, (2010) High-Temperature Erosion-Corrosion Performance of High-Velocity Oxy-Fuel Sprayed Ni-20Cr Coating in Actual Boiler Environment, Metallurgical and Materials Transactions A, 42(7) 1836-1846

Khan F. F., G. Bae, K. Kang, H. Na, J. Kim, T. Jeong, and C. Lee, (2011) Evaluation of Die-Soldering and Erosion Resistance of High Velocity Oxy-Fuel Sprayed MoB-Based Cermet Coatings, Journal of Thermal Spray Technology 20(5) 1022-1034

Knuuttila J., P. Sorsa1, T. Mäntylä, J. Knuuttila and P. Sorsa, (1999) Sealing of thermal spray coatings by impregnation, Journal of Thermal Spray Technology 8(2) 249-25

Lebedev A. S. and S. V. Kostennikov, (2008) Trends in Increasing Gas-Turbine Units Efficiency, Thermal Engineering 55(6) 461–468

Lee S.-H., N. J. Themelis and M. J. Castaldi, (2007) High-Temperature Corrosion in Waste-to-Energy Boilers, Journal of Thermal Spray Technology16(1) 104-110

Leivo E., T. Wilenius, T. Kinos, P. Vuoristo, T. Mäntylä, (2004) Properties of thermally sprayed fluoropolymer PVDF, ECTFE, PFA and FEP coatings, Progress in Organic Coatings 49 69–73

Leivo E.M., M.S. Vippola, P.P.A.Sorsa, P.M.& Vuoristo, and T.A.Mäntylä, (1997) Wear and Corrosion Properties of Plasma Sprayed$Al_2O_3$and Cr_2O_3Coatings Sealed by Aluminum Phosphates, Journal of Thermal Spray Technology 6(2) 205-210

Lenling W.J., M.F. Smith and J.A. Henfling, (1991) Beneficial effects of austempering post-treatment on tungsten carbide based wear coatings, in Thermal Spray Research and Applications (ed.) T.F. Bernecki (pub.) ASM Int. Materials Park, OH, USA 227-232

Li L., N. Hitchman, and J. Knapp, (2010) Failure of Thermal Barrier Coatings Subjected to CMAS Attack, Journal of Thermal Spray Technology 19(1-2) 148-155

Lima R.S. and B.R. Marple, (2007) Thermal Spray Coatings Engineered from Nanostructured Ceramic Agglomerated Powders for Structural, Thermal Barrier and Biomedical Applications: A Review, Journal of Thermal Spray Technology 16(1) 40-63

Lima R.S. and B.R. Marple, (2005) Superior Performance of High-Velocity Oxyfuel-Sprayed Nanostructured TiO2 in Comparison to Air Plasma-Sprayed Conventional Al_2O_3-13TiO_2, Journal of Thermal Spray Technology 14(3) 397-404

Lin L. and K.Han, (1998) Optimization of surface properties by flame spray coating and boriding, Surface and Coatings Technology 106 100–105

Liu Z., J. Cabrero, S. Niang, Z.Y. Al-Taha, (2007) Improving corrosion and wear performance of HVOF-sprayed Inconel 625 and WC-Inconel 625 coatings by high power diode laser treatments, Surface & Coatings Technology 201, 7149–7158

Ma X., A. Matthews, (2009) Evaluation of abradable seal coating mechanical properties, Wear 267, 1501–1510

Maranho O., D. Rodrigues, M. Boccalini, and A. Sinatora, (2009) Bond Strength of Multicomponent White Cast Iron Coatings Applied by HVOF Thermal Spray Process, Journal of Thermal Spray Technology 18(4) 708-713

Markocsan N., P. Nylén, J. Wigren, X.-H. Li, and A. Tricoire, (2009) Effect of Thermal Aging on Microstructure and Functional Properties of Zirconia-Base Thermal Barrier Coatings, Journal of Thermal Spray Technology 18(2) 201-208

Meng H., (2010) The performance of different WC-based cermet coatings in oil and gas applications-A comparison, in ITSC 2010 Thermal spray: global solutions, future applications (pub.) DVS Düsseldorf, Germany e-proc.

Miller R.A., (1997) Thermal Barrier Coatings for Aircraft Engines: History and Directions, Journal of Thermal Spray Technology 6(1) 35-42

Mizuno H. and Junya Kitamura, (2007) MoB/CoCr Cermet Coatings by HVOF Spraying against Erosion by Molten Al-Zn Alloy, Journal of Thermal Spray Technology 16(3) 404-413

Moskowitz L. N., (1993) Application of HVOF thermal spraying to solve corrosion problems in the petroleum industry—an industrial note, Journal of Thermal Spray Technology 2(1) 21-29

Murakami K.and M. Shimada, (2009) Development of Thermal Spray Coatings with Corrosion Protection and Antifouling Properties, in Thermal Spray 2009: Proceedings of the International Thermal Spray Conference (eds.) B.R. Marple, M.M. Hyland, Y.-C. Lau, C.-J. Li, R.S. Lima, G. Montavon (pub.) ASM Int. Materials Park, OH,USA 1041-1044

Nagai M., S. Shigemura, A. Yoshiya, (2009) Thermal-Sprayed CFRP Roll with Resistant to Thermal Shock and Wear - For Papermaking Machine - in Thermal Spray 2009: Proceedings of the International Thermal Spray Conference (eds.) B.R. Marple,

M.M. Hyland, Y.-C. Lau, C.-J. Li, R.S. Lima, G. Montavon (pub.) ASM Int., Materials Park, OH,USA 607-611

Notomi A., N. Sakakibara, (2009) Recent Application of Thermal Spray to Thermal Power Plants, in Thermal Spray 2009: Proceedings of the International Thermal Spray Conference (eds.) B.R. Marple, M.M. Hyland, Y.-C. Lau, C.-J. Li, R.S. Lima, G. Montavon (pub.) ASM Int.Materials Park, OH, USA 1106-1111

Pacheo da Silva C. et al, (1991) 2nd Plasma Technik Symposium 1 363-373 (Pub.) Plasma Technik Wohlen, CH

Pawlowski L., (1996) Technology of thermally sprayed anilox rolls: State of art, problems, and perspectives, Journal of Thermal Spray Technology, 5(3) 317-334

Petrovicova E. and L. S. Schadler, (2002) Thermal Spraying of Polymers, Int. Mater. Rev., 47(4) 169-190

Pint B.A., J.A. Haynes, Y. Zhang, (2010) Effect of superalloy substrate and bond coating on TBC lifetime Surface & Coatings Technology 205 1236–1240

Pomeroy M.J., (2005) Coatings for gas turbine materials and long term stability issues, Materials and Design 26, 223–231

Racek O., (2010) The Effect of HVOF Particle-Substrate Interactions on Local Variations in the Coating Microstructure and the Corrosion Resistance, Journal of Thermal Spray Technology 19(5) 841-851

Richer P., M. Yandouzi, L. Beauvais, B. Jodoin, (2010) Oxidation behavior of CoNiCrAlY bond coats produced by plasma, HVOF and cold gas dynamic spraying, Surface & Coatings Technology 204, 3962–3974

Sanz A., (2001) Tribological behavior of coatings for continuous casting of steel, Surface and Coatings Technology 146 -147, 55-64

Schulz U., O. Bernardi, A. Ebach-Stahl, R. Vassen, D. Sebold, (2008) Improvement of EB-PVD thermal barrier coatings by treatments of a vacuum plasma-sprayed bond coat, Surface & Coatings Technology 203, 160–170

Scrivani A., S. Ianelli, A. Rossi, R. Groppetti, F. Casadei, G. Rizzi, (2001) A contribution to the surface analysis and characterisation of HVOF coatings for petrochemical application, Wear 250, 107–113

Seong B.-G., J.-H. Kim, J.-H. Ahn, K.-H. Baik, (2009) A Case Study of Arc-spray Tooling Process for Production of Sheet Metal Forming Dies, in Thermal Spray 2009: Proceedings of the International Thermal Spray Conference (eds.) B.R. Marple, M.M. Hyland, Y.-C. Lau, C.-J. Li, R.S. Lima, G. Montavon, (pub.) ASM Int., Materials Park, OH, USA, e-proc. 562-566

Seong B.G., S.Y. Hwanga, M.C. Kima, K.Y. Kimb, (2001) Reaction of WC-Co coating with molten zinc in a zinc pot of a continuous galvanizing line, Surface and Coatings Technology 138, 101-110

Sidhu H.S., B.S. Sidhu, and S. Prakash, (2007) Hot Corrosion Behavior of HVOF Sprayed Coatings on ASTM SA213-T11 Steel, Journal of Thermal Spray Technology16(3) 349-354

Sidhu T.S., A. Malik, S. Prakash, and R.D. Agrawal, (2007) Oxidation and Hot Corrosion Resistance of HVOF WC-NiCrFeSiB Coatingon Ni- and Fe-based Superalloys at 800 °C, Journal of Thermal Spray Technology16(5-6) 844-849

Sidhu B. S., Prakash S., (2006) Erosion-corrosion of plasma as sprayed and laser remelted Stellite-6 coatings in a coal fired boiler, Wear 260, 1035–1044

Sidhu H. S., Sidhu B. S., and S. Prakash, (2006) Comparative Characteristic and Erosion Behavior of NiCr Coatings Deposited by Various High-Velocity Oxyfuel Spray Processes, Journal of Materials Engineering and Performance 5(6) 699-704

Sidhu T.S., S. Prakash, and R.D. Agrawal, (2006) Characterizations and Hot Corrosion Resistance of Cr3C2-NiCr Coating on Ni-Base Superalloys in an Aggressive Environment, Journal of Thermal Spray Technology 15(4) 811-816

Sidhu T. S., S. Prakash, and R. D. Agrawal, (2005) Studies on the Properties of High-Velocity Oxy-Fuel Thermal Spray Coatings for Higher Temperature Applications, Materials Science 41(6) 805-823

Souza V.A.D. and A. Neville, (2006) Mechanisms and Kinetics of WC-Co-CrHigh Velocity Oxy-Fuel Thermal Spray Coating Degradation in Corrosive Environments, Journal of Thermal Spray Technology 15(1) 106-117

Souza V.A.D., A. Neville, (2003) Linking electrochemical corrosion behavior and corrosion mechanisms of thermal spray cermet coatings (WC/CrNi and WC/CrC/CoCr), Materials Science and Engineering A352 202-211

Soveja A., S. Costil, H. Liao, P. Sallamand, and C. Coddet, (2010) Remelting of Flame Spraying PEEK Coating Using Lasers, Journal of Thermal Spray Technology 19(1-2) 439-447

Toma D., W. Brandl, G. Marginean, (2001) Wear and corrosion behaviour of thermally sprayed cermet coatings, Surface and Coatings Technology 138, 149-158

Toscano J., R. Vassen, A. Gil, M. Subanovic, D. Naumenko, L. Singheiser, W.J. Quadakkers, (2006) Parameters affecting TGO growth and adherence on MCrAlY-bond coats for TBC's Surface & Coatings Technology 201, 3906 -3910

Tuominen J., P. Vuoristo, T. Mäntylä, M. Kylmälahti, J. Vihinen, and P.H. Andersson, (2000) Improving Corrosion Properties of High-VelocityOxy-Fuel Sprayed Inconel 625 by Using a High-Power Continuous Wave Neodymium-DopedYttrium Aluminum Garnet Laser, Journal of Thermal Spray Technology 9(4) 513-519

Uozato S., K. Nakata, (2005) M. Ushio, Evaluation of ferrous powder thermal spray coatings on diesel engine cylinder bores,Surface & Coatings Technology 200 2580 - 2586

Uusitalo M.A., P.M.J. Vuoristo, and T.A. Mantyla, (2002) ElevatedTemperature Erosion-Corrosion of Thermal Sprayed Coatings in Chlorine Containing Environments, Wear 252(7-8) 586-594

Vaßen R., M. O. Jarligo, T. Steinke, D. E. Mack, D. Stöver, (2010) Overview on advanced thermal barrier coatings, Surface & Coatings Technology 205 938–942

Vaßen R., S. Giesen, and D. Stöver, (2009)Lifetime of Plasma-Sprayed Thermal Barrier Coatings: Comparison of Numerical and Experimental Results, Journal of Thermal Spray Technology 18(5-6) 835-845

Vassen R., A. Stuke, and D. Stöver, (2009) Recent Developments in the Field of Thermal Barrier Coatings, Journal of Thermal Spray Technology 18(2) 181-186

Vuoristo P.and P. Nylen, (2007) Industrial and Research Activities in Thermal Spray Technology in the Nordic Region of Europe Journal of Thermal Spray Technology 16(4) 466-471

Wang B.-Q., A. Verstak, (1999) Elevated temperature erosion of HVOF Cr3C2/TiC- NiCrMo cermet coating, Wear 233-235 342 - 351

Wang B., (1996) Erosion-corrosion of thermal sprayed coatings in FBC boilers, Wear 199 24-32

Weiss L. E., D. G. Thuel, L. Schultz and F. B. Prinz, (1994) Arc-sprayed steel-faced tooling, Journal of Thermal Spray Technology 3(3) 275-281

Wilson S., D. Sporer, and M. R. Dorfman, (2008) Technology advances in compressor and turbine abradables, in Thermal Spray Conference: Crossing Borders (ed.) E. Lugsheider (pub.) DVS, Düsseldorf, Germany (2008) e-proc

Yoshiya A., S. Shigemura, M. Nagai, M. Yamanaka, (2009) Advances of Thermal Sprayed Carbon Roller in Paper Industry, in Thermal Spray 2009: Proceedings of the International Thermal Spray Conference (eds.) B.R. Marple, M.M. Hyland, Y.-C. Lau, C.-J. Li, R.S. Lima, G. Montavon (pub.) ASM Int., Materials Park, OH,USA 601-606

Zhang C., G. Zhang, V. JI, H. Liao, S. Costil, C. Coddet, (2009) Microstructure and mechanical properties of flame-sprayed PEEK coating remelted by laser process, Progress in Organic Coatings 66, 248–253

Zhang J., Z. Wang, P. Lin, W. Lu, Z. Zhou, and S. Jiang, (2011) Effect of Sealing Treatment on Corrosion Resistance of Plasma-Sprayed NiCrAl/Cr2O3-8 wt.%TiO2 Coating, Journal of Thermal Spray Technology 20(3) 508-513

Zhang T., D.T. Gawne, Y. Bao, (1997) The influence of process parameters on the degradation of thermally sprayed polymer coatings, Surface and Coatings Technology 96 337-344

Zeng Z., N. Sakoda, T. Tajiri, S. Kuroda, (2008) Structure and corrosion behavior of 316L stainless steel coatings formed by HVAF spraying with and without sealing, Surface & Coatings Technology 203 284–290

2

The Influence of Dry Particle Coating Parameters on Thermal Coatings Properties

Ricardo Cuenca-Alvarez[1], Carmen Monterrubio-Badillo[2], Fernando Juarez-Lopez[1], Hélène Ageorges[3] and Pierre Fauchais[3]
[1]*Instituto Politécnico Nacional, CIITEC*
[2]*Instituto Politécnico Nacional, CMP+L*
[3]*SPCTS-UMR 6638, University of Limoges*
[1,2]*Mexico*
[3]*France*

1. Introduction

The physical properties of coatings elaborated by plasma spraying, especially the mechanical properties are strongly influenced by some fifty operating parameters of the spraying process. Several studies have been conducted to correlate these operating parameters with the coating microstructure, via the behavior of molten particles in flight to be impacted against the surface substrate, well known as splats. Then, it is expected to build coatings with tailored properties for mechanical and even thermal applications (Fauchais & Vardelle, 2000).

Simultaneously to the operating parameters of plasma spraying, characteristics of raw powder play an important role in the coating elaboration (Vaidya et al, 2001). Depending on the production process, particles feature different characteristics concerning shape, size, specific density, purity, etc. This has a significant influence on the resulting coating properties (Sampath et al, 1996). Consequently, it becomes mandatory to have an intensive knowledge about the powder characteristics in order to better control the behavior of in-flight particles and, thus obtaining coatings with the expected performance.

For the elaboration of composite coatings, it is commonly to use composite powders. However, different characteristics of powders are obtained from the variety of processes nowadays available for powder production, even for powders with the same chemical composition! (Kubel, 2000) Kubel has compared powders produced from different techniques for plasma spraying (atomization, agglomeration by spray-drying, melting and grinding, wet particle coating; sintering). A variety of powder characteristics is found for which the operating parameters for plasma spraying must be adapted to obtain deposits featuring the desired properties. From this, certain components or materials are fabricated by some of these methods or exclusively just one.

For example, when the particle shape is different, a change in powder flowability is induced. If the more spherical particles are, then powder flows much better. Consequently, the resulting properties of coatings obtained by plasma spraying of these powders are so different, even if the projection conditions, particle size and mass flow of the powders used, are constant. This is due to the difference in behavior of particle during injection and in-flight. A powder that flows with difficulty causes a blockage in the pipe injection, resulting in a decrease in the rate of deposition. Then, the overheating of substrate and detaching of coating are expected. Similarly, the injection of fine particles in a plasma jet has been a major difficulty in thermal spraying.

All these inconvenient are critical for the elaboration of composite coatings due to the great differences in physical and chemical properties between metals and ceramics. For example, when co-spraying is performed, the deposition of different phases into the coating microstructure is heterogeneous, becoming the undesired goal. For this, agglomeration of particles is the solution used to spray fine particles of metals, carbides or ceramics. Spray-drying, granulation or compression methods are used to join fine particles each other resulting in bigger agglomerates featuring higher specific area and lower density values. In certain manner, these agglomerates still retain some properties from fine particles. The result is a low rate of deposition and a porous microstructure coating because of incipient fusion of agglomerates during their passage through the plasma jet. However, particle agglomeration can easily help to prepare the metal-ceramic composite particles and get coatings composed of a metal matrix reinforced by ceramic grains. The agglomerates can be eventually, densified by sintering or calcinating and, then crushing before their thermal spraying.

To find ways of avoiding the heterogeneous deposit of different phases, coating of particles is a promising method to deposit simultaneously, metallic and ceramic phases. A variety of processes is nowadays available by including wet and dry routes. In the wet route, also known as the chemical route, a liquid phase is used to disperse organic binders in order to attach two or more different materials such as aluminum coated with nickel, or titanium carbide coated with graphite. This method seems to be used less because of the difficulties posed by the process itself, mainly by the use of organic binders, often regarded as environmental pollutants, the heterogeneity of the coating layer and the cost of production. Changes in quantity of coating material on the particles and the loss of it during spraying induce a heterogeneous distribution of phases and mechanical properties of deposits.

A new trend in the production of composite powders is actually required. This technology must be able to meet the industrial needs for manufacturing composite coatings taking into account the reproducibility of results, problems related to environmental pollution, costs and the feasibility of powders production.

In the mid-80's, Yokoyama developed the process, called mechanofusion for the production of PMMA particles coated by alumina (Yokoyama et al, 1987). Later, powders based on nickel and aluminum, were prepared by Ito (Ito, 1991). This technology allows the production of composite powders in a dry route with no need to add a binder, or sintering for attaching to the particles coating the surface of host particles. Another feature of the Mechanofusion process is the obtaining of particles with a nearly spherical shape.

Technologies for dry particle coating are relatively new and are still under research and development stages, but have a high industrial interest. In comparison with other methods for producing coated particles, the dry technique is considered "clean" since it does not use solvents or organic binder, and even water is avoided. Therefore, the cost and time of production is considerably reduced, if only by avoiding the step of drying powders.

By giving the proposal for using of two different techniques, the mechanofusion and plasma spraying, this chapter is aimed to describe how the particles characteristics play an important role in order to adapt the mechanofusion process for producing composite powders and their influence on the coatings building by plasma spraying.

2. Raw material: The particles processing

Currently, the materials used in the preparation of composite coatings can be obtained from different techniques; however, the selection of this technique will depend on the plasma spraying technique used in order to obtain the appropriate treatment of the particles in the jet and then the desired deposit. Next, it will be summarized a brief classification of composite coatings under the only criterion of the production technique of powder.

2.1 Wet coated particles

The coating of particles in a wet route is most often used for the protection of carbide powders, which decompose rapidly during spraying. This decomposition or loss of carbon causes either the oxidation of released elements or the formation of intermediate phases which degrade the deposits properties such as their oxidation resistance, hardness or wear resistance. Then, it is though that could be useful the adding a protective layer on the particles sensitive. For example tungsten carbide (WC) was coated with cobalt (Co), whose content varies from 12 to 17 wt% (Vinayo et al, 1985; Kim et al, 1997; Jacobs, 1998). Other examples, TiC can be coated with carbon or graphite (Moreau, 1990), and chromium oxide by cobalt CrO_2/Co (Lugscheider, 1992). Sol-gel is a technique for the production of composite particles at the nanoscale such as Al_2O_3/SiC system that allows obtaining deposits with some metastable phases of Al_2O_3 into the stable phase α-SiC (Jiansirisomboon, 2003).

2.2 Self-propagating High-temperature Synthesis (SHS)

The SHS process (Self-propagating High-temperature Synthesis) is part of the family of combustion reactions involving the metal reducing and oxidizing (oxygen is the oxidizing agent the most common). For the synthesis of materials by direct reaction, self-combustion is established by the exothermicity of the reaction and converts the reactants into products that are still in solid form. This does not necessarily mean the involvement of oxygen. The SHS is used for the production of composite powders containing titanium carbide (TiC), considered the best replacement for the tungsten carbide (WC) traditionally used in applications that require good wear resistance. The obtained composites coatings consist of TiC phase dispersed within a metal matrix formed by the NiCr alloy (Bartuli & Smith, 1996). In other applications, the $MoSi_2$ compound prepared by SHS is used to form a protective layer resistant to corrosion at high temperatures, such as casting nozzles in the glass industry (Bartuli et al, 1997; Gras, 2000).

2.3 Plasma spheroidized powder

Spheroidization of powders by plasma is primarily to heat and melt the particles while holding in a plasma jet. The raw materials are often milled and sintered powders with poor flowability. The spherical droplets that form are then cooled and solidified gradually. The wollastonite mineral ($CaSiO_3$) are very popular in the field of cement and ceramics, including one of its metastable phase called wollastonite TC (triclinic structure) has a great success in medical applications. Obviously, the control of chemical composition and impurities becomes mandatory. However, the irregular morphology of minerals leads to difficulties in marketing. That's why a spherical morphology of particles is desired. Work on this subject have lead to encouraging results using the plasma spraying in water, whose particle shape is spherical then this improves the flowability characteristics and the quality of the deposit (Liu & Ding. 2002; Liu & Ding. 2003). By the same principle, it is possible to improve the flow characteristics of particles initially irregulars; mixtures of milled powders of NiCoCrAlY and ZrO_2-Y_2O_3 were densified and spheroidized by plasma spraying in distilled water (Khor et al, 2000). On an industrial scale, companies like Tekna Plasma Systems Inc. produce a wide variety of powders spheroidized by plasma radiofrequency induction including powders such as YSZ/ZrO_2, Al_2TiO_6, Cr/Fe/C, SiO_2, Re/Mo, Re, WC, CaF_2, TiN (Boulos, 2011).

2.4 Atomization

As already described, spraying of particles is the most used method for the production of alloys of iron, cobalt, nickel, or aluminium (Rautioaho et al, 1996; Wang et al, 2006; Krajnikov, 2003; Kelly, 1999). But the resulting particles have no the same shape depending of the atomization media, particles may show a spherical (gas atomization) or an irregular (water atomization) shape. The deposits obtained with these powders feature an homogeneous distribution of phases. This is due to the excellent flowing characteristics of particles in the plasma and the absence of metastable phases of the compounds prepared by this technique (Sordelet, 1998; Zhao et al, 2003; Zhao & Lugscheider, 2002).

2.5 Mechanical alloying

The main objective of the plasma spraying of powders obtained by mechanical alloying is to obtain homogeneous microstructure but also very fine. Since mechanical alloying can lead to intermetallic phases that are often difficult to form even at high temperatures, the plasma spraying of powders prepared by high energy milling is then an excellent alternative for the formation of deposits of this type of composite phases. The versatility of the mechanical alloying allows processing systems such as HA hydroxyapatite reinforced with zirconia stabilized with yttria, Cu/Al_2O_3 and Ti/Al/Si_3N_4 (Fukumoto & Okane. 1992). In the case of systems with explosive materials such as aluminium powder with a very fine particle size (< 3 μm), the short-term mechanical alloying reduces the reactivity of this powder due to either the inclusion of particles of a hard phase (Al_2O_3 and/or SiC) within the Al particles or the bonding of small particles of Al_2O_3 and SiC at surface of Al (Bach et al, 2000).

2.6 Reactive plasma spraying

To reduce costs, several authors propose the use of particles capable of reacting with the environment by forming new compounds due to the reactivity of the in-flight particles and get a more homogeneous distribution of phases. Depending on the working atmosphere, the resulting species may be of oxide, nitride or carbide. Examples of this kind of coatings are those obtained from spraying of materials such as $FeTiO_3$, whose resulting deposits are composed of Fe/TiC-Ti_3O_5; and titanium Ti to obtain the deposition TiN, Ti_2N, TiN_{1-x}, TiC or TiC_{1-x} (Ananthapadmanabhan & Taylor, 1999; Valente & Galliano, 2000; Lugscheider et al, 1997).

2.7 Melted/sintered and milled powders

The melted/sintered and milled powders are commercially the most popular due to their relative simplicity of the production process. The main difference between the two processes is the temperature of production, which induces some differences in the properties of powders. In the case of WC/Co, sintered particles are more porous than melted ones, and the appearance of intermediate phases such as W_2C and Co_3W_3C or even only tungsten, is far more important in the melted powders than in sintered. Regarding particle morphology, melted powders are the most irregular, since the fracture is created along the crystal planes and twins while in the sintered particles, the fracture propagates between defects and grain boundaries. Obviously, resulting deposits feature different characteristics even if is the same material, such as fracture toughness and modulus are higher for deposits prepared with powders milled and blended than those sintered and crushed (Khor, 2000; De Villiers, 1998; Jacobs et al, 1998). The process of melting and milling the powder can also be used to vary the chemical composition, particle size distribution and homogeneity of the powder system (Ananthapadmanabhan, 2003).

2.8 Agglomerated powders

Apart from the more conventional technology of all, the mechanical mixing of powders [,],, the technique of preparing agglomerated powders is the most used in the field of plasma spraying of composite powders. The spray drying, commonly known as agglomeration of the powder is used to form spherical agglomerates followed by a sintering treatment in a controlled atmosphere to prevent their destruction during their penetration into the plasma jet. Different systems of powders are produced including: WC/Co, WC/CoCr, WC/NiMoCr, Ni/SiC (Wang et al, 2006; Krajnikov, 2003; Kelly, 1999; Vinayo et al, 1985; Bach, 2000; Khan & Clyne, 1996; Zimmermann et al, 2003; Wielage et al, 2001).

2.9 Other techniques

The study of development of composite coatings is not only related to the pre-processing of powders but also to different spraying protocols. For example the formation of multilayer deposits is becoming popular in applications such as deposition of thermal barrier to reduce the problem of cracking of the deposit or detaching due to thermal shock. This is also valid for the development of deposits with a combination of the properties of wear resistance and lubrication (Ramaswany et al, 1997; Gadow, & Scherer, 2001). However, if the deposit should keep just certain homogeneity of phase distribution, the co-spraying of powders allows the deposition of powders having different densities without the need for binders or pre-mix powder (Trice, 1999; Denoirjean, 2003). Another possibility for development of composite coatings is the co-precipitation of phases by melting and tempering of materials often immiscible each other (Colaizzi, 2000).

2.10 Mechanofusion

The elaboration of composite coatings using mechanofused powders was proposed in the 90's by the inventors of the system (Yokoyama et al, 1987). After that, a limited number of studies was presented in the literature. The system Ni/Al was investigated first by H. Ito et al. The system Ni/Al was investigated first by H. Ito et al. with interesting results encouraging for the industrial use of mechanofusion process as a new alternative of powder preparation for plasma spraying (Ito et al, 1991). Mechanofused powders exhibit improved flowability as compared to raw powders because of the spherical shape, which facilitates injection. Consequently, deposits are built with a homogeneous distribution of phases and the appearance of intermetallic phases formed during spraying (Ito et al, 1991; Kim, 1997; Jacobs, 1998). Several authors have evaluated different configurations of powdered systems for plasma spraying including: NiAl/TiC/ZrO_2 (Herman et al, 1992a; Herman et al, 1992b), AlCuFe and AlCuCo (Csanády, 1997), NiAl or NiCrAl-TiC-ZrO_2 (Bernard, 1994), and 316L stainless steel - α-Al_2O_3 (Ageorges & Fauchais, 2000; Cuenca-Alvarez et al, 2003a, 2003b).

2.11 An example of application

In the following sections, the influence of main parameters of mechanofusion processing, henceforth called MF, firstly on deformation of metallic particles and, secondly the particle coating will be described. The powder system is selected by considering a review of the previous bibliography oriented towards a wear resistance application.

2.11.1 Powder characteristics

Stainless steel (SS) is specified as the host particles, whereas alumina as the guest ones. The last is sustained by the increase of wear resistance by combining toughness of metals with hardness of ceramics. Physical characteristics and SEM micrographs of raw powders are given in table I. Since dry particle coating depends on the particle size distributions (PSD) of host and guest particles, PSD must be different each other at least in an order of 2 as confirmed by laser granulometry. Commercial gas atomized 316L stainless steel is provided by Sultzer Metco with two particle size distributions whereas finer α-alumina is from Baikowzki, France.

Preparation of composite particles is performed by using an in-house designed MF set-up, consisting of a cylindrical chamber rotating on the vertical axis at 1400 rpm, with a concentric joint of compression hammers and scraper blades remaining static. The gap between the inner wall of the chamber and the compression hammer is adjustable. Due to centrifugal forces and, depending on the compression gap, the powder is forced against the chamber wall and dynamically compressed through the gap. Consequently, particles bed is intensively mixed and subjected to different phenomena such as compression, attrition, frictional shearing or rolling. Then, mechanical energy input, plus the generated heat can lead to mechanical alloying, homogeneous mixing, or deformation of metallic particles.

When two different types of particles, in terms of chemical composition and particle size distribution, are MF processed; the finest particles (secondary) are attached on the coarser particle surface (host) without needing to use binders (Yokoyama et al, 1987). There are several operating parameters affecting the performance of the MF device (Cuenca-Alvarez, 2003c). However, once the characteristics of host and guest particles are specified, the key parameters are the rotation speed, processing time as a function of the powder input rate, compression gap and the mass ratio of host to guest particles. Then, a compression rate (τ) is defined by the relation between the powder bed thickness formed over the inner wall (EC) and the spacing of compression gap (EF):

$$\tau = \frac{EC - EF}{EC} \tag{1}$$

As mentioned above, this work analyzes firstly the influence of compression rate affecting the stainless steel particle shape, followed by the study of feasibility of the MF device to coat stainless steel host particles by pure alumina in function of the powder charges and the powder rate input. The corresponding variations in the operating parameters are given in table II.

Function	Characteristics	Morphology
Host Designation: *SS*	316L Stainless Steel Mean Particle size: 142 μm Specific mass 7960 kgm^{-3}	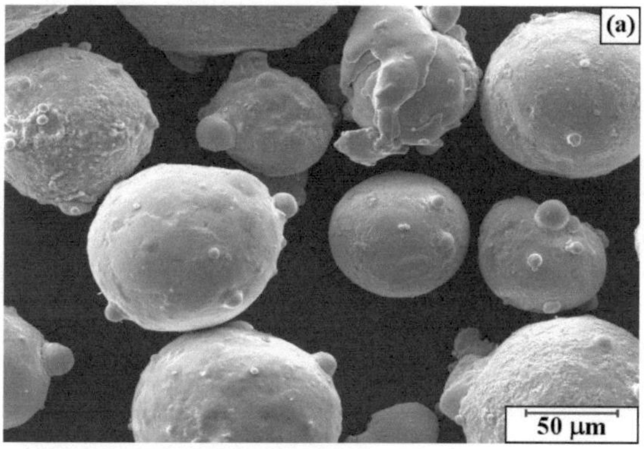
Guest Designation: *AL*	Alumina (α- phase) Mean Particle size: 1.5 μm Specific mass 3900 kgm^{-3}	

Table 1. Powder characteristics

Parameter	Value
Stainless steel charge [g]	150
Compression rate [τ]	25, 15, 5
Mass ratio of Host/Guest particle	15, 7.5, 3
Processing time [h]	1, 2, 3, 4, 5

Table 2. MF operating parameters.

2.11.2 Deformation of metallic particles

Compression rate plays an important role on the particle shape. When τ = 25, metallic particles are welded to the internal chamber wall due to the overheating generated by the friction between the compression hammer and the powder bed (figure 1a). However, friction decreases rapidly at lower values of τ (15) where deformed particles are obtained as shown in figure 1b. For τ=5, the compression gap is widely spaced to induce a moderate deformation of particles with a tendency to spheroidize them (figure 1c).

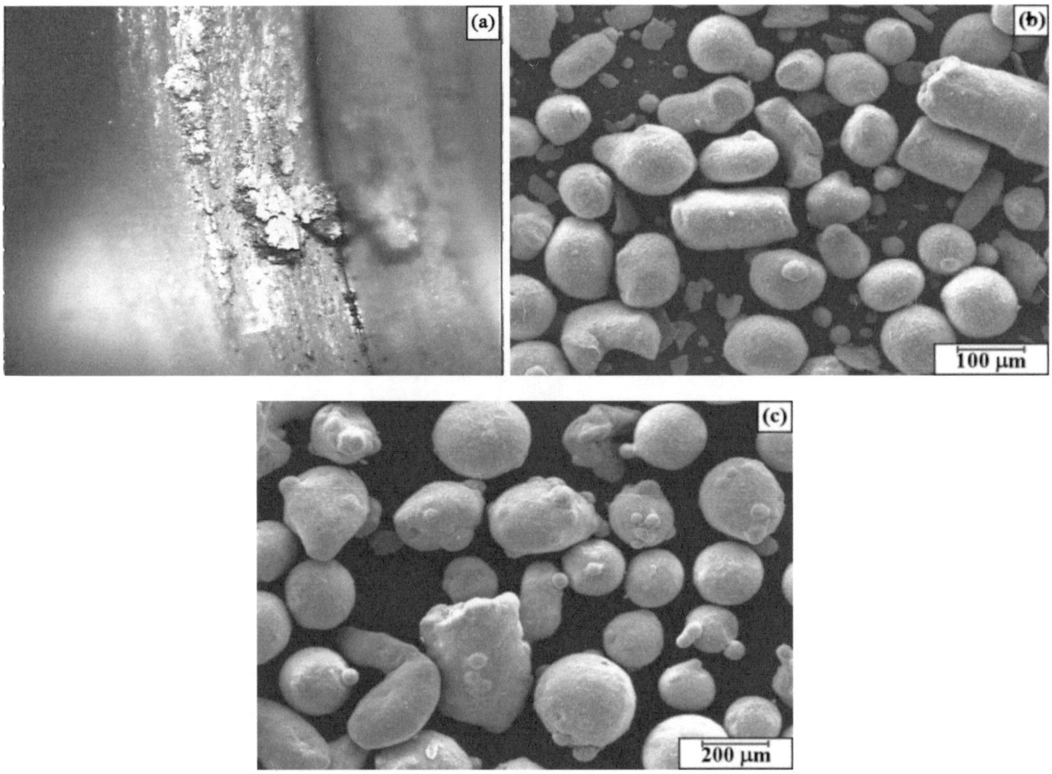

Fig. 1. Appearance of mechanofused particles at different τ: (a) 25 (b) 15 and (c) 5.

Nevertheless, fine particles (~1 μm) appear as a result of the abrasion effect taking place into the wide gap formed by the geometry of scraper blades. Thus, in order to reduce this effect, scraper blades geometry is modified to recover more efficiently the agglomerated powder from the wall surface with an incipient abrasion effect.

2.11.3 Coating of stainless steel particles by Al_2O_3

By using τ = 5, milling and overheating of particles is avoided but a rolling effect is still present, particle coating is strongly influenced by the behaviour of alumina particles into the powder bed. When alumina content is evaluated by means of processing powder at values of mass ratio of host to guest particles of 3 and 7.5, alumina particles are segregated onto the chamber wall surface as shown in Fig. 2a. This behaviour allows just the coverage of some particles featuring a heterogeneous surface coating (Fig. 2b). However, if MF process is performed with smaller amounts of fine guest particles, surface coverage is more uniform. This phenomenon applies for a mass ratio of host/guest particles of 15.0 and, is explained by a better dispersing of alumina particles within the bed of particles, avoiding their segregation.

Fig. 2. (a) Agglomeration of alumina particles onto chamber components. (b) SEM micrographs of mechanofused particles at higher alumina contents.

By considering the latest, processing time is investigated as a function of the powder input rate by introducing alumina particles at 0.05 g/min in order to ensure a well dispersion of both phases into the powdered bed. Samples are taken by intervals of 1 h up to 5 h. A comparison in particle size distributions of mixtures processed at different periods (Fig. 3a), shows a slight difference in the main peak centered at 105 μm. It is likely that, even though the compression gap is widely spaced, a strong rolling effect is still induced, thus attrition of coarser metallic particles take place in the early stages of MF processing, reducing the size of metallic particles. However, a small peak is observed in the range of 0.3 to 1 μm for the samples processed up to 4 and 5 h. This phenomenon suggests that guest alumina particles, which previously have been attached to the surface of host particles, now are detached due to their successive passing throughout the compression gap. In Fig. 3b, the corresponding XRD patterns reveal an increase in size of Al_2O_3-α peak as more alumina particles are introduced. Nevertheless, a slight oxidation of stainless steel particles is detected on 47° 2θ in all cases as a result of attrition taking place in the early stages of processing described before. Then, material not oxidized is renewed at the surface of metallic particles, but oxidation does not continue because of attaching of alumina particles onto that metallic surface preventing its wearing.

Attrition and deformation effects, described above, lead the composite particles to adopt a spherical shape after 4 h of processing, achieving a shape factor of 1.25 (1.0 corresponds to the perfect sphere). Morphology and cross-sections from the resulting composite particles (fig. 4) reveal the formation of a uniform coating of alumina onto the surface of stainless steel particles, attaining up to 5.4 μm of thickness.

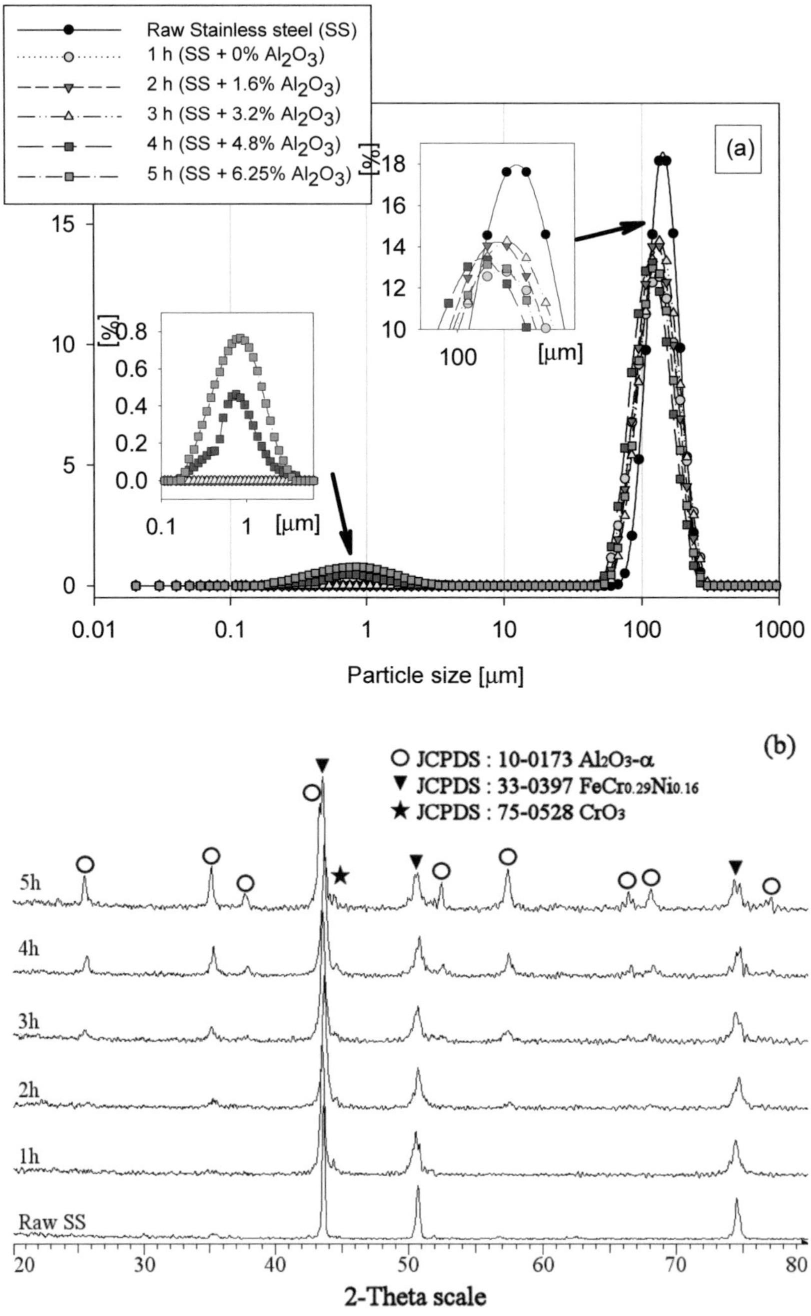

Fig. 3. Evolution of (a) particle size distribution and (b) XRD patterns during MF process of stainless steel SS plus alumina at different processing times: 1, 2, 3, 4, and 5 h.

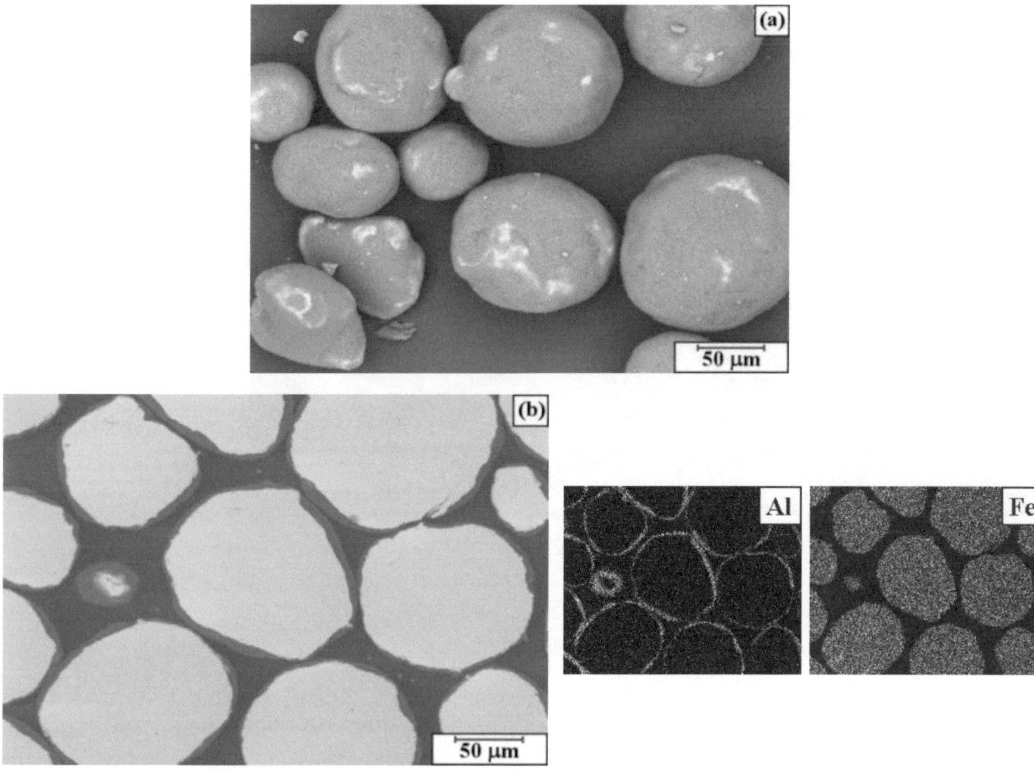

Fig. 4. Evolution of (a) PSD and (b) XRD patterns during MF process of stainless steel SS plus alumina at different processing times: 1, 2, 3, 4, and 5 h.

2.11.4 Coating of stainless steel particles by Al_2O_3 and SiC

The study for preparing a metal/oxide/carbide composite powder was performed by working with the same operating parameters indicated on table 2 but adding up to of 4 wt % alumina and 1.6 wt % silicon carbide. After 6 h of processing, the resulting particles were mesh sieved between 40 - 200 µm.

Samples of SS/AL/SiC mechanofused-composite powders consist also of a stainless steel core uniformly coated by a ceramic shell composed by a mixture of Al_2O_3/SiC. Typical morphologies and cross sections of these powders are shown in Fig. 5. All composite powders are found to be nearly spherical with a mean shape factor of 1.05 and the ceramic shell thickness attains 3.6 µm in thickness. No phase transformation or contamination was detected after the mechanofusion processing as confirmed by XRD analysis.

3. Plasma spraying: The operating parameters

Metal, ceramic or composites coatings, produced by plasma spraying are formed via the stacking of impacted particles at a very high speed (100 to 350 m / s), then flattened due to a molten or plastic state, over the surface of the substrate to be coated. The microstructure of these deposits depends on the particle behaviour in-flight into the plasma and at the impact against the substrate which was prepared previously to certain characteristics.

Simultaneously, this behavior is majorly controlled by the spraying conditions and the thermophysical properties of plasma gas. In the following, it will be presented the resulting composite coatings from the previously described powders processed by mechanofusion considering the main operating parameters of the spraying process.

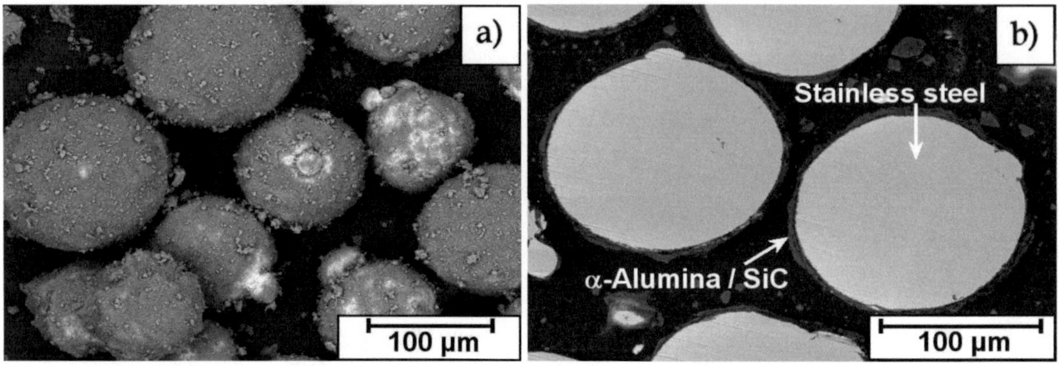

Fig. 5. SEM micrographs of (a) morphology and (b) cross section of Stainless steel/Al_2O_3/SiC mechanofused-composite powders.

3.1 Generation of plasma jet

The plasma jet is formed from a high voltage discharge (5 - 10kV) and high frequency (few MHz) between the tip of a thoriated tungsten cathode (2% Wt Thoria) and the wall of a nozzle-anode electrolytic copper (concentric to the cathode). Thus, a plasma jet flows trough out of the nozzle at high speed (between 1000 and 2500 m/s) and a temperature between 8000 and 14000 K, for an enthalpy about 100 kWh/m3. The plasma gas at the nozzle exit, have a low density (1/30th of the density of cold gas) and viscosity at 10 000 K could be ten times higher than the same mixtures at room temperature. For this case, air plasma spraying (APS) was performed with a conventional d.c. plasma torch under the parameters listed on Table 3.

3.2 The convective motion inside the particles

Inside the plasma jet, a strong movement is induced at the interface liquid-gas due to a significant difference in velocity between the fluid and the molten particles, forcing the displacement of material within of droplet. This is evidenced by the appearance of waves on the surface of the particles and oxide nodules in their core, after passing through the plasma jet (the Reynolds number is 20 to 40) (Espié, 2000). Figure 6 shows the morphology of particles collected at 100 mm downstream of the nozzle exit. Three types of behaviours, according to the state of heating of composite particles, can be observed: (Fig. 6a) those corresponding to particles just over the melting temperature where the alumina shell is broken due to the large difference of expansion coefficient between both materials (for stainless steel is 17 x 10-6K-1 and 8 x 10-6K-1 for alumina); (Fig. 6b) those more heated than in the preceding case but where the molten alumina shell was not entrained to the tail of the in flight particle, then the alumina shell is already consolidated but the host particle is burst into pieces. The third case (Fig. 6c) corresponds to the overheated particles where the light alumina shell at the surface of the molten stainless steel droplet is entrained either to the front or the back edge of the moving droplet.

Anode nozzle i.d. [mm]	7
Arc current [A]	550
Voltage [V]	57
Argon flow rate [slm]	45
Hydrogen flow rate [slm]	15
Gun thermal efficiency [%]	56
Injector external position [mm]	x = 7.5 z = 3.0*
Injector i.d. [mm]	1.8
Spray distance [mm]	100

Table 3. Plasma jet parameters

Fig. 6. Aspect of different types of particles collected in mid-flight: (a) semi-solid composite particle, (b) ceramic shell still remaining consolidated and (c) completely molten composite particle.

3.3 Chemical reactions of particles in-flight: oxidation and/or decomposition

If thermal spraying is performed at atmospheric pressure in the open air, the plasma jet is mixed with the entrained air, having different physicochemical properties, especially density is 30 or 40 times higher, inducing the formation of vortex rings. Thus, coalescence of these vortices is expected to form large amplitudes, showing difficulty to mixing with the air jet, as 'dense particles', until the plasma is correlatively cooled by heat exchange. Consequently, changes in the distributions of temperature, velocity and composition of the plasma conduce to a heterogeneous treatment of particles and, in particular it is likely to react with the entrained oxygen. The reaction rate depends mainly on the oxygen content in their neighbourhood, the exposure time of metal particles to oxygen and the particles temperature (Vardelle et al, 1995).

The oxidation reaction is developed through two mechanisms (Espié, 2000):

- By diffusion of oxygen from the surface to the core of particles (very slow, thickness about one hundred nanometers to a few microns for metals such as pure iron) that represents 1 to 2% by weight oxide;
- By convection from the particle surface towards within. In a continuous circulation, this induces the introduction of metal oxidized and dissolved oxygen inside the core of particle and the refreshing of metal to the surface. Then, the formation of oxide nodules is expected with much higher weight percentage of oxide (12 to 14 wt% for iron, for example), as compared to that obtained by diffusion.

Obviously, oxidation phenomenon inside the plasma have a significantly influence on the composition, microstructure, properties and performance of the deposits obtained. Typically, this is responsible for the appearance of defects in lamellae cohesion; chemical differences on the surface and on the coefficients of dilatation that eventually degrade the mechanical and thermal properties of deposits. The only way to prevent or slow down their occurrence is the isolating of spraying process in vacuum chambers or in controlled atmosphere, but their use remains limited to applications that justify the significantly higher installation cost (by a ratio 10 to 25).

XRD analysis (Fig. 7) reveals a slight oxidation rate of the metallic phase when spraying SS/Al_2O_3 composite powder. It is worth noting that ferrochromium oxides promote fractures and cracking when coatings are subjected to compressive stress (Volenik, 1998).

3.4 Coating building

The deposit is built by a series of successive passes that allow the deposition of particles in a melted or plastic state. The stacking of particles begins on the substrate surface and then continues on particles already deposited and, generally, solidified. Consequently, the contact conditions between lamellae/substrate and lamellae/lamellae are critical for the final properties of coatings (Bianchi, 1995; Branland, 2002). The time between two successive passes must be also considered because, for small size parts is about a few seconds while for larger parts (15 m long) this time can reach even several tens of hours.

Obviously, the final properties of coatings are directly controlled by factors concerning particles (kinetics, viscosity, chemical reactivity of droplets, temperature) and substrate (chemical asset, temperature, roughness). This is explained from a variety of studies that is found in references (Léger et al, 1996; Sampath et al, 1996; Fauchais et al, 2004; Pech, 1999).

3.4.1 Substrate temperature

From all operating parameters, the substrate temperature seems to play the most important role in the formation of lamellae. For a smooth substrate (Ra<0.05 µm), below of a substrate temperature, so-called "transition temperature, TT", the droplet breaks into interconnected pieces. The lower the temperature of the substrate, the morphology of lamellae is more irregular splash-shaped. However, over the TT temperature, the morphology of the lamellae is rather cylindrical disk-shaped with higher contact area and stronger adhesion to the substrate. It should be noted that TT depends on the sprayed and substrate materials. For example, for a zirconia or alumina on stainless steel 316L, TT is about 200 ° C.

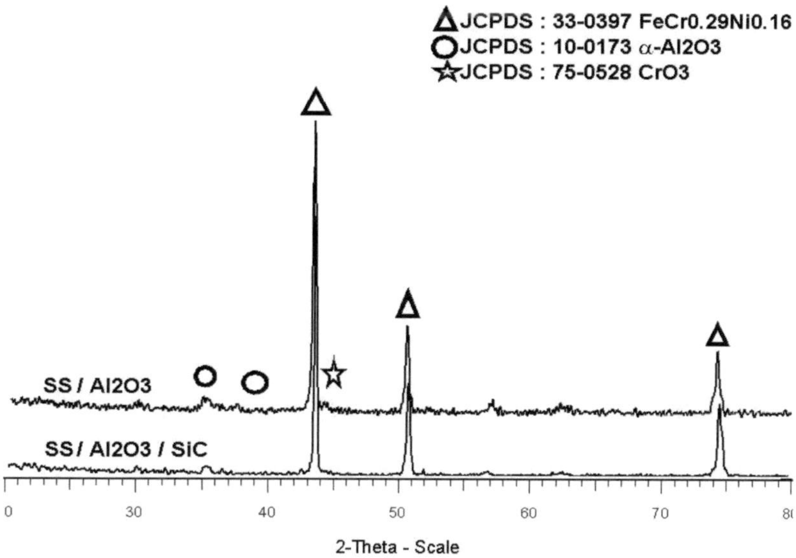

Fig. 7. XRD spectra of different plasma sprayed coatings from SS/Al_2O_3 and SS/Al_2O_3/SiC.

By considering wettability, TT also depends either on the oxidation state of the substrate and in-flight particles. If substrate is heated too long at too high temperature, an oxide layer is formed changing the characteristics of substrate surface in terms of nature, thickness, morphology and roughness of the oxide formed. Although the substrate is at a temperature greater than TT, lamellae will show a weak adhesion or even will not attach to it (Fauchais et al, 2004).

The best contact conditions observed on smooth substrates at a temperature greater than TT, also applies to rough substrates (Ra > 1 µm) and deposit adhesion is greatly increased (by a factor of 3 to 4). In addition, the morphology of lamellae also governs the size and distribution of pores, residual stresses and microstructure of the deposit.

The morphology of splats of SS/Al_2O_3 depends on substrate surface temperature. On cold substrates (TS<100°C), alumina has splashed all around the fingered stainless steel splat, as shown in Fig. 8.a, whereas a nearly disk-shaped splat, is obtained on a substrate preheated to 300°C where the aluminum is placed either over (Fig. 8b) or under (Fig. 8c) the stainless steel splat according to the host particle size. This phenomenon was explained concerning about the alumina cap position relatively to the stainless steel droplet: for the particles smaller than 100 µm the alumina cap is behind the stainless steel droplet at impact on the substrate while with particles bigger than 100 µm it is in front of stainless steel (Cuenca-Alvarez, 2003c).

For another type of splats, corresponding to particles shown in Fig 6a., alumina is scattered in small pieces over the splat surface (Fig. 8d). In-flight, the alumina pieces are either solid or close to their melting temperature i.e. very viscous. Upon flattening the stainless steel which has a high momentum pushes away the alumina pieces or maybe those beneath the flattening particle in its rim where the contact with the substrate is poor. Thus, alumina pieces are distributed evenly at the top of the splat and more regularly in its rims. It occurs whatever may be the preheating temperature of the substrate.

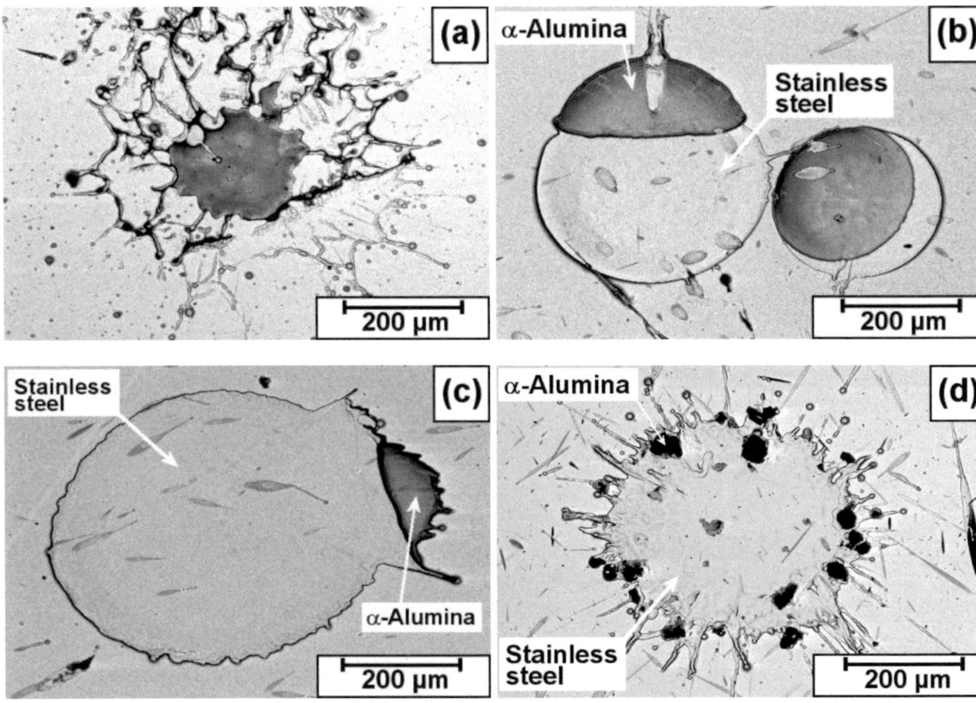

Fig. 8. Splats of stainless steel/alumina composite particles collected on (a) cold and (b,c,d) 350 °C preheated substrates.

The powdered system SS/Al_2O_3/SiC develops particular splat morphology (Fig. 9). It consist of an alumina-mesh net interconnected by fine SiC grains and distributed over the stainless steel splat as confirmed by the EDS analysis presented in white color.

3.4.2 Phases distribution and hardness properties

Typical microstructures of the resulting plasma sprayed coatings of SS/Al_2O_3 and SS/Al_2O_3/SiC mechanofused powders are shown in the Fig. 10. Both coatings exhibit a dense lamellar structure with randomly distributed hard phases within the stainless steel matrix. However the alumina distribution is coarser when spraying SS/Al_2O_3 powder due to the higher alumina content. Homogeneous distribution of ceramics is then expected either by adding a lower content of hard phase or using a smaller core particle size.

These microstructural characteristics of coatings influence their hardness properties. A comparison between the different coatings developed, illustrated in Fig. 11, shows that higher hardness is obtained with both SS/Al_2O_3 (HV5 843 MPa ± 63) while with SS/Al_2O_3/SiC is lower (HV5 756 MPa ± 38). However, the resulting hardness of composite coatings is in both cases higher than that obtained with pure stainless steel deposits (HV5

747 MPa ± 44). Two possible reasons can explain these observations: the uniformly distributed alumina within the coating and the formation of ferrochromium oxides increasing its hardness by dispersion strengthening of hard phases.

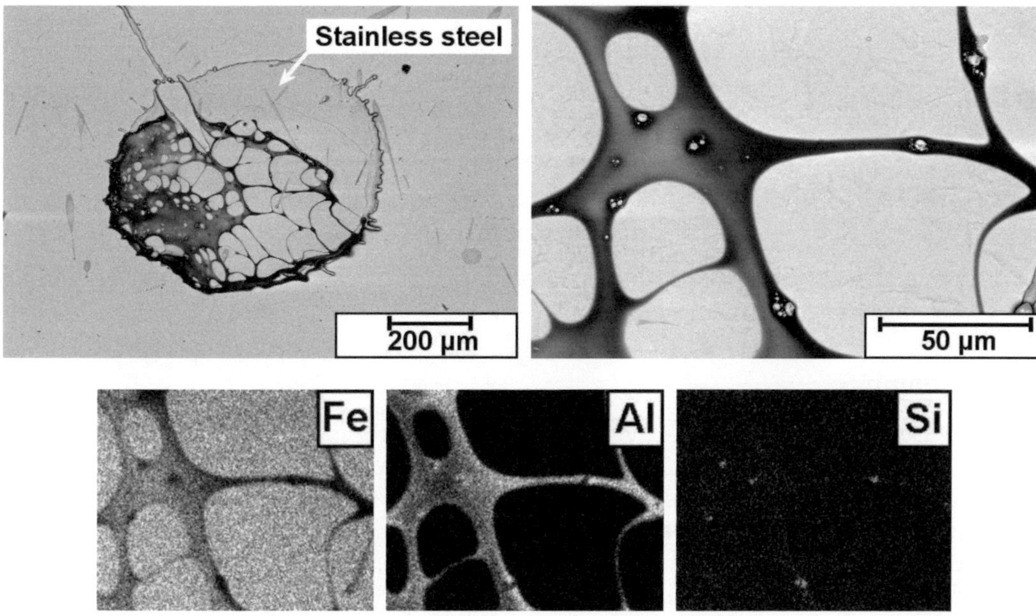

Fig. 9. Particular morphology of SS/Al_2O_3/SiC splats showing the ceramic-mesh net on the stainless steel splat with their corresponding EDS analysis in white color.

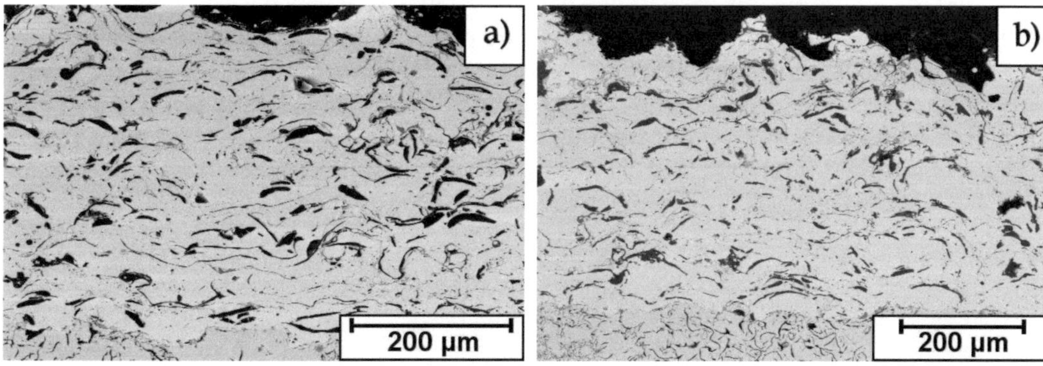

Fig. 10. Typical microstructures of plasma sprayed coatings from (a) SS/Al_2O_3, and (b) SS/Al_2O_3 SiC

By comparing with SS/Al_2O_3, hardness of SS/Al_2O_3/SiC coating is lower due to the incomplete melting of particles limiting the oxide formation. It is most likely that mainly coarse host particle size and a thermal barrier effect of the alumina shell promote this state of incomplete fusion. But also, the hardness attains a value similar to that of pure stainless steel deposits. Nevertheless, no oxide formation is detected by XRD analysis with this type of composite coating. This suggests that coatings' strengthening is mainly governed by the formation of a fine ceramic-mesh net as described above in Figure 9.

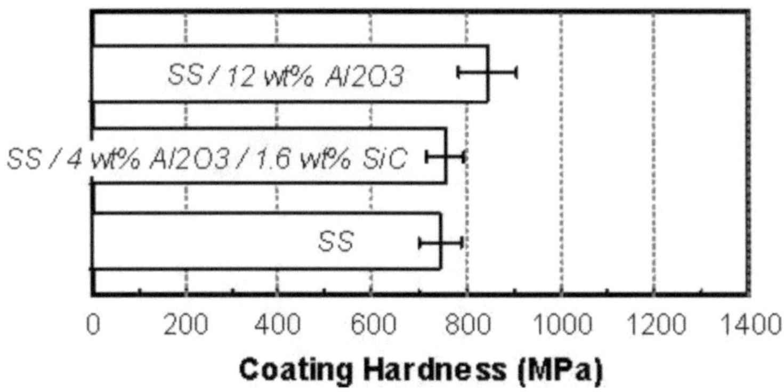

Fig. 11. Comparison of hardness between different plasma sprayed coatings developed from either pure stainless steel or composites powders.

4. Conclusions

Mechanofusion process is an effective means to prepare composite powders to be thermal sprayed and, consequently to control the plasma spray deposit microstructure. The high energy input of the mechanofusion process is directed towards the creation of particle interfaces via agglomeration of particles with a very fine size, in this case alumina (0.6 μm) and silicon carbide (3 μm), coated on stainless steel particles (-90 μm +45 μm). It is likely that agglomeration of fine alumina and silicon carbide particles on stainless steel particles is governed by the large difference in particle size distributions.

When spraying these composite powders, alumina and silicon carbide particles are found embedded and uniformly distributed in a dense steel matrix enhancing hardness properties. The final hardness is according to the kind of composite but it could be considered that the responsible for increasing the coating hardness, is mainly the uniformly distributed ceramic hard phase within the metallic matrix. Actually, the formation of ferrochromium oxides is not an option to increase coating hardness, because coarser particles (100-140 μm) are not completely melted during their passing through the plasma jet, so oxidation is still diffusion controlled

By spraying a ternary composite powder (stainless steel/alumina/silicon carbide), coating hardness is slightly higher than that of pure stainless steel. These composite coatings exhibit a particular mechanism of strengthening consisting of the formation of an alumina-mesh net interconnected by fine SiC grains and distributed over the surface of the stainless steel splat. This allows to joint directly metal splats, retaining the hard phase between lamellae.

Finally it is likely that oxidation of stainless steel particles is limited or almost stopped by their coarse particle size and a molten Al_2O_3 and SiC layer.

5. Acknowledgment

Authors would like to thank CONACYT-Mexico, SFERE-France, and National Polytechnic Institute through CIITEC, EDI & COFAA for their financial support.

R. Cuenca-Alvarez acknowledges to Teotihuacan Group for their kind help on this work.

6. References

Ageorges, H. & Fauchais, P. (2000). Plasma Spraying of Stainless-Steel Particles Coated with an Alumina Shell, *Thin Solid Films*, (370) p. 213-222, ISSN: 0040-6090.

Ananthapadmanabhan, P.V. & Taylor, P.R. (1999). Titanium Carbide-Iron Composite Coatings by Reactive Plasma Spraying of Ilmenite, *Journal of Alloys. & Compounds*, (287), p. 121-125, ISSN: 0925-8388.

Ananthapadmanabhan, P.V.; Thiyagarajan, T.K.; Sreekumar, K.P.; Satpute, R.U.; Venkatramani, N. & Ramachandran, K. (2003). Co-spraying of Alumina-Titania: Correlation of Coating Composition and Properties with Particle Behaviour in the Plasma Jet, *Surfaces & Coatings Technology*, (168), p. 231-240, ISSN: 0257-8972.

Bach, Fr.-W.; Duda, T.; Babiak, Z.; Bohling, F. & Formanek, B. (2000). Characterization of Al_2O_3 and SiC Particles Reinforced Al Powders and Plasma Sprayed Wear Resistance Coatings, *in Thermal Spray : Surface Engineering via Applied Research*, Ohio-USA, 2000, p. 299-302, ISBN: 978-0-87170-680-5.

Bartuli, C. & Smith, R.W. (1996). Comparison between Ni-Cr-40 vol% TiC Wear-Resistant Plasma Sprayed Coatings Produced from Self-Propagating High-Temperature Synthesis and Plasma Densified Powders, *Journal of Thermal Spray Technology*, p. 335-342, ISSN: 1059-9630.

Bartuli, C.; Smith, R.W. & Shtessel, E. (1997). SHS Powders for Thermal Applications, *Ceramics International*, (23), p. 61-68, ISSN: 0272-8842.

Bernard, D.; Yokota, O.; Grimaud, A.; Fauchais, P.; Usmani, S.; Chen, Z.J.; Berndt, C.C. & Herman, H. (1994). Mechanofused Metal-Carbide-Oxide Cermet Powders for Thermal Spraying, *in Thermal Spray: Industrial Applications*, Ohio-USA, 1994, p. 171-178, ISBN: 0-87170-509-5.

Bianchi, L. (1995). Arc Plasma and Inductive plasma Spraying of ceramic coatings: Influence of Mechanism Formation of first layer on coating Properties, PhD Thesis, University of Limoges, France.

Boulos M. (August 2011). Powder Densification and Spheroidization Using Induction Plasma Technology, in: Tekna Plasma Inc.,. Available from: http://www.tekna.com/index.php?module=CMS&id=1&newlang=eng

Branland, N. (2002). Plasma Spraying of Titanium Dioxide Coatings: Contribution to study their microstructures and electric properties, PhD Thesis, University of Limoges, France.

Colaizzi, J.; Kear, B.H.; Mayo, W.E.; Shropshire, R.; Rigney, R.W. & Bunhouse, S. (2000). Micro and Nano-Scaled Composites via Decomposition of Plasma Sprayed Ceramics, *in Thermal Spray: Surface Engineering via Applied Research*, Ohio-USA, p.813-820, ISBN: 978-0-87170-680-5.

Csanády, A.; Csordás-Pintér, A.; Varga, L.; Tóth, L. & Vincze, G. (1997). Solid State Reactions in Al Based Composites Made by Mechanofusion, Mikrochimika Acta, (125), p. 53-62, ISSN: 1436-5073.

Cuenca-Alvarez, R.; Ageorges, H.; Fauchais, P.; Fournier, P. & Smith, A. (2003a). The Effect of Mechanofusion Process and Planetary-Milling on Composite Powder Preparation: Agglomeration and Fragmentation, *Materials Science Forum*, Vol. 442, (2003), pp. 67-72, ISBN 0-87849-930-x.

Cuenca-Alvarez, R.; Ageorges, H. & Fauchais, P. (2003b). Plasma Spraying of Mechanofused Carbide-Oxide and Carbide Metal Powders: The influence of Chemical Composition of Protective Shell, *Materials Science Forum*, Vol. 442, (2003), pp. 73-78, ISBN 0-87849-930-x.

Cuenca-Alvarez, R. (2003c). In French: Contribution to the elaboration of composite coatings by arc plasma spraying of powders prepared by mechanofusion, PhD Thesis, University of Limoges, France.

De Villiers, H.L. (1998). Powder/Processing/Structure Relationships in WC-Co Thermal Spray Coatings: A Review of the Published Literature, *Journal of Thermal Spray Technology*, vol. 7, (3), p. 357-373, ISSN: 1059-9630.

Denoirjean, P.; Syed, A.A.; Cuenca-Alvarez, R.; Denoirjean, A.; Ageorges, H.; Labbe, J.C. & Fauchais, P. (2004). Comparison of Stainless Steel-Alumina Coatings Plasma Sprayed in Air by Two Different Techniques", *Physical and chemical news*, 2004, vol. 20 pp. 21-26, ISSN 1114-3800.

Espié, G. (2000) Oxydation of iron particles into an air plasma jet: its influence on coating properties, PhD Thesis, University of Limoges, France.

Fauchais, P. & Vardelle, A. (2000). Heat, Mass and Momentum Transfer in Coating Formation by Plasma Spraying, *International Journal of Thermal Science*, (39), p. 852-870, ISSN 12900729.

Fauchais, P.; Fukumoto, M.; Vardelle, A. & Vardelle, M. (2004). Knowlegde Concerning Splat Formation, *Journal of Thermal Spray Technology*, Volume 13(3) p. 337-360, ISSN: 1059-9630.

Fukumoto, M. & Okane, I. (1992). Application of Mechanically Alloyed Composite Powders to Thermal Plasma Spraying, *in Thermal Spray: International Advances in Coatings Technology*, Ohio-USA, 1992, p.595-600, ISBN: 0-87170-443-9.

Gadow, R. & Scherer, D. (2001). Ceramic and Metallurgical Composite Coatings with Advanced Tribological Properties under Dry Sliding Conditions, *in Thermal Spray 2001: New Surfaces for a New Millenium*, Ohio-USA, 2001, p.1069-1074, ISBN: 0-87170-737-3.

Gras, C. (2000). Réactivité et thermodynamique dans le procédé MASHS. Application aux systèmes Mo-Si et Fe-Si (in French: Reactivity and thermodynamics of MASHS process applicated on Mo-Si et Fe-Si systems), University of Bourgogne, PhD. Thesis, France.

Herman, H.; Chen, Z.J.; Huang C.C.; Cohen, R. & Tiwari, R. (1992a). Vacuum Plasma Sprayed Mechanofused Ni-Al Composite Powders and Their Intermetallics, *in Thermal Spray: International Advances in Coatings Technology*, Ohio-USA, 1992, p. 355-361, ISBN: 0-87170-443-9.

Herman, H.; Chen, Z.J.; Huang, C.C. & Cohen, R. (1992b). Mechanofused Powders for Thermal Spray, *Journal of Thermal Spray Technology*, 1992, (12), p.129-135, ISSN: 1059-9630.

Ito, H.; Umakoshi, M.; Nakamura, R.; Yokoyama, T.; Urayama, K. & Kato, M., (1991). Characterization of Ni-Al Composite Powders Formed by the Mechanofusion Process and Their Sprayed Coatings, *in Thermal Spray Coatings: Properties, Processes and Applications*, p. 405-410, Ohio-USA, ISBN-10: 0871704374.

Jacobs, L.; Hyland, M.M. & De Bonte, M. (1998). Comparative Study of WC-Cermet Coatings Sprayed via the HVOF and the HVAF Process, *Journal of Thermal Spray Technology*, vol. 7, (2), p. 213-218, ISSN: 1059-9630.

Jacobs, L.; Hyland, M.M. & de Bonte, M., (1998). Comparative Study of WC-Cermet Coatings Sprayed via the HVOF and the HVAF Process, *Journal of Thermal Spray Technology*, vol. 7, (2), p.213-218, ISSN: 1059-9630.

Jiansirisomboon, S.; MacKenzie, K.J.D.; Roberts, S.G. & Grant, P.S., (2003). Low Pressure Plasma-Sprayed Al2O3 and Al2O3/SiC Nanocomposite Coatings from different Feedstock Powders, *Journal of European Ceramics Society*, (23), p. 961-976, ISSN: 0955-2219.

Kelly, T.F.; Larson, D.J.; Miller, M.K. & Flinn, J.E. (1999). Three Dimensional Atom Probe Investigation of Vanadium Nitride Precipitates and the Role of Oxygen and Boron in Rapidly Solidified 316 Stainless Steel, *Materials Science & Engineering*. A, (A270), p.19-26, ISSN: 0921-5093.

Khan, M.S.A. & Clyne, T.W. (1996). Microstructure and Abrasion resistance of Plasma Sprayed Cermet Coatings, *in Thermal Spray : Practical Solutions for Engineering Problems*, Ohio-USA, 1996, p. 113-122, ISBN-10: 0871705834.

Khor, K.A.; Dong, Z.L. & Gu, Y.W. (2000). Influence of Oxide Mixtures on Mechanical Properties of Plasma Sprayed Functionally Graded Coating, *Thin Solid Films*, (368), p. 86-92, ISSN: 0040-6090.

Khor, K.A.; Fu, L.; Lim, V.J.P. & Cheang, P. (2000). The Effects of ZrO2 on the Phase Compositions of Plasma Sprayed HA/YSZ Composite Coatings, *Materials Science & Engineering A*, (A276), p. 160-166, ISSN: 0921-5093.

Kim, M.C.; Kim, S.B. & Hong, J.W. (1997). Effect of Powder Types on Mechanical Properties of D-Gun Coatings, *in Thermal Spray: A United Forum for Scientific and Technological Advances*, p. 791-795 , Ohio-USA, ISBN-10: 0871706180.

Krajnikov, A. V.; Likutin, V. V. & Thompson, G. E. (2003). Comparative Study of Morphology and Surface Composition of Al–Cr–Fe Alloy Powders Produced by Water and Gas Atomisation Technologies, *Applied. Surface Science*, Vol. 210, (3-4), p. 318-328, ISSN: 0169-4332.

Kubel Jr., E.J., (1990). Powders Dictate Thermal-Spray-Coating Properties, *Advanced Materials Processing*, (12), p. 12-32, ISSN: 0882-7958.

Léger, A.C.; Vardelle, M.; Vardelle, A.; Fauchais, P.; Sampath, S.; Berndt, C.C. & Herman, H. (1996). Plasma Sprayed Zirconia: Relationships Between Particle Parameters, Splat Formation and Deposit Generation - Part 1: Microstructure and Solidification", *in Thermal Spray: Practical Solutions for Engineering Problems*, Ohio-USA, 1996, p. 623-628, ISBN-10: 0871705834.

Liu, X. & Ding, C. (2002). Characterization of Plasma Sprayed Wollastonite Powder and Coatings, *Surfaces & Coatings Technology*, (153), p. 173-177, ISSN: 0257-8972.

Liu, X. & Ding, C. (2003). "Plasma Sprayed Wollastonite 2M/ZrO2 Composite Coating, *Surfaces & Coatings Technology*, (172), p. 270-278, ISSN: 0257-8972.

Lugscheider, E.; Jungklaus, H.; Zhao, L. & Reymann, H. (1997). Reactive Plasma Spraying of Coatings Containing In-Situ Synthesized Titanium Hard Phases, *International Journal of Refractory Metals and Hard Materials*, (15), p. 311-315, ISSN: 0263-4368.

Lugscheider, E.F.; Loch, M. & Suk, H.G. (1992). Powder Technology-State of the Art, *in Thermal Spray: International Advances in Coatings Technology*, p. 555-559, Ohio-USA, 1992, ISBN: 0-87170-443-9.

Moreau, C. & Dallaire, S., (1990). Plasma Spraying of Carbon-Coated TiC Powders in Air and Inert Atmosphere, *in Thermal Spray: Research and Applications*, p. 747-752, Ohio-USA, ISBN-10: 0871703920.

Pech, J. (1999). Pre-oxiydation generated by arc plasma spraying: Relations between surface, oxidation and coating adhesion, PhD Thesis, University of Rouen, France.

Ramaswany, P.; Seetharamu, S.; Varma, K.B.R. & Rao, K.J. (1997). Al_2O_3-ZrO_2 Composite Coatings for Thermal-Barrier Applications, *Composites Science & Technology*, (57), p. 81-89, ISSN: 0266-3538.

Rautioaho, R.; Riipinen, M.M.; Saven, T. & Tamminen, A. (1996). Ni_3Al and Ni_3Si-Based Intermetallics Produced by the Osprey Process, *Intermetallics*, (4), p. 99-109, ISSN: 0966-9795.

Sampath, S.; Matejicek, J.; Berndt, C.C. ; Herman, H. ; Léger, A.C.; Vardelle, M.; Vardelle, A. & Fauchais, P. (1996). Plasma Sprayed Zirconia: Relationships Among Particle Parameters, Splat Formation and Deposit Generation - Part II: Microstucture and Properties, *in Thermal Spray: Practical Solutions for Engineering Problems*, p. 629-636, Ohio-USA, ISBN-10: 0871705834.

Sampath, S.; Matejicek, J.; Berndt, C.C.; Herman, H.; Léger, A.C.; Vardelle, M.; Vardelle, A. & Fauchais, P. (1996). Plasma Sprayed Zirconia: Relationships Among Particle Parameters, Splat Formation and Deposit Generation - Part II: Microstucture and Properties", *in Thermal Spray: Practical Solutions for Engineering Problems*, Ohio-USA, p. 629-636, ISBN-10: 0871705834.

Sordelet, D.J.; Besser, M.F. & Logsdon, J.L. (1998). Abrasive Wear Behavior of Al-Cu-Fe Quasycrystalline Composite Coatings", *Materials Science & Engineering A*, (A255), p. 54-65, ISSN: 0921-5093.

Trice, R.W.; Jennifer Su, Y.; Faber, K.T.; Wang, H. & Porter, W. (1999). The Role of NZP Additions in Plasma-Sprayed YSZ: Microstructure, Thermal Conductivity and Phase Stability Effects", *Materials Science & Engineering A*, (A272), p. 284-291, ISSN: 0921-5093.

Vaidya, A.; Bancke, G.; Sampath, S. & Herman, H. (2001). Influences of Process Variables on the Plasma Sprayed Coatings : An Integrated Study, *in Thermal Spray : New Surfaces for a New Millenium*, p. 1345-1349. Ohio-USA, 2001, ISBN: 0-87170-737-3.

Valente, T. & Galliano, F.P. (2000). Corrosion Resistant Properties of Reactive Plasma-Sprayed Titanium Composite Coatings, *Surfaces & Coatings Technology*, (127), p. 86-91, ISSN: 0257-8972.

Vardelle, A.; Fauchais, P. & Themelis, N.J. (1995). Oxidation of Metal Droplets in Plasma Sprays, *in Thermal Spray: Science & Technology*, Ohio-USA, 1995, p. 175-180, ISBN: 10-0871705419.

Vinayo, M.E.; Kassabji, F.; Guyonnet, J. & Fauchais, P., (1985). Plasma Sprayed WC-Co Coatings: Influence of Spray Conditions (Atmospheric and Low Pressure Plasma Spraying) on the Crystal Structure, Porosity and Hardness, *Journal of Vacuum Science Technology A*, A3 (6), p. 2483-2489, ISSN: 0003-6951.

Volenik, K.; Novak, V.; Dubsky, J.; Chraska, P. & Neufuss, K. (1998). Compressive Behaviour of Plasma Sprayed High-Alloy Steels, *in Thermal Spray: Meeting the Challenges of the 21th Century*, Ohio-USA, (1998), pp. 671/675, ISBN-10: 0871706598.

Wang, F.; Yang, B.; Duan, X. J.; Xiong, B. Q. & Zhang, J. S. (2003). The Microstructure and Mechanical Properties of Spray-Deposited Hypereutectic Al–Si–Fe Alloy, *Journal of Materials Processing Technology*, vol. 137, (1-3), p. 191-194, ISSN: 0924-0136.

Wielage, B.; Wilden, J. & Schnick, T. (2001). Manufacture of SiC Composite Coatings by HVOF, *in Thermal Spray 2001: New Surface for a New Millenium*, Ohio-USA, 2001, p. 251-258, ISBN: 0-87170-737-3.

Yokoyama, T.; Urayama, K.; Naito, M. & Kato, M., The Angmill Mechanofusion System and Its Applications (1987). *KONA*, (5), p. 59-68, ISSN: 0288-4534.

Zhao, L. & Lugscheider, E. (2002). Influence of the Spraying Processes on the Properties of 316L Stainless Steel Coatings, *Surfaces & Coatings Technology*, (162), p. 6–10, ISSN: 0257-8972.

Zhao, L.; Maurer, M. & Lugscheider, E. (2003). Thermal Spraying of Nitrogen Alloyed Austenitic Steel, *Thin Solid Films*, (424), p. 213-218, ISSN: 0040-6090.

Zimmermann, S.; Keller, H. & Schwier, G. (2003). New Carbide Based Materials for HVOF Spraying, *in Thermal Spray 2003: Advancing the Science & Applying the Technology*, Ohio-USA, 2003, p. 227-232, ISBN: 978-0-87170-785-7.

3

Isothermal Oxidation Behavior of Plasma Sprayed MCrAlY Coatings

Dowon Seo[1] and Kazuhiro Ogawa[2]
[1]*Dept. of Mechanical Eng., Toyohashi University of Technology, Toyohashi,*
[2]*Fracture & Reliability Research Institute, Tohoku University, Sendai,*
Japan

1. Introduction

Thermal spray coatings are deposited in an ambient atmosphere or vacuum chamber. Although vacuum plasma spray (VPS) coating is deposited inside vacuum, oxygen can penetrate into the flame during spraying process, as in high velocity oxygen-fuel (HVOF) spraying (C.J. Li & W.Y. Li, 2003). This causes the spray materials to be exposed directly to an oxidizing atmosphere. This oxidation significantly influences the phase composition, microstructure, properties and performance of the sprayed coatings. Metal oxides are grown on the lamellar interface. The oxides are brittle and have different thermal expansion coefficients than that of the metal, the inclusion of which may cause the spalling of the coating (Neiser et al., 1998). Moreover, the inclusion of oxides in the MCrAlYs (where M is the alloy base metal; typically nickel, cobalt, or combination of these two), the coatings will degrade the resistance to sulfur and vanadium, etc., under high temperature corrosion. The presence of the oxides in steel coating also affects its mechanical properties (Volenik et al., 1997). However, some coating properties can be improved by metal oxides in sprayed coatings. A typical example is the improvment of wear resistance (Neiser et al., 1998). The deposited oxides increase also the hardness of the coating (Dobler et al., 2000). Therefore, it is important to understand the oxidizing behavior of spray materials at spraying.

The oxide content in the as-sprayed coating depends on the spraying technique, spraying parameters and starting material compositions (Dobler et al., 2000). Espie et al. reported that the oxygen content in plasma sprayed low carbon steel particles increased with spray distance (Espie et al., 2001). Fukushima and Kuroda presented similar effect for plasma sprayed Ni-20Cr coating, but they also reported that the oxygen content in the thermal sprayed coating was decreased with decreasing the stand-off distance (Fukushima and Kuroda, 2001). From the results in the study, it can be found that the oxygen content reached up to 10wt.% in plasma sprayed coatings, and others reported an oxygen intake up to 3-6 wt.% in the coating deposited by HVOF process. The detailed examination in these studies revealed that there exists a clear difference in the grain size of powders. Therefore, it can be considered that such a difference in grain size of spray powder may be a cause responsible for the notable difference in coating oxygen content and oxidation behavior. From such a result, it may be considered that the properties of VPS process also might be affected by particle size, even though a vacuum process can better suppress oxidation of metallic coatings during spraying than the air plasma spray (APS) or HVOF process can.

The oxidation of metals depends on the rates at which anion or cation transport can occur through the crystal lattice or along grain boundaries in the oxide (Pomeroy, 2005). In alloys, which oxide is the most stable depends on the oxide dissociation pressure which is lowest for Al and Cr compared to the typical base elements Fe, Co and Ni. For Ni-Cr alloys containing more than about 10 at.% Cr, a continuous protective chromia layer is formed. A higher Cr level (25 at.%) is required for Co-based alloys because Cr diffuses more slowly in Co and so cannot form a continuous chromia layer at lower concentrations (Pomeroy, 2005). Chromium oxide can itself oxidize at temperatures greater than about 1123 K to a volatile CrO_3 compound. Because of this, the use of aluminium additions for oxidation resistance is preferred at this temperature and above for key components such as those used in gas turbines. There is added advantage, because the rate of formation of aluminium oxide is slower than for chromium oxide at the same temperature. Addition levels of Al to Ni for the formation of a continuous alumina scale are as for chromia formation (10 at.%). Alumina formation at lower Al contents can be induced by mixed Cr+Al additions since Cr getters oxygen allowing alumina to form at lower aluminium activities (contents) (Pomeroy, 2005). The aluminium and chromium contents referred to above apply to isothermal oxidation conditions. However, when thermal cycling conditions prevail, oxide scales can spall from the substrate surface due to thermally induced stresses.

Therefore, in this work, the influence of particle size and exposure time on the mechanical and oxidation behavior of selected VPS coatings was investigated. The particle and coating morphology, the oxygen contents of coatings, porosity, surface roughness, the growth rate of thermally grown oxide (TGO) thickness and mass gain, and the oxidation behavior were examined.

2. Materials and experimental procedures

Five commercially available CoNi- and CoCrAlY powders of different size ranges, from several micrometers to over 45 μm, were used as starting materials. Table 1 shows the nominal compositions of the five starting powders and the respective as-sprayed coating thickness, referred as (a) AMDRY-9951, (b) CO-210-1, (c) UCT-195, (d) CO-110, and (e) UCT-1348. All powders were manufactured through gas atomization process. In order to obtain a reliable relationship between particle size and oxygen content, the particle size of the powders was measured statistically from scanning electron microscopy (SEM, Hitachi S-4700, Japan) images. The Inconel 718 alloy (MA718, Mitsubishi Materials Co., Japan) was used as substrate to deposit the coating, which is widely used for gas turbine components. The chemical composition of the substrate material was 19Cr-19Fe-5.1Nb-3Mo-0.9Ti-0.5Al-balance Ni. The VPS processes were carried out with the A-2000V VPS system (Plasma Technik, Germany), under the following conditions: preheating temperature is 843 K during transferred arc treatment, voltage 56-57 V, current 580-590 A, spraying distance 310 mm, argon gas atmosphere 8 kPa. The single face of substrate, which was 80×70×5 mm after pretreatment of blasting, was sprayed with 0.28 mm in average thickness.

To characterize the oxidation behavior, static oxidation experiments were carried out in air under isothermal condition at 1273 K up to 1000 h. After eight-different exposure times, the specimens were characterized microstructurally using light optical and scanning electron microscopy (SEM; Hitachi S-4700, Japan). Each exposure condition consisted of high heating rate (about 32 K/min) in a kanthal muffle furnace, and about 3 K/min cool-down to ambient temperature. The specimens were kept in alumina boats and then the boats were inserted into the furnace. The aim of this oxidation procedure was to create accelerated conditions for testing. The studies were performed on uncoated as well as plasma sprayed specimens for the purpose of comparison. Weight change measurements were made at the end of each exposure period with the help of an electronic balance (Sartorius LA120S, Germany) with a sensitivity of 0.1 mg. The coatings were removed from substrates and polished by 0.5 μm alumina media before the weight gain measurements, to exclude the effect of surface roughness on the oxidation. The spalled scale also was included at the time of measurements of weight change to determine total rate of oxidation. The coated samples were degreased using ethanol and then subjected to optical microscopy, SEM and energy dispersive X-ray analysis (EDX; Oxford 6841, USA) to characterize the surface morphology. Electron spectroscopy for chemical analysis (ESCA; PHI Quantum 2000, USA) was also used to obtain surface profiles information in the first few layers, especially oxygen concentration profile. The porosity measurements were made with an image analyzer software (ImageJ, NIMH, USA), which is based on ASTM B276. The image was obtained through the SEM. After preliminary characterization, the samples were cross-sectioned, mounted in a conductive, phenolic resin with carbon filler and subjected to mirror polishing to identify the cross-sectional details. The coating thickness was measured by taking a back scattered electron image (BSEI) with the SEM, which was attached to a Robinson back scattered detector (RBSD). The surface roughness of the coatings was measured by the contactless measuring system (Mitaka NH-3T, Japan).

Type	Designation	Chemical composition* (wt.%)		Coat thick.
		Main	Trace	(μm)
CoNiCrAlY	(a) AMDRY-9951	32Ni-21Cr-8Al-0.5Y-Bal.Co	-	269.15
	(b) CO-210-1	32Ni-21Cr-8Al-0.5Y-Bal.Co	Fe, O, C, P, Se, N, H, S	247.85
	(c) UCT-195	33Ni-21Cr-8Al-0.4Y-Bal.Co	Si, Fe, C, P, S, O, N, H	308.54
CoCrAlY	(d) CO-110	23Cr-13Al-0.7Y-Bal.Co	P, C, S	266.23
	(e) UCT-1348	23Cr-13Al-0.6Y-Bal.Co	C, H, S, P	316.38

Table 1. Nominal compositions and coating thickness of the as-sprayed coatings

3. Results and discussion

3.1 Morphology and oxygen contents in deposits

Fig. 1 shows the typical morphology of spray powders screened into different sizes. It can be found that all powders have a spherical shape. The measurement of particle size using SEM images yielded the results shown in Table 2 and Fig. 2. From the results, it is clear that the measured mean particle size was larger than the mean value of the opening sizes of the sieves specified by suppliers. The AMDRY-9951 and CO-110 powders show the smallest and narrowest ranges than others.

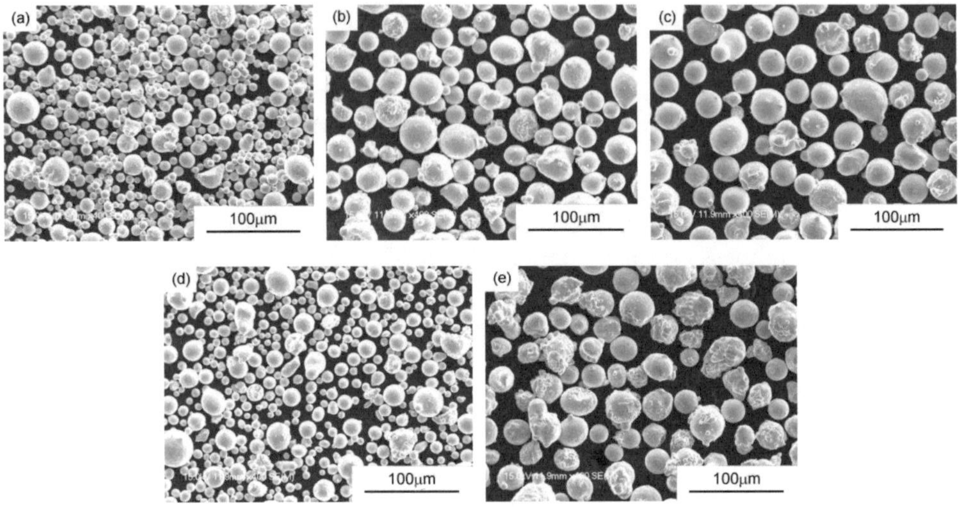

Fig. 1. Morphology of the spray powders: (a) AMDRY-9951; (b) CO-210-1; (c) UCT-195; (d) CO-110; (e) UCT-1348

Designation	Sieved size	Analyzed size	Range	Mean	Median	S.D.	Variance	Error
(a) AMDRY-9951	-35 +5	-34.5 +2.5	32.06	9.40	8.375	4.22	17.80	0.16
(b) CO-210-1	-45 +10	-43.5 +3.6	39.86	20.51	20.35	8.36	69.95	0.75
(c) UCT-195	-45 +22	-40.3 +6.2	34.1	24.84	26.34	7.59	57.55	0.84
(d) CO-110	-44 +5	-38.4 +2.0	36.39	10.08	8.58	4.79	22.99	0.22
(e) UCT-1348	-45 +15	-44.7 +12.6	32.06	25.71	25.04	6.51	42.33	0.72

Table 2. Particle size ranges and descriptive statistics of powders (μm)

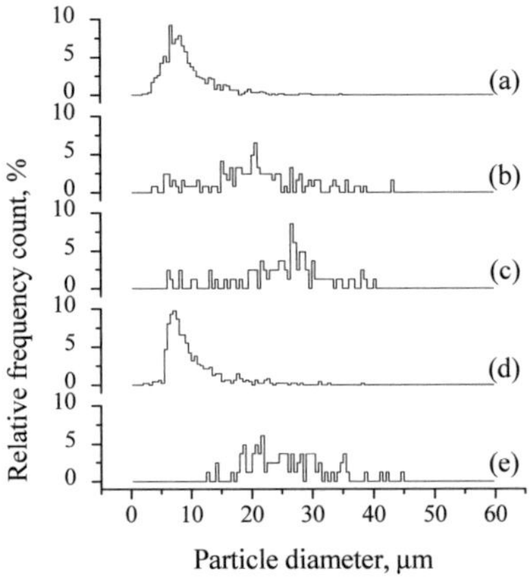

Fig. 2. Particle size distributions of the spray powders: (a) AMDRY-9951; (b) CO-210-1; (c) UCT-195; (d) CO-110; (e) UCT-1348

Fig. 3 shows a cross-section of a collected particle of approximately 18 µm in diameter and EDX analysis results obtained at three typical points marked as 1, 2 and 3. It was found that there was a thin-splat-shell covering over the particle. The analysis revealed that this covering had oxide film formed during in-flight (point 3, Fig. 3a). Moreover, no evidence of oxygen was found inside the particle as shown in the point 1 of Fig. 3b. This result suggested that there is an interface between the oxide surface covering and the non-oxidized inner fraction of the particle. Therefore, in spite of VPS, the oxidation of an in-flight particle proceeds from particle surface towards the inside. As the result of referred research, Li et al. (C.J. Li & W.Y. Li, 2003) showed an in-flight particle of HVOF spraying and its EDX analysis. The results of their analysis revealed that the thin-film-shaped covering was an oxide film formed during in-flight. During particle flight, with the decrease in the particle size, the oxygen content in the powders increase. When the mean particle size of powders was reduced to approximately 30µm, the rapid increase in oxygen content in the in-flight powders was reported (C.J. Li & W.Y. Li, 2003). It is clear that the particle size has a significant effect on the oxidization of the in-flight particle and consequently the oxygen content in spraying powders.

Generally, the oxidation of an alloy particle in a flame occurs on the particle surface. The oxygen content in the particle will increase with time because of the diffusion of oxygen from the surface towards the inside of the particle. Based on the relation between the oxygen content in an in-flight particle and the distance as reported by Espie et al. for plasma spraying, it can be considered that the oxygen content in the particle would follow the parabolic growth law, although the effect of internal flow in a well melted particle may occur and intensify the oxidation by exposing the fresh metal surface to the flame as pointed out by Neiser et al. and Espie et al. (Neiser et al., 1998; Espie et al., 2001). Therefore, at the initial stage, the oxygen content will be increased rapidly. With the progressing of the oxidation process, the increase in oxygen content will become less intensive. Moreover, after the particle flies off the high temperature zone of the flame, oxidation becomes slower. Due to the high stiffness of the flame, it was difficult to collect the powders at short spray distance. However, the very similar oxygen contents in both the coatings and in-flight particles provided evidence that the oxidation takes place mainly in the flame.

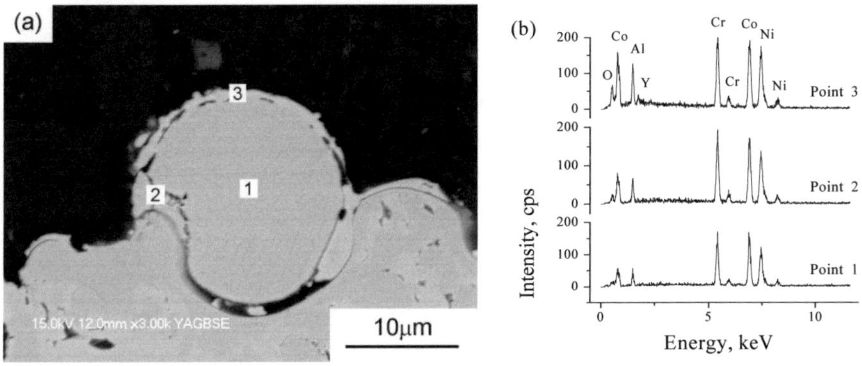

Fig. 3. Cross-sectional microstructure of a deposited particle using (a) CO-210-1 powder and (b) EDX spectrum analysis results at the center (point 1), near periphery (point 2) of un-melted particle, and fully melted splat (point 3)

Fig. 4 shows the oxygen contents in the coatings versus mean particle size of the spray powder. It can be found that the oxygen contents in (a) AMDRY-9951 and (d) CO-110 coatings were larger than those in the other coatings. Table 3 shows the results of the oxygen contents in three kinds of starting powders and deposited coatings. It can be clearly observed that a remarkable increase in the oxygen content in the coating occurred with a decrease in the mean particle size. When the mean particle size is approximately 10 µm, the oxygen content was increased by a half order of magnitude with regard to the coating deposited by the powders larger than 20 µm. This implies that the decrease of mean powder size to a half will result in an increase in oxygen content by a half order of magnitude. It is clear that the particle size of spray powder has a substantial effect on the oxygen content in the deposited coating. This fact explains why the coatings deposited by the same type of material may significantly differ in the oxygen contents. Therefore, when the spray materials with grain size of less than 40 µm are used and the oxidation is involved, the particle size and the distribution of the size as well should be taken into account essentially.

But after thermal exposure, the difference of the oxygen contents between the smaller and larger particle coatings was decreased with increasing exposure time. Generally, the oxide scale growth in isothermal condition initiates on the free surface of coatings. The oxygen content in the exposed coating will increase with exposure time because of the diffusion of oxygen from the surface towards the inside of the coating. Therefore, at the initial stage (1 hour), the oxygen content was increased rapidly. With the progressing of the oxidation process, the increase in oxygen content will become less intensive. So the influence of particle size on the oxygen content in the aged coating also becomes less intensive as shown in Fig. 4 and Table 3. The oxides in the as-sprayed coatings result from powder oxidation.

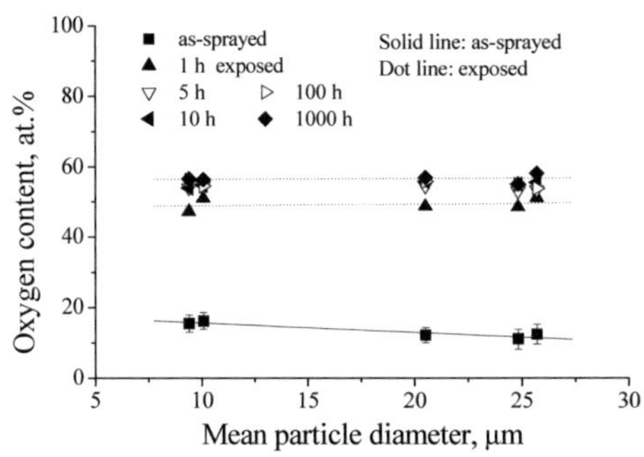

Fig. 4. Effect of particle size on the oxygen content in CoNi-/CoCrAlY coatings

Designation	Powder* (wt.%)	Coating (at.%)						
		In micro-level depth**						Nano-level***
		As-sprayed	1 h	5 h	10 h	100 h	1000 h	As-sprayed
AMDRY-9951	-	15.50	47.13	54.05	54.04	56.22	56.60	58.55
CO-210-1	0.028	12.18	48.70	54.58	55.89	55.90	56.97	59.93
UCT-195	0.040	11.12	48.52	53.16	55.04	54.86	55.16	59.14

* Specified by supplier. ** Measured using EDX. *** Measured using ESCA.

Table 3. Oxygen contents in starting powders and coatings

3.2 Effect of particle distribution on the porosity of coatings

Fig. 5 shows the distribution of the cross-sectional porosity in the coating about 100 µm from the coating surface versus particle size. The amount of porosity increased normally with increasing particle size. This, of course, could be due to reduced melting efficiency of the coarser particles in the plasma plume. The coating from the finer powder shows well-adhered splats, while the interlamellar pores are more prominent in the medium powder coating. The unmelted and poorly adhered particles can be seen in the coarse powder coating. Earlier studies (Kulkarni et al., 2003) have shown that the morphology of the splats changes from a contiguous disk-like shape to a fragmented shape with increasing particle size. These fragmented splats lead to poor splat-splat contact and to the formation of pores. But the distribution of particle size plays the important role of porosity than only the particle size. In some cases the unmelted particles press the splats and make the poorly bonded splat-splat contact closed as shown in Fig. 6. The (b) CO-210-1 coating shows the lowest porosity, this could be due to the good compacting of the small, medium and larger particles. This powder has the widest range and sample variance as shown in Table 2, so these lead to low porosity inside the coating during the deposition process.

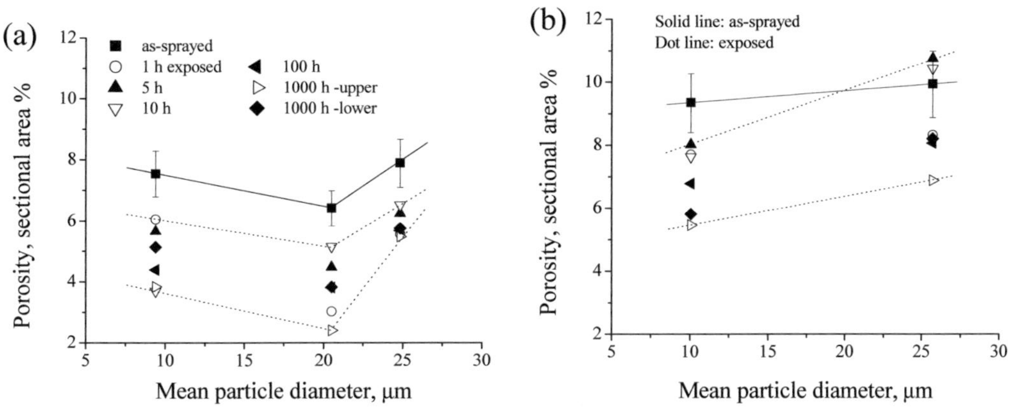

Fig. 5. Effect of particle size on the porosity of (a) CoNiCrAlY and (b) CoCrAlY coatings

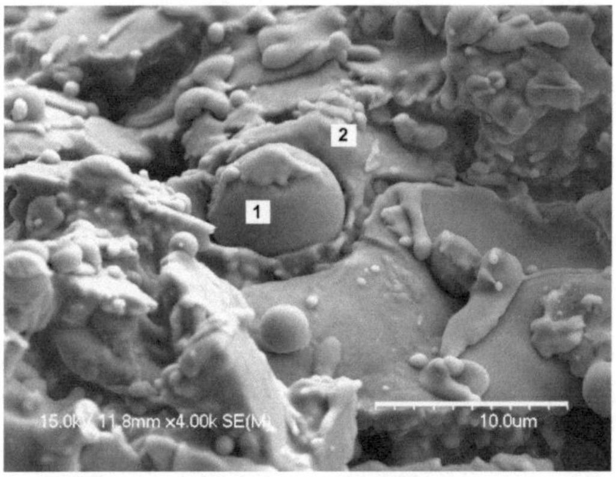

Fig. 6. Compressing phenomenon of the as-sprayed CoNiCrAlY (UCT-195) coatings: (1) an unmelted particle; (2) the pressed splat by the particle

As shown in Fig. 5, with increasing exposure time, the amount of porosity decreased gradually in all coatings. Upon detail review of the CoNiCrAlY coatings, the porosity decreased at Stage I (1-10 h of (a) AMDRY-9951, and 0-1 h of (b) CO-210-1 and (c) UCT-195), and increased at Stage II (10-100 h of (a) AMDRY-9951, and 1-10 h of (b) CO-210-1 and (c) UCT-195), and then finally decreased again beyond Stage II. Stage I represents the sintering effect. These sharp changes in porosity after short sintering times have also been observed by Thompson and Clyne (Thompson & Clyne, 2001). It might be expected that short exposure (Stage I) to temperatures is sufficient to heal cracks and lock splats together leading to decreased porosity inside of the MCrAlY coatings. Inside pores concentrated on the lower part of the coating over the Stage II.

From the cross-sectional morphology of the interfaces between the MCrAlY coatings and the substrates after 1000 hours of exposure, the intergranular voids present in the scale which may be formed. In the absence of the effective shrinkage for vacancies, vacancies generated by outward diffusion of alloying elements such as Ni or Co condense to form cavities. One of the most effective methods to develop oxidation resistance in alloys and coatings at temperatures above about 1223 K is to form continuous scales of α-Al_2O_3 via selective oxidation. As the oxidation time increase, transient oxidation stage is represented by the formation of the protective alumina oxide layer, followed by Al-depletion, Ni or Co outward diffusion and solid state reaction with pre-existing alumina to form spinel phase, and the Cr_2O_3 formation at the oxide-to-metal interface. These processes highly depend on the oxidation temperature and microstructures of the VPS MCrAlY coatings.

Authors tried to predict the porosity of as-sprayed coatings, using bi- or tri-models which are composed with two or three type of particle diameters as shown in Table 4. But, real porosity values are lower than proposed theoretical values. It is believed that most particles are round and spherical, but small particles change to flat shape. Generally, packing of spheres leads to a higher density than other shapes. To predict more exactly, splat-reflected model is required, and more useful. The apparent density of the metal powder is influenced by many factors, such as particle size, shape, size distribution, inter-particle friction, surface chemistry, agglomeration and packing type, etc. In general, with the decrease of the particle size, the packing density decreases because of the higher inter particle friction. The greater the surface roughness or the more irregular the particle shape, the lower the packing density is. A higher relative apparent density can be achieved by mixing different sizes of powders. The principle involves using finer particles to fill the voids formed by the larger powders.

Fig. 7 shows four possible occurrences in the mixture with different sizes of powders, where L, M and S represent the diameters of the large, middle and small spheres, respectively. From these models, it can be found that using the tri-modal arrangement one can obtain the highest apparent density (Fig. 7b), and using bi-modal, i.e. mono-sized small particles with one middle-sized particle yields the lowest apparent density (Fig. 7a). The quantitative analysis has been conducted as shown in Table 4. For instance, an ideal situation is shown for plain tri-modal particles in Fig. 7b where the small size disk just touches the large neighboring disks. In fact, the composition of a mixture according to the geometrical relationship is not feasible. This is caused by two reasons. The first reason is that metal powders consisting of purely mono-sized particles are impossible to attain. In spraying process, the powder mixture always consists of several powders. Another is that the size of the powder in spraying is restricted by many factors, such as layer thickness, flowability, deposit characteristic and cost.

Designation	Dia. group*	Frequency count ratio	Mean dia. ratio	Mean dia. (μm)	Porosity (area %) Theoretical	Measured
(a) AMDRY-9951	S	14.69	1.00	8.60	22.92	6.87
	M	1.00	2.38	20.50		
(b) CO-210-1	S	2.43	1.00	10.64	14.42	4.96
	M	5.36	2.08	22.15		
	L	1.00	3.36	35.71		
(c) UCT-195	S	0.79	1.00	10.30	28.73	6.19
	M	4.07	2.44	25.14		
	L	1.00	3.40	35.02		
(d) CO-110	S	1.00	1.00	5.62	30.44	7.47
	M	7.36	1.86	10.46		
(e) UCT-1348	M	4.27	1.00	23.85	29.93	7.91
	L	1.00	1.51	36.03		

* S, M and L represent the groups of the small, middle and large particles. Range of S, M and L are -16.3+2.0, -30.6+16.3, and -45.0+30.6, respectively.

Table 4. Theoretical analysis of porosity and particle distribution of as-sprayed coatings

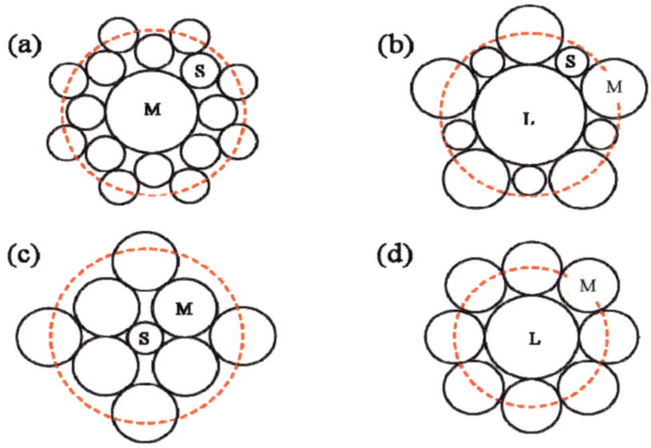

Fig. 7. Plain bi- and tri-model of powder deposition for: (a) AMDRY-9951; (b) CO-210-1 and UCT-195; (c) CO-110; (d) UCT-1348

3.3 Effect of exposure time on the porosity of coatings

Fig. 8 shows the distribution of the cross-sectional porosity in the coating about 100 μm depth from the free surface of coatings according to the increase of exposure time. The amount of porosity increased normally with increasing the particle size. This, of course, could be due to reduced melting efficiency of coarser particles in the plasma plume. A coating from fine powder shows well-adhered splats, while the interlamellar pores are more prominent in medium powder coating. The unmelted and poorly adhered particles can be seen in the coarse powder coating. Earlier studies (Kulkarni et al., 2003) have shown that the morphology of the splats changes from a contiguous disk-like shape to a fragmented shape with increasing particle size. These fragmented splats lead to poor splat-splat contact and to the formation of pores. But the distribution of particle size performs the important role of porosity than only the particle size. Some cases the unmelted particles press the splats and make the poor splat-splat contact close. The CO-210-1 coating shows the lowest porosity, this could be due to the good compounding the small, medium and larger particles. This powder has the widest range and sample variance as shown in Table 2, so these lead to make lower porosity inside coatings during the deposition process.

As shown in Fig. 8, with increasing exposure time, the amount of porosity decreased gradually in all coatings. But in view of details, the porosity decreased at Stage I_P (0-10 h for AMDRY-9951 coating; 0-1 h for CO-210-1 and UCT-195 coatings), and increased at Stage II_P (10-100 h for AMDRY-9951 coating; 1-10 h for CO-210-1 and UCT-195 coatings), and then finally decreased again over Stage II_P. Stage I_P represents the sintering effect as shown in Fig. 9. These sharp changes in porosity after short sintering times have also been observed (Thompson & Clyne, 2001). It might be expected that short exposure (Stage I_P) to high temperatures sufficient to heal cracks and lock splats together would lead to decrease the porosity inside of CoNiCrAlY coatings. Internal pores concentrated on the lower part of coating (over 100 μm from coating surface) over the Stage II_P as shown in Fig. 10. From the cross-sectional morphology of the interfaces between CoNiCrAlY coatings and substrates after 1000 h exposure, the intergranular voids present in the scale which may be formed, as described by Choi (Choi et al., 2002). In the absence of the effective sinks for vacancies, vacancies generated by outward diffusion of alloying elements such as Ni or Co condense to form cavities. One of the most effective methods to develop oxidation resistance in alloys and coatings at temperatures above about 1223 K is to form continuous scales of α-Al_2O_3 via selective oxidation. As the oxidation time increase, transient oxidation stage is represented, and then protective alumina oxide layer formation, Al-depletion, Ni or Co outward diffusion and solid state reaction with pre-existing alumina to form spinel phase, Cr_2O_3 formation at the oxide-to-metal interface, these processes highly depend on the oxidation temperature and microstructures of the MCrAlY coatings.

It reveals that the substrate interface possessed a lot of processing defects like pores (3.5-6% porosity) in 1000 h heat exposed specimens. Further, it was evident that the metallic MCrAlY overlay coating also possessed a lot of porosities as defects incorporated from plasma spraying. An Al-depleted zone at the overlay coat/substrate interface as well as just below the TGO along the TGO/MCrAlY interface was observed in the heat exposed specimens. Similar observations were also made by previous investigators in studying the oxidation behavior of sprayed MCrAlY coatings in air (Brandl et al., 1996; Ray & Steinbrech, 1999; Ray et al., 2001). They have observed Al depleted zone below the MCrAlY coating and also below the TGO scale after a long time oxidation of 500 h. The MCrAlY coat/substrate interface revealed the presence of porosity and carbide precipitations (Ray & Steinbrech, 1999; Ray et al., 2001) in heat exposed specimens. The metal matrix shows grain growth with carbide precipitation within the grains. Possibly secondary carbides precipitate out along the grain boundaries and primary carbides in the center of the grains as described in some references (Ray & Steinbrech, 1999; Ray et al., 2001). Due to the high oxygen conductivity of scales at high temperature and the presence of porosity in the TGO and in the MCrAlY coating, the Al in the MCrAlY coat oxidizes to Al_2O_3 at the TGO/MCrAlY coat interface. Therefore following this interface, there is an Al depleted zone in the MCrAlY coat. It appears that a fraction of Al in the MCrAlY coating also diffuses into substrate. Therefore an Al depleted zone also occurs between the interface of the MCrAlY/substrate and the

MCrAlY coat. Previous authors have also reported similar observations in such MCrAlY overlaid substrates (Brandl et al., 1996; Ray & Steinbrech, 1999; Ray et al., 2001).

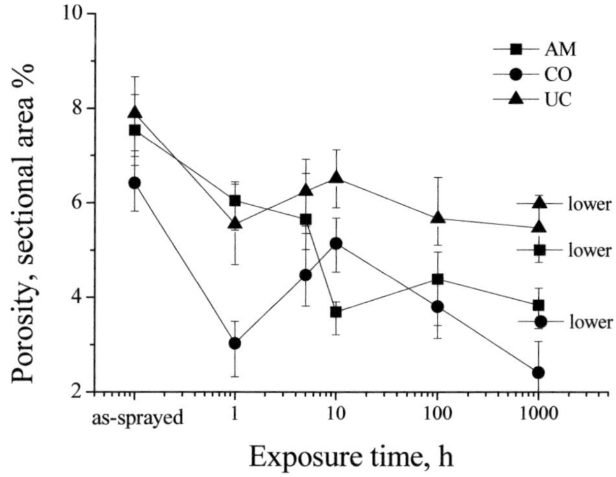

Fig. 8. Effect of exposure time on the porosity of CoNiCrAlY coatings with heat exposure

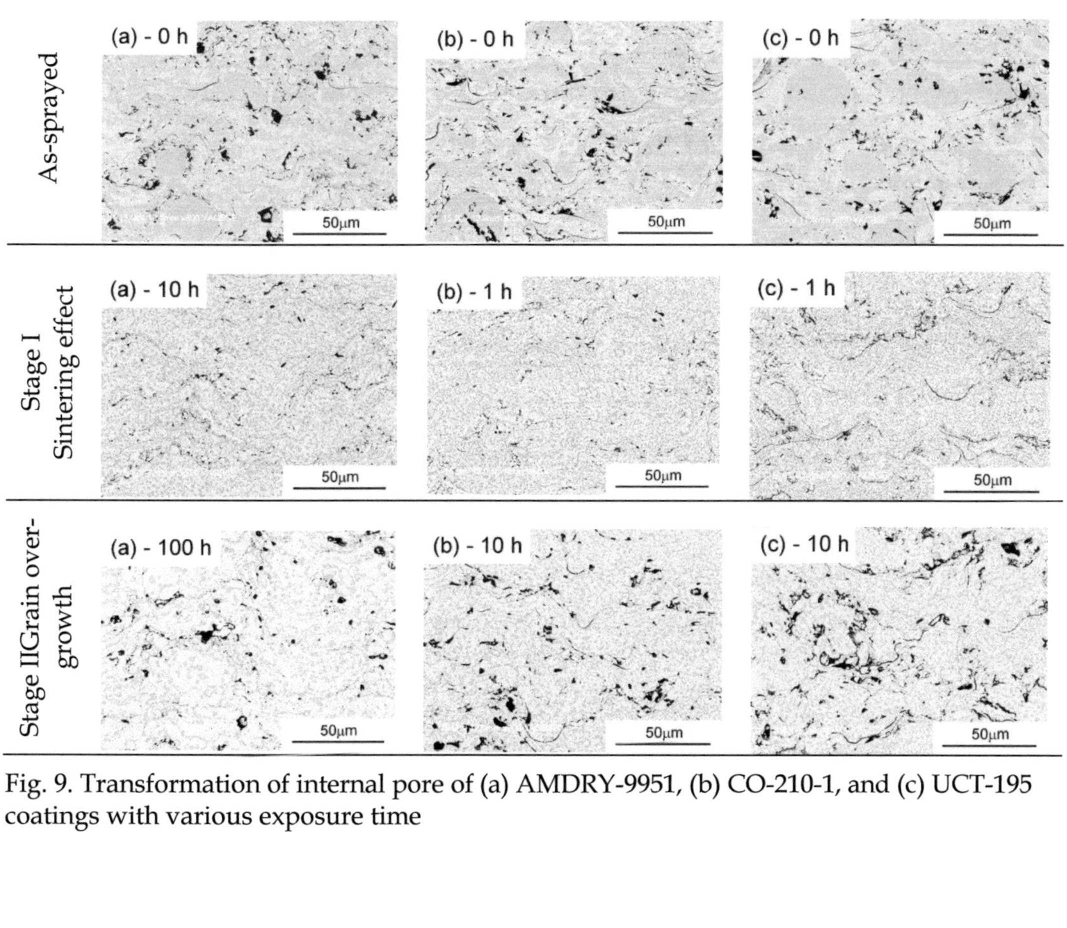

Fig. 9. Transformation of internal pore of (a) AMDRY-9951, (b) CO-210-1, and (c) UCT-195 coatings with various exposure time

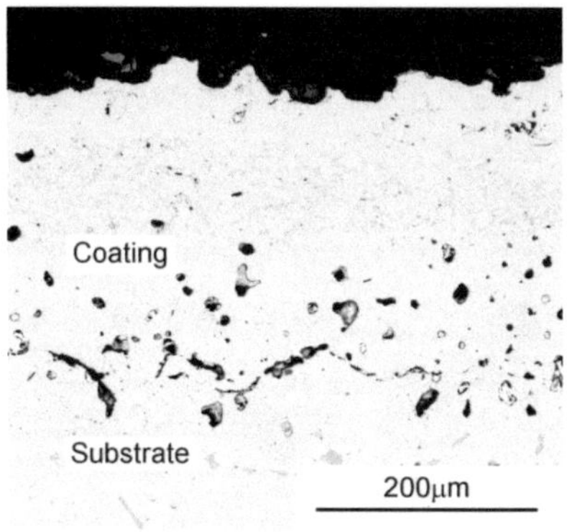

Fig. 10. Internal pore microstructure of the 1000 h-exposed CO-210-1 coating

3.4 Effect of exposure time on the surface roughness and TGO

Fig. 11 shows the surface roughness of as-sprayed and exposed coatings with increasing exposure time. The surface roughness increased with increasing the particle size all over. The surface roughness decreased up to Stage I_R (0-1 h for AMDRY-9951 coating; 0-5 h for CO-210-1 and UCT-195 coatings) with increasing exposure time, but over the time of Stage I_R, surface roughness was kept in relatively fixed value or increased slightly. This could be due to the partial sintering effect of inter-splat pores at the time of Stage I_R. And another possibility is due to the TGO, i.e., the growth of spinel-type Al_2O_3 on the surface as shown in Fig. 12(a). The open pores of surface could be filled with these oxides as shown in Fig. 13. Tang et al. (Tang et al., 2004) also showed that the oxide scale grown on the coating consisted primarily of spinel-type oxides after 1 h exposure at 1273 K. After exposure at 1273 K for longer time, beside the spinel-type oxides, α-alumina and Cr_2O_3 were also identified (will be discussed at section 3.5). The TGO scale was concentrated on the valley of surface from initial 1 h exposure to heat, as shown in Fig. 13 (point 1). At next stage, TGO scales also grew on the convex surface (hill-side), so the surface roughness increased slightly (refer to Fig. 11). And then the roughness decreased again or kept in relatively fixed value. The crystal size of grown scales is relatively small than the concavo-convex as-sprayed surfaces, so the heat-exposed surface becomes smooth and fine-grained gradually.

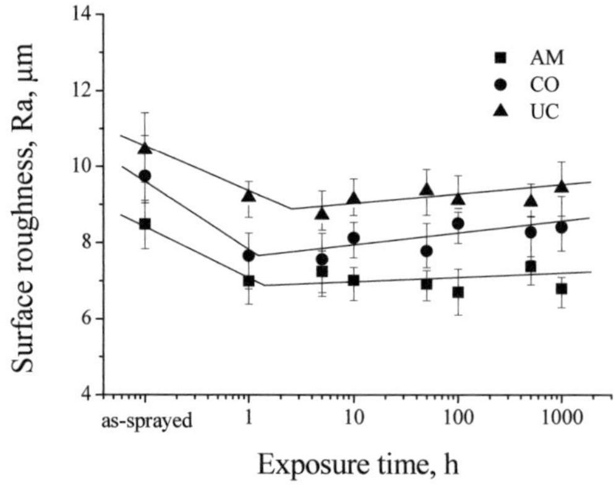

Fig. 11. Effect of exposure time on the surface roughness of CoNiCrAlY coatings

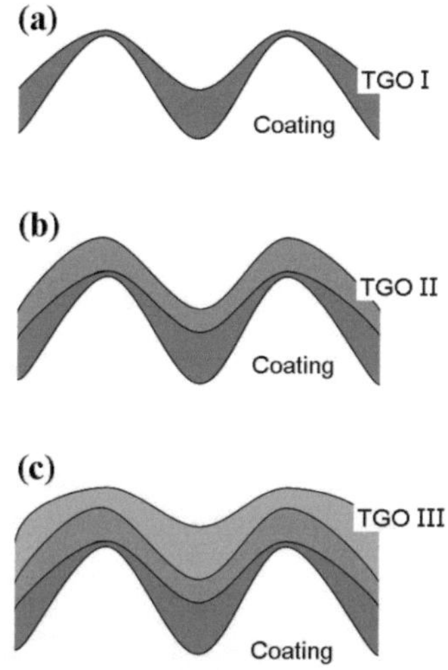

Fig. 12. Schematics of the surface roughness transition with TGO growth

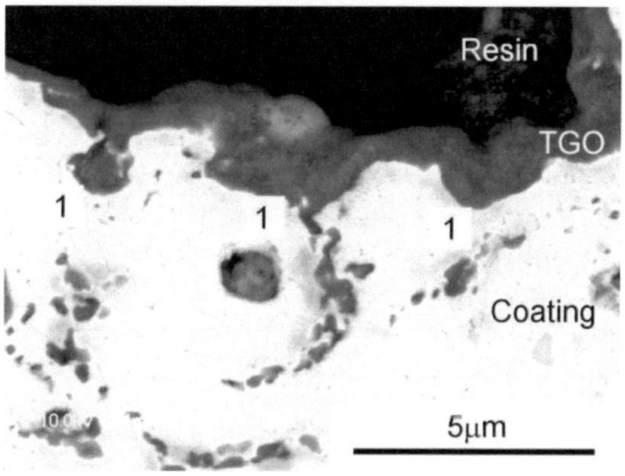

Fig. 13. Transition of surface roughness after heat exposure of 1h (AMDRY-9951 coating); TGO scales are filled into the concave surfaces (point 1) at first step, and then grow on the convex surfaces

The TGO thickness and weight gain plots for the three coatings and substrate in static air at 1273 K are shown in Fig. 14(a) and (b) respectively. The square value of TGO thickness was proportional to the square root of exposure time, and the square value of weight gain was proportional to the heat exposure time. The oxidation can be divided into two stages (Choi et al., 2002). One is the transient oxidation stage which shows initial relatively rapid oxidation rate and the other is the steady-state oxidation stage which shows a relatively slow oxidation rate. The transition time from the initial transient oxidation to the steady-state oxidation can be defined as the time of intersection of the extrapolation lines of the initial rapid oxidation and that of the later slow oxidation. The transition times for all the coatings which show this transition were below 23 h. UCT-195 coating showed a faster oxidation rate than other coatings at both plots, which might be due to the bigger particle size and higher surface roughness shown in Fig. 11. AMDRY-9951 coating shows the lowest rate of TGO thickness growth and weight gain, but UCT-195 coating shows the highest.

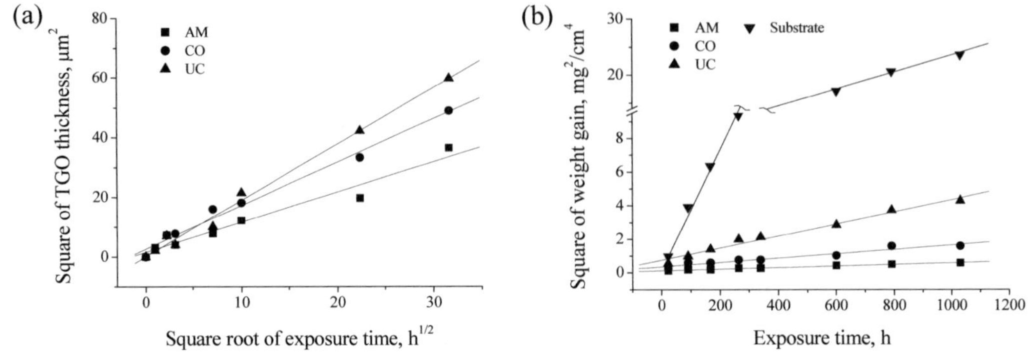

Fig. 14. Effect of exposure time on the (a) oxide scale thickness and (b) weight gain of heat exposed coatings

All the coatings exhibited a tendency to have uniform TGO thickness and weight gain after an initially higher rate of oxidation, i.e. they followed the parabolic rate law of oxidation. The parabolic oxidation rate constant is generally defined by the equation $(\Delta W/A)^2 = k_p t$ where ΔW is the weight change at any time, A is the surface area, t is time and k_p is the constant for the parabolic rate law (Tang et al., 2004; Briks & Meier, 1983). The parabolic rate constants are 4.65×10^{-4}, 1.3×10^{-3}, and 3.6×10^{-3} $mg^2cm^{-4}h^{-1}$ for AMDRY-9951, CO-210-1, and UCT-195 coatings, respectively. Thus, the rate constant of AMDRY-9951 coating is a factor of eight smaller than that of UCT-195 coating. The weight gain of AMDRY-9951 coating (0.5 mg/cm^2) also showed over sextuple lower value than that of uncoated substrate (3.4 mg/cm^2) at 340 h. But substrate Inconel 718 does not follow an exact parabolic rate law, this may be because inhomogeneous oxides locally formed and grew rapidly below 340 h (Tang et al., 2004). The transient oxidation stage which shows a rapid oxidation rate can not be explained by Wagner's model (Briks & Meier, 1983). From the gradients of the plots, approximate values for the parabolic rate constant, k_p, were calculated for the transient and steady-state oxidation range of Fig. 14b. These were 3.32×10^{-2} and 1.74×10^{-2} $mg^2cm^{-4}h^{-1}$ for the transient and steady-state range, respectively. Decrease of gradient occurs after oxidation time of 340 h. These calculated k_p values cannot be compared directly with the data of coatings owing to the differences in compositions. However, they are higher than the corresponding value for the coatings. In particular, in the case of steady-state oxidation range, the rate constant of UCT-195 coating is a factor of five smaller than that of substrate.

The rates of TGO thickness growth and weight gain increase with increasing the particle size of starting powder. The oxide scales of AMDRY-9951 coating exhibited excellent adhesion to the overlay coating relatively, although minor cracks were observed during the isothermal oxidation as shown in Fig. 15a. CoNiCrAlY coatings had two important and inherent discontinuities that could act as non-uniform oxide formation sites (Tang et al., 2004). One is the oxide stringers that result from an oxidation reaction between scale-forming elements (Al or Y) in a molten droplet and oxygen entrapped from the surrounding atmosphere. This phase implies pre-consumption of Al, a very important scale-forming element while there is poor contact at the matrix and oxide stringer interface. The other is the improperly flattened zones that are due to the deposition of partially melted particles resulting from insufficient heat transport during flight and splashing of spreading particles impacting on rough surface. These zones have high densities of open pores that are easy penetration paths for oxygen, leading to internal oxidation. These two features have influenced the oxidation behavior in both the transient and steady-state oxidation stage. Owing to above reasons, AMDRY-9951 coating, which has the lowest surface roughness as shown in Fig. 11, presented the smaller TGO thickness growth rate and parabolic oxidation rate constant than other coatings.

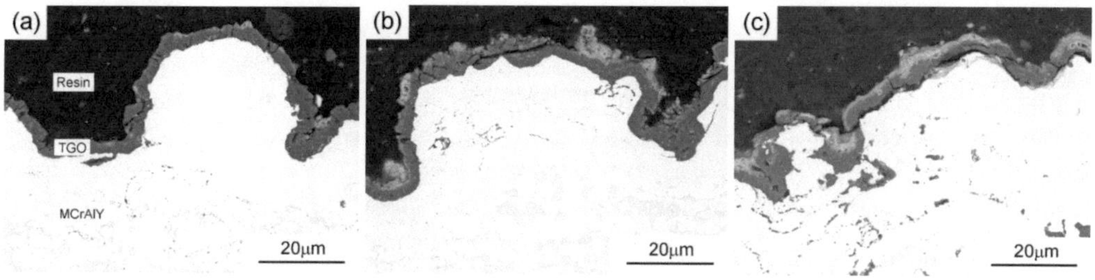

Fig. 15. Formation of the TGO on the surface of the CoNiCrAlY coatings after 100 h exposure: (a) AMDRY-9951; (b) CO-210-1; (c) UCT-195

3.5 Effect of chemical composition on the oxidation of coatings

Fig. 16 shows the results of EDX analysis of the CoNiCrAlY coatings according to the heat exposed time. There were metallic phases composed mainly of Cr, Co and Ni for as-sprayed coatings. The absence of peaks corresponding to the oxide phases in the as-sprayed coatings is due to the relatively small thickness of the oxide scale, in spite of the presence of the oxide layer on the surface of the as-sprayed coatings. In the case of the exposed coatings at 1273 K, all three coatings showed similar oxidation behaviors especially from the initial oxidation stages. The major metallic elements, i.e., Cr, Co and Ni decreased, but Al and O increased gradually according to increasing the exposure time. This phenomenon represents the growth of Al_2O_3 oxides. But over 100 h exposure time for the CO-210-1 and UCT-195 coating, Cr, Co and Ni elements have become much richer than Al relatively. These Cr-Co-Ni-O systems represent the development of typical mixed oxides at the latter thermal exposure stage. In the case of AMDRY-9951 coating, this phenomenon starts from 1000 h, and the peaks of Cr, Co and Ni were relatively lower than other coatings. It may be inferred from the above that AMDRY-9951 coating which has a relatively narrow range of particles has less mixed oxides left on the exposed surfaces than other coatings.

In order to understanding the development of corrosion protective coatings, it is necessary to appreciate the processes by which oxidation and corrosion occur and how the mechanisms by which they occur depends on environmental and temperature. Three accelerated degradative processes occur depending on temperature and these can be defined in order of increasing temperature as: Type II hot corrosion, Type I hot corrosion and oxidation (see Fig. 17) (Briks & Meier, 1983). Type II hot corrosion occurs at temperatures in the range 600-800 °C and involves the formation of base metal (nickel or cobalt) sulphates which require a certain partial pressure of sulphur trioxide for their stabilization. These sulphates react with alkali metal sulphates to form low melting point compounds which prevent a protective oxide forming. Indeed, the oxide formed as a striated one has shown that Ni-Cr-Al materials resist this form of corrosion most effectively (Viswanathan, 2001; Nicholls et al., 2002). Type I hot corrosion involves the transport of sulphur from a sulphatic deposit (generally Na_2So_4) across a preformed oxide into the metallic material with the formation of the most stable sulphides. Once stable sulphides formers (e.g. Cr) are fully reacted with the sulphur moving across the scale, then base metal sulphides can form with catastrophic consequences as they are molten at the temperatures at which Type I hot corrosion is observed (800-950 °C) (Viswanathan, 2001; Eliaz et al., 2002). Thus, the formation of NiS_2 (molten at 645 °C) and Co_xS_y (lowest liquidus ~840 °C) can cause degradation levels which are serious enough to cause major component degradation. The most suitable materials which can resist Type I hot corrosion are $PtAl_2$-(Ni-Pt-Al) coatings and M-CrAlY coatings containing up to 25 wt.% Cr and 6 wt.% Al (Goward, 1998). The oxidation resistance under such conditions can be markedly improved by the addition of so-called reactive elements (Y, Hf, Ce) to alloys and coatings. An exhaustive search for the reasons behind this effect has finally shown that yttrium and rare earth metals segregate to grain boundaries within alumina scales causing a reduction in Al and O transport rates through the oxide and thus reduced oxidation rates (Pint, 1996). Furthermore, reactive elements combine with sulphur and phosphorus impurities in metallic materials and coatings with the result that these impurities cannot selectively diffuse to the surface and contaminate the oxide-metallic interface. This gives extremely good scale adherence and categorically explains why minor yttrium or cerium (<0.8 %) additions to oxidation resistant coatings and alloys greatly improve oxidation resistance.

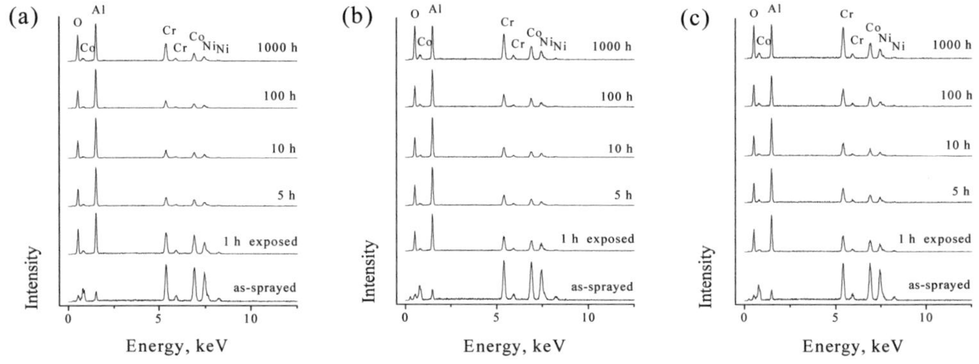

Fig. 16. EDX spectrum analysis results of (a) AMDRY-9951, (b) CO-210-1 and (c) UCT-195 coatings according to exposure time

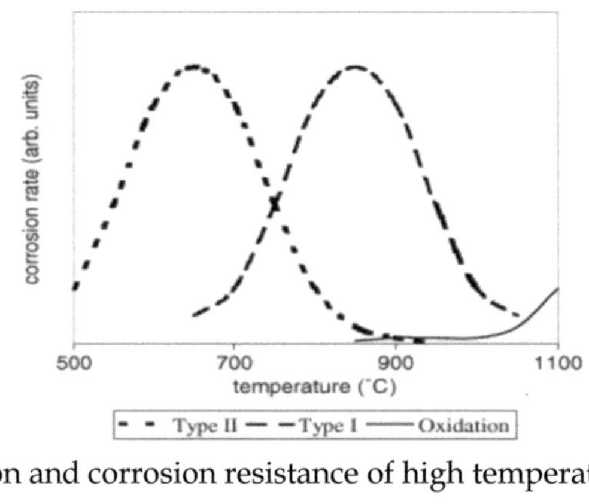

Fig. 17. Relative oxidation and corrosion resistance of high temperature coating systems.

Hot corrosion problems (type I and type II hot corrosion, vanadic corrosion) are a direct result of salt contaminants such as Na_2SO_4, NaCl and V_2O_5 which, in combination, produce low melting point deposits which dissolve the protective surface oxides. A number of fluxing mechanisms has been proposed to account for the different corrosion morphologies that are observed (Briks & Meier, 1983 ; Khanna, 2004) and this has resulted in the general classification of high temperature (type I) hot corrosion, low temperature (type II) hot corrosion and vanadic corrosion (535-950°C). These corrosion processes can be separated into an initiation and propagation stage. During the initiation stage the corrosion rate is comparatively low as breakdown of the surface oxide occurs. However, once this has happened and repair of the oxide is no longer possible, then the propagation phase results in the rapid consumption of the alloy. Since the coating provides for the repair of the protective surface oxide scales, the initiation stage can be extended, ideally for the design life of the component. However, once coating penetration occurs, the propagation stage often results in catastrophic corrosion rates.

As a background to the development of protection coating technology, the corrosion performance of a wide range of diffusion and overlay coatings, under high temperature oxidation and type I and type II hot corrosion conditions have been extensively reviewed (Nicholls et al., 1990; Nicholls, 2000). The platinum-modified aluminides performed exceptionally well under high temperature oxidation conditions and in type I hot corrosion environments but performed less well under type II hot corrosion conditions, although out-performing conventional aluminides (Nicholls et al., 1990). Of the other diffusion coatings, the silicon-containing diffusion aluminides (for example Sermetel 1515) perform well under type II hot corrosion conditions. Chromised and chrome-aluminised coatings also offer protection under type II corrosion conditions. Thus, silicon containing or chromium-rich diffusion coatings offer improved corrosion resistance at the lower temperatures that are often encountered within utility turbine environments (Nicholls et al., 1990; Luthra & LeBlanc, 1987). Overlay coatings of classic design, with 18-22% Cr and 8-12% Al, generally perform better at higher temperatures where oxidation is the dominant failure mode (above 900°C) reflecting the good adherence of the thin alumina scales, which is promoted by the presence of active elements such as yttrium. Generally under this high temperature oxidizing conditions NiCrAlYs and NiCoCrAlYs out-perform the cobalt-based systems. However, at low temperatures where type II hot corrosion predominates, corrosion rates for the NiCrAlY and NiCoCrAlY overlay coatings can be relatively high. CoCrAlYs generally outperform NiCrAlY-based systems, with the high chromium containing CoCrAlYs showing best performance (Nicholls et al., 1990; Nicholls, 2000; Novak, 1994). This is illustrated schematically in Fig. 18 reproduced from a paper by Novak (Novak, 1994). Several methods have been investigated to improve the traditional MCrAlY coatings by use of a platinum underlayer and overlayers (Nicholls et al., 2002). Other additions such as Ti, Zr, Hf, Si and Ta have been examined (Saunders & Nicholls, 1984). Surface modification results in the formation of a duplex coating structure and this can result in improved performance, for example, a pulse-aluminized CoNiCrAlY coating exhibits superior corrosion resistance at 750 and 850 °C compared to its plasma sprayed counterpart (Nicholls et al., 1990; Nicholls, 2000). Some overlay coating is formed by a gas phase diffusion treatment of a plasma sprayed MCrAlY coating. Silicon modifications to the surface of CoCrAlY coatings have also been proposed and improve the resistance to low temperature hot corrosion (Nicoll, 1984).

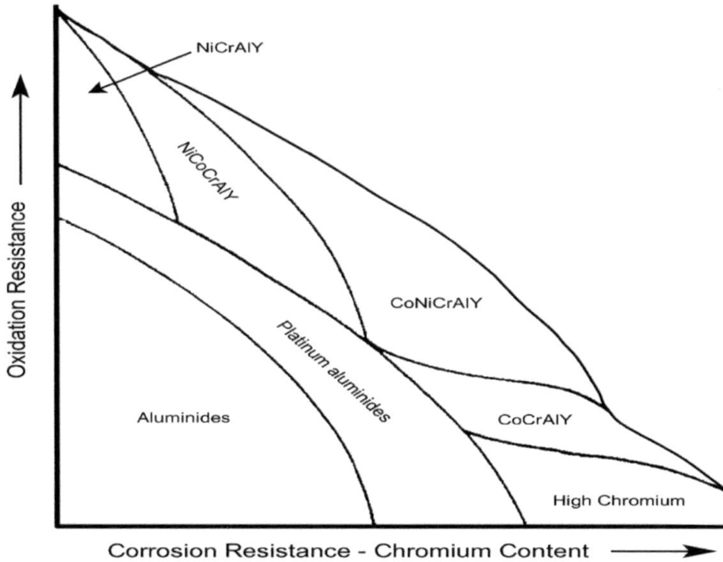

Fig. 18. Schematic representation of rate-temperature curves for Type II hot corrosion, Type I hot corrosion and oxidation (alumina former).

Two categories of coatings arise: diffusion coatings and overlay coatings. Overlay coating involves the application of coatings to substrates using physical deposition processes. Typical application methods include: thermal spraying, physical vapor deposition (PVD), electron beam physical vapor deposition (EBPVD), ion plating/sputtering and electroplating (Stringer, 1987). Overlay coatings typically comprise β+γ' aluminide in a γ matrix (Sims et al., 1987) and are of typical composition (Ni, Co)-15-28 wt.% Cr, 4-18 wt.% Al, 0.5-0.8 wt.% Y (Goward, 1998). The relative amounts of Ni and Co depend upon: (a) coating ductility requirements (>Ni) and (b) corrosion resistance (>Co). The excellent oxidation and corrosion resistance of the coatings is afforded by the formation of highly adherent alumina scales which grow slowly as a result of the reactive element effect attributable to yttrium. In addition, their high Cr contents make them useful protective coatings against Type II hot corrosion. These overlay coatings are deposited using thermal spray (Ar shrouded plasma (APS) or low pressure plasma (LPPS)) techniques or EBPVD). The EBPVD technique is typically preferred for high quality coatings since a certain amount of oxidation of the coating particles typically occurs during thermal spraying giving rise to nanoscale oxide particles at splat boundaries. However, the thermal spray processes are frequently used in practice.

Overlay coatings show interdiffusion effects with alloy substrates and Itoh and Tamura indicate the rate of interdiffusion decreases in the order NiCrAlY > CoCrAlY > NiCoCrAlY > CoNiCrAlY, where NiCo represents higher Ni contents compared to Co and CoNi higher Co contents (Itoh & Tamura, 1999). These results, collected using vacuum plasma sprayed coatings were independent of the three substrates used in the experimental program. The most recent technological advance in overlay coating technology is the development of so-called smart coatings (Nicholls et al., 2002). These coatings attempt to address the problems associated with the differences in temperature over the surface of an airfoil. Temperatures vary from a maximum of the order of 1100°C at leading and trailing edges to about 650 °C in the center of the airfoil surfaces and near airfoil roots. Because of this, the nature of environmental degradation varies from oxidation through Type I hot corrosion to Type II hot corrosion as indicated in Fig. 17. A practical example of this variation in corrosion type with temperature is graphically portrayed by Viswanathan in his analysis of the failure of airfoils (Viswanathan, 2001). Following extensive laboratory studies employing alkali-metal sulphates in simulated gaseous environments, Nicholls et al. have been able to very carefully define optimized chromium and aluminium contents for oxidation and corrosion protection in the temperature ranges referred to above (Nicholls et al., 2002). As might be expected, high Cr contents (>40 wt.%) and low Al contents (6-8 wt.%) are most suitable for protection from Type II hot corrosion. For protection from Type I hot corrosion, roughly equal amounts of Cr and Al represent optimized compositions. Oxidation protection (1100°C), is best afforded by coatings containing 25 wt.% Cr and 14 wt.% Al. The final coating developed by Nicholls et al. comprises a commercial base coating (Co-32Ni-21Cr-8Al-0.5Y) adjacent to the substrate, a Cr enriched layer of variable composition from Ni-60Cr-20Al to Ni-35Cr-40Al and a surface layer of composition Ni-15Cr-32Al (Nicholls et al., 2002). These multi-layer coatings have been shown to outperform typical a commercial Pt modified aluminide and an Al enriched version of the base coat at 700-800 °C. Whilst this technology appears a step forward, issues related to coating ductility and additional improvements afforded by cobalt with respect to corrosion resistance, may be of additional importance.

4. Conclusions

In this work, the thermal degradation behavior of selected CoNiCrAlY coatings has been studied. Isothermal oxidation behaviors of the VPS MCrAlY metallic overlay coatings were observed for comparative study. Coatings were oxidized isothermally at a 1273 K for different time periods in order to form the thermally grown oxide layer. Particular focus is set on the influence of particle size, chemical composition, and exposure time on the oxygen content in coatings, porosity, surface roughness, and TGO growth.

The particle size has a significant effect on the oxidization of the in-flight particle and consequently the oxygen content in the as-sprayed coatings. But after thermal exposure, the difference of the oxygen contents between the smaller and larger particle coatings was decreased with increasing exposure time. The distribution of particle size plays the important role of porosity than only the particle size. It can be found that using the tri-modal arrangement in the mixture with different diameters of the large, middle and small spheres can obtain the highest deposition density. In some cases, the unmelted particles press the splats and make the poorly bonded splat-splat contact closed.

The isothermal degradation of coatings was considerably influenced by the heat exposure time. The porosity decreased gradually in all coatings with increasing exposure time, but in view of details, the porosity decreased at Stage I_P (0-10 or 0-1 h), and increased at Stage II_P (10-100 or 1-10 h), and then finally decreased again over Stage II_P. These could be due to the sintering effect at Stage I_P, which leads to heal cracks and lock splats together. The surface roughness decreased up to Stage I_R with increasing exposure time, but over Stage I_R, surface roughness was kept in relatively fixed value or increased slightly. The TGO thickness growth rate and parabolic oxidation rate constant are influenced by the surface roughness of coatings with increasing heat exposure time.

5. Acknowledgment

The authors would like to express thanks to Mr. S. Murata at Murata Boring Technology Research Co., Japan for the financial support of this work. Authors also thank Mr. M. Tanno for assisting in coating characterization.

6. References

Birks, N. & Meier, G.H. (1983). *Introduction to High Temperature Oxidation of Metals*, Edward Arnold, ISBN 978-0713134643, London, UK

Brandl, W., Grabke, H.J., Toma, D., & Kruger, J. (1996). The Oxidation Behaviour of Sprayed MCrAlY Coatings. *Surf. Coat. Technol.*, Vols.86-87, No.1, (December 1996), pp. 41-47, ISSN 0257-8972

Choi, H., Yoon, B., Kim, H., & Lee, C. (2002). Isothermal Oxidation of Air Plasma Spray NiCrAlY Bond Coatings. *Surf. Coat. Technol.*, Vol.150, Nos.2-3, (February 2002), pp. 297-308, ISSN 0257-8972

Dobler, K., Kreye, H., & Schwetzke, R. (2000). Oxidation of Stainless Steel in the High Velocity Oxy-Fuel Process. *J. Therm. Spray Technol.*, Vol.9, No.3 (September 2000), pp. 407-413, ISSN 1059-9630

Eliaz, N., Shemesh, G., Latanision, R.M. (2002). Hot Corrosion in Gas Turbine Components. *Eng. Fail. Anal.*, Vol.9, No.1, (February 2002), pp. 31-43, ISSN 1350-6307

Espie, G., Fauchais, P., Labbe, J.C., Vardelle, A., & Hannoyer, B. (2001). Oxidation of Iron Particles during APS: Effect of the Process on Formed Oxide Wetting of Droplets on Ceramics Substrates. *Proceedings of ITSC 2001: New Surfaces for a New Millennium*, pp. 821-827, ISBN 978-0871707376, Singapore, May 28-30, 2001

Fukushima, T. & Kuroda, S. (2001). Oxidation of HVOF Sprayed Alloy Coatings and Its Control by a Gas Shroud. *Proceedings of ITSC 2001: New Surfaces for a New Millennium*, pp. 527-532, ISBN 978-0871707376, Singapore, May 28-30, 2001

Goward, G.W. (1998). Progress in Coatings for Gas Turbine Airfoils. *Surf. Coat. Technol.*, Vols.108-109, (October 1998), pp. 73-79, ISSN 0257-8972

Itoh, Y., Tamura, M. (1999). Reaction Diffusion Behaviors for Interface Between Ni-Based Super Alloys and Vacuum Plasma Sprayed MCrAlY Coatings. *J. Eng. Gas. Turb. Power*, Vol.121, No.3, (July 1999), pp. 476-483, ISSN 0742-4795

Khanna, A.S. (2004). *Introduction to High Temperature Oxidation and Corrosion*, ASM International, ISBN 0-87170-762-4, Materials Park, OH, USA

Kulkarni, A., Vaidya, A., Goland, A., Sampath, S., & Herman, H. (2003). Processing Effects on Porosity-property Correlations in Plasma Sprayed Yttria-stabilized Zirconia Coatings. *Mater. Sci. Eng. A*, Vol.359, Nos.1-2, (October 2003), pp. 100-111, ISSN 0921-5093

Li, C.J. & Li, W.Y. (2003). Effect of Sprayed Powder Particle Size on the Oxidation Behavior of MCrAlY Materials during High Velocity Oxygen-fuel Deposition. *Surf. Coat. Technol.*, Vol.162, No.1, (January 2003), pp. 31-41, ISSN 0257-8972

Luthra, K.L. & LeBlanc, O.H. (1987). Low temperature hot corrosion of CoCrAl alloys. *Mater. Sci. Eng.*, Vol.87, No.C, (March 1987), pp. 329-335, ISSN 0025-5416

Neiser, R.A., Smith, M.F., & Dykhuizen, R.C. (1998). Oxidation in Wire HVOF-Sprayed Steel. *J. Therm. Spray Technol.*, Vol.7, No.4, (December 1998), pp. 537-545, ISSN 1059-9630

Nicholls, J.R. & Saunders, S.R.J. (1990). *High Temperature Materials for Power Engineering*, Kluwer, ISBN 978-0792309277, Dordrecht, Netherlands

Nicholls, J.R. (2000). Designing Oxidation-Resistant Coatings. *JOM*, Vol.52, No.1, (January 2000), pp. 28-35, ISSN 1543- 1851

Nicholls, J.R., Simms, N.J., Chan, W.Y., Evans, H.E. (2002). Smart overlay coatings - concept and practice. *Surf. Coat. Technol.*, Vol.149, Nos.2-3, (January 2002), pp. 236-244, ISSN 0257-8972

Nicoll, A.R. (1984). *Coatings and Surface Treatment for Corrosion and Wear Resistance*, Ellis Horwood, ISBN 978-0131400962, Chichester, UK

Novak, R.C. (1994). *Coatings Development and Use: Case Studies, Presentation to the Committee on Coatings for High-Temperature Structural Materials*, National Materials Advisory Board, National Research Council, Irvine, California, US

Pint, B.A. (1996). Experimental Observations in Support of the Dynamic-segregation Theory to Explain the Reactive-element Effect. *Oxid. Met.*, Vol.45, Nos.1-2, (February 1996), pp. 1-37, ISSN 1573-4889

Pomeroy, M.J. (2005). Coatings for Gas Turbine Materials and Long Term Stability Issues. *Mater. Des.*, Vol.26, No.3, (May 2005), pp. 223-231, ISSN 0264-1275

Ray, A.K., Roy, N., & Godiwalla, K.M. (2001). Crack Propagation Studies and Bond Coat Properties in Thermal Barrier Coatings under Bending. *Bull. Mater. Sci.* , Vol.24, No.2, (April 2001), pp. 203-209, ISSN 0250-4707

Ray, A.K. & Steinbrech, R.W. (1999). Crack Propagation Studies of Thermal Barrier Coatings under Bending. *J. Eur. Ceram. Soc.*, Vol.19, No.12, (October 1999), pp. 2097-2109, ISSN 0955-2219

Saunders, S.R.J. & Nicholls, J.R. (1984). Hot Salt Corrosion Test Procedures and Coating Evaluation. *Thin Solid Films*, Vol.119, No.3, (September 1984), pp. 247-269, ISSN 0040-6090

Sims, C.T., Stoloff, N.S., Hagel, W.C. (1987). *Superalloys II*, Wiley, ISBN 978-0471011477, New York, USA

Stringer, J. (1987). Role of Coatings in Energy-producing Systems: an Overview. *Mater. Sci. Eng.*, Vol.87, (March 1987), pp. 1-10, ISSN 0025-5416

Tang, F., Ajdelsztajn, L., Kim, G.E., Provenzano, V., & Schoenung, J.M. (2004). Effects of Surface Oxidation during HVOF Processing on the Primary Stage Oxidation of a CoNiCrAlY Coating. *Surf. Coat. Technol.*, Vol.185, No.2-3, (July 2004), pp. 228-233, ISSN 0257-8972

Thompson, J.A. & Clyne, T.W. (2001). The Effect of Heat Treatment on the Stiffness of Zirconia Top Coats in Plasma-Sprayed TBCs. *Acta Mater.*, Vol.49, No.9, (May 2001), pp. 1565-1575, ISSN 1359-6454

Viswanathan, R. (2001). An Investigation of Blade Failure in Combustion Turbines. *Eng. Fail. Anal.*, Vol.8, No.5, (October 2001), pp. 493-511, ISSN 1350-6307

Volenik, K., Novak, V., Dubsky, J., Chraska, P., & Neufuss, K. (1997) Properties of Alloy Steel Coatings Oxidized during Plasma Spraying, *Mater. Sci. Eng. A*, Vols.234-236, No.30, (August 1997), pp. 493-496, ISSN 0921-5093

4

Analysis of Experimental Results of Plasma Spray Coatings Using Statistical Techniques

S.C. Mishra
Department of Metallurgical and Materials Engineering, National Institute of Technology, Rourkela, India

1. Introduction

Surface modification is a generic term applied to a large field of diverse technologies that can be gainfully harnessed to achieve increased reliability and enhanced performance of industrial components. The incessant quest for higher efficiency and productivity across the entire spectrum of manufacturing and engineering industries has ensured that most modern-day components are subjected to increasingly harsh environments during routine operation. Critical industrial components are, therefore, prone to more rapid degradation as the parts fail to withstand the rigors of aggressive operating conditions and this has been taking a heavy toll of industry's economy. In an overwhelmingly large number of cases, the accelerated deterioration of parts and their eventual failure has been traced to material damage brought about by hostile environments and also by high relative motion between mating surfaces, corrosive media, extreme temperatures and cyclic stresses. Simultaneously, research efforts focused on the development of new materials for fabrication are beginning to yield diminishing returns and it appears unlikely that any significant advances in terms of component performance and durability can be made only through development of new alloys.

As a result of the above, the concept of incorporating engineered surfaces capable of combating the accompanying degradation phenomena like wear, corrosion and fatigue to improve component performance, reliability and durability has gained increasing acceptance in recent years. The recognition that a vast majority of engineering components fail catastrophically in service through surface related phenomena has further fuelled this approach and led to the development of the broad interdisciplinary area of surface modifications. A protective coating deposited to act as a barrier between the surfaces of the component and the aggressive environment that it is exposed to during operation is now globally acknowledged to be an attractive means to significantly reduce/suppress damage to the actual component by acting as the first line of defense. Coating is a layer of material formed naturally or synthetically or deposited artificially on the surface of an object made of another material with the aim of obtaining required technical or decorative properties.

The increasing utility and industrial adoption of surface engineering is a consequence of the significant recent advances in the field. Very rapid strides have been made on all fronts of science, processing, control, modeling, application developments etc. and this has made it an invaluable tool that is now being increasingly considered to be an integral part of component design. Surface modification today is best defined as "the design of substrate and surface together as a system to give a cost effective performance enhancement, of which neither is capable on its own". The development of a suitable high performance coating on a component fabricated using an appropriate high mechanical strength metal/alloy offers a promising method of meeting both the bulk and surface property requirements of virtually all imagined applications. The newer surfacing techniques, along with the traditional ones, are eminently suited to modify a wide range of engineering properties. The properties that can be modified by adopting the surface engineering approach include tribological, mechanical, thermo-mechanical, electrochemical, optical, electrical, electronic, magnetic/acoustic and biocompatible properties.

The development of surface engineering has been dynamic largely on account of the fact that it is a discipline of science and technology that is being increasingly relied upon to meet all the key modern day technological requirements: material savings, enhanced efficiencies, environmental friendliness etc. The overall utility of the surface engineering approach is further augmented by the fact that modifications to the component surface can be metallurgical, mechanical, chemical or physical. At the same time, the engineered surface can span at least five orders of magnitude in thickness and three orders of magnitude in hardness.

Driven by technological need and fuelled by exciting possibilities, novel methods for applying coatings, improvements in existing methods and new applications have proliferated in recent years. Surface modification technologies have grown rapidly, both in terms of finding better solutions and in the number of technology variants available, to offer a wide range of quality and cost. The significant increase in the availability of coating process of wide ranging complexity that are capable of depositing a plethora of coatings and handling components of diverse geometry today, ensures that components of all imaginable shape and size can be coated economically.

Although there are different techniques available for the deposition of materials on suitable substrates, thermal spraying process is being widely used for depositing thick coatings for various industrial applications. The type of thermal spraying depends on the type of heat source employed and consequently flame spraying (FS), high velocity oxy-fuel spraying (HVOF), plasma spraying (PS) etc. come under the umbrella of thermal spraying. Plasma spraying utilizes the exotic properties of the plasma medium to impart new functional properties to conventional and non-conventional materials and is considered as one highly versatile and technologically sophisticated thermal spraying technique instead of having relatively high price of the sprayable consumables.

Plasma spraying, one of the thermal spraying processes, is increasingly popular owing to its versatility in spraying a large number of materials and is being researched well. It is a very large industry with applications in corrosion, abrasion and temperature resistant coatings and the production of monolithic and near net shapes [**1**]. The process can be applied to coat on variety of substrates of complicated shape and size using metallic, ceramic and /or polymeric consumables. The production rate of the process is very high and the coating adhesion is also adequate. Since the process is almost material independent, it has a very wide range of applicability, e.g., as thermal barrier coating, wear resistant coating etc. Thermal barrier coatings are provided to protect the base material, e.g., internal combustion engines, gas turbines etc. at elevated temperatures. Zirconia (ZrO_2) is a conventional thermal barrier coating material used as the top coat, over a bond coat. As the name suggests, wear resistant coatings are used to combat wear especially in cylinder liners, pistons, valves, spindles, textile mill rollers etc. alumina (Al_2O_3), titania (TiO_2) and zirconia (ZrO_2) are the some of the conventional wear resistant coating materials [**2**].

Plasma spraying is a surface modification technique that combines particle melting, rapid solidification and consolidation in a single process. Because of their higher strength-to-weight ratio and superior wear-resistant properties, ceramics are preferred in most tribological applications. The ceramic materials can be applied for the overlay coating due to the higher gas enthalpy of the thermal plasma jet. The suitability of a ceramic coating on metal substrates depends on (i) the adherence strength at coating-substrate interface, and (ii) stability at operating conditions.

Critical components in high-tech industries operate under extremely hostile conditions of temperature, gas flow, heat flux and corrosive media, which severely limit their service life. This problem can be minimized by using composite structures consisting of the core material to with stand the load and with a suitable surface coating to improve the component life span at operating environment. Plasma spray technology, the process of preparing overlay coating on any surface, is one of the most widely used techniques to prepare such complex structural parts with improved properties and increased life span [**3**].

Alumina–titania coating, which is one of the material largely manufactured, used the atmospheric plasma sprayed (APS) process. This material is known for its wear, corrosion and erosion resistance applications. These types of coatings can be prepared by blending the matrix powder with reinforcement and by plasma spraying [**4, 5**]. The coating process is based on the creation of a plasma jet to melt a feedstock powder [**3**]. Powder particles are injected with the aid of a carrier gas; they gain their velocity and temperature by thermal and momentum transfers from the plasma jet. At the surface of the substrate, particles flatten and solidify rapidly forming a stack of lamellae.

The use of the composite in preference to pure aluminum oxide has certain advantages. TiO_2 is a commonly used additive in plasma sprayable alumina powder. TiO_2 has a relatively low melting point and it effectively binds the alumina grains leading to higher density and wear resistance coating [**6**]. However, a success of an Al_2O_3 - TiO_2 coating depends upon a judicious selection of the arc current, which can melt the powders effectively. This results in a good coating adhesion along with high wear resistance [**7**]. Al_2O_3 with low wt. %. of TiO_2 coatings provide high electric resistance and are suitable where good insulating properties and high electric strength are required [**8**]. But the coatings of mixtures with high wt. %. TiO_2 possess good electrical conductivity due to its manufacturing process of powder and preparation of coatings [**9**].

A qualitative analysis of the experimental results with regard to erosion wear rate using statistical techniques is presented. The analysis is aimed at identifying the operating variables/factors significantly influencing the erosion wear rate of alumina titania on metals. Factors are identified in accordance to their influence on the coating erosion wear rate. A prediction model based on artificial neural network is also presented considering the significant factors. Neural computation is used since plasma spraying is a complex process that has many variables and multilateral interactions. This technique involves construction of a database, training, and validation and then provides a set of predicted results related to the coating adhesion strength and erosion wear rate at various operating parameters.

During plasma spraying, various operating parameters are determined mostly based on past experience. It therefore does not provide the optimal set of parameters for a particular objective. In order to obtain the best result with regard to any specific coating quality characteristic, accurate identification of significant control parameters is essential. Solid particle erosion is considered as a non-linear process with respect to its variables: either materials or operating conditions. To obtain the best functional output coatings exhibiting selected in-service properties and the right combinations of operating parameters are to be known. These combinations normally differ by their influence on the erosion wear rate or/i.e. coating mass loss. In order to control the wear loss in such a process one of the challenges is to recognize parameter interdependencies, co-relations and there individual effects on wear. This chapter is devoted to analyze the experimental results of the erosion wear behavior of alumina titania coatings made at different operational conditions on mild steel and cupper substrates. For this purpose, a statistical technique i.e. Taguchi experimental design is used. Factors are identified according to their influence on the coating erosion rate. The most significant parameter is found. A prediction model using artificial neural network (ANN) is presented considering the significant factors. Beside that, this analysis is made taking into account of training and test procedure to predict the dependence of coating adhesion strength of the coatings made at different operating power levels on different substrate materials.

2. Taguchi experimental design

Taguchi method of experimental design is a simple, efficient and systematic approach to optimize designs for performance and cost effectiveness [10]. In the present work, this method is applied to the process of plasma spraying for identifying the significant process variables/interactions influencing coating erosion wear rate. The levels of these factors are also found out so that the process variables can be optimized within the test range.

Experiments are carried out to investigate the influence of the four selected control parameters. The code and levels of control parameters are shown in table 1. This table shows that the experimental plan has two levels. A standard Taguchi experimental plan with notation **L16 (2^{15})** is chosen as outlined in table 2. In this method, experimental results are transformed into a signal-to-noise (S/N) ratio. It uses the S/N ratio as a measure the quality characteristics deviating from or nearing to the desired values. There are three categories of quality characteristics in the analysis of the S/N ratio, i.e. the lower-the-better, the higher-the-better, and the nominal-the-better. To obtain optimal spraying parameters, the lower-the-better quality characteristic for erosion wear rate is taken.

2.1 Analysis of control factor

Table 2 shows experimental lay out and results with calculated S/N ratios for erosion wear rate of the coatings made at 18Kw power level. Analysis of the influence of each control factor on the coating efficiency is made with a signal-to-noise (S/N) response table, using

MINITAB computer package. The response data of the testing process is presented in table 3. The S/N response graph for coating erosion wear rate is shown in Fig.1.The influence of interactions between control factors is also analyzed in the response table. The control factor with the strongest influence is determined by differences values. The higher the difference, the more influential is the control factor or an interaction of two controls. The strongest influence on coating erosion wear rate is found out to be of impact angle (A) followed by impact velocity (B) and stand off distance (C) then size of the erodent (D) respectively.

Parameter	**Code**	**Level 1**	**Level 2**
Impact Angle(Degree)	A	30	90
Impact Velocity(m/sec)	B	32	58
Stand Off Distance(mm)	C	100	150
Erodent Size(μm)	D	200	400

Table 1. Control factors and selected test levels.

Exp. No.	**A**	**B**	**C**	**D**	**Coating erosion wear rate**	**S/N Ratio**
1	1	1	1	1	10.00	-20.0000
2	1	1	1	2	11.00	-20.8279
3	1	1	2	1	11.5	-21.2140
4	1	1	2	2	12.20	-21.7272
5	1	2	1	1	14.40	. -23.1672
6	1	2	1	2	12.50	-21.9382
7	1	2	2	1	18.10	-25.1536
8	1	2	2	2	19.80	-25.9333
9	2	1	1	1	.6	4.4370
10	2	1	1	2	.8	1.9382
11	2	1	2	1	2.10	-6.4444
12	2	1	2	2	2.41	-7.6403
13	2	2	1	1	6.10	-15.7066
14	2	2	1	2	8.00	-18.0618
15	2	2	2	1	17.10	-24.6599
16	2	2	2	2	17.74	-24.9791

Table 2. Experimental lay out and results with calculated S/N ratios for coating erosion wear rate.

Level	A	B	C	D
1	-11.39	-11.43	-14.17	-16.49
2	-22.50	-22.45	-19.72	-17.40
Diff.	11.11	11.02	5.5	0.91
Rank	1	2	3	4

Table 3. The S/N response table for coating erosion wear rate.

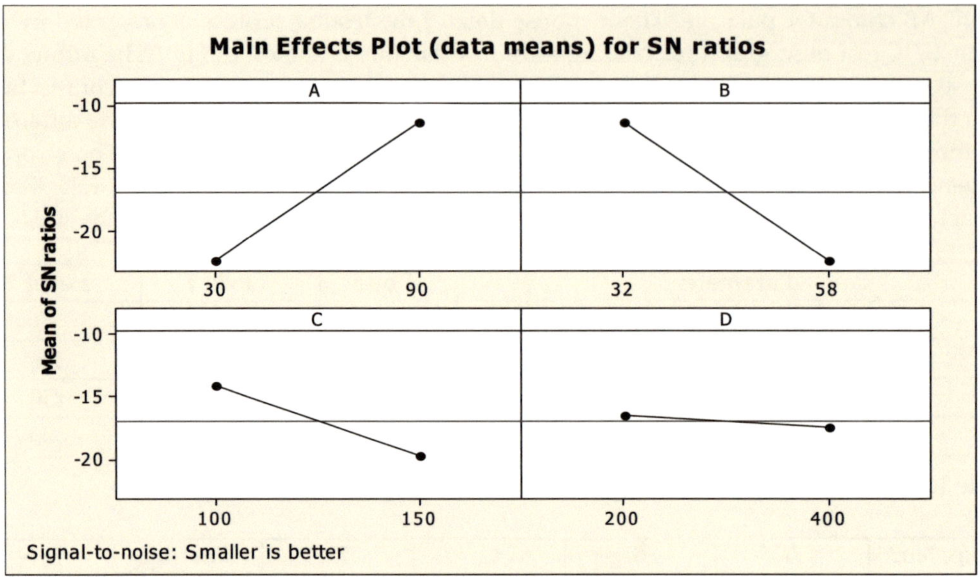

Fig. 1. The S/N response graph for coating erosion wear rate.

It is interesting to note that the Taguchi experimental design method identified impact angle and impact velocity as the most powerful factor influencing the erosion wear rate of the alumina titania coatings. The stand off distance, size of the erodent emerge as the other significant factors affecting the coating erosion wear rate. The impact angle, thus is a significant process variable and in this work, is rightly taken as the basis for studying its effect on the coating erosion wear characteristics.

3. Artificial neural network (ANN) analysis

Plasma spraying is considered as a non-linear problem with respect to its variables: either materials or operating conditions. To obtain functional coatings exhibiting selected in-service properties, combinations of processing parameters have to be planned. These combinations differ by their influence on the coating properties and characteristics. In order to control the spraying process, one of the challenges nowadays is to recognize parameter interdependencies, correlations and individual effects on coating characteristics. Therefore a robust methodology is needed to study these interrelated effects. In this work, a statistical method, responding to the previous constraints, is implemented to correlate the processing parameters to the coating properties. This methodology is based on artificial neural networks (ANN), which is a technique that involves database training to predict property-parameter evolutions. This section presents the database construction, implementation protocol and a set of predicted results related to the coating erosion wear. ANNs are excellent tools for complex processes that have many variables and complex interactions. The analysis is made taking into account training and test procedure to predict the dependence of erosion wear behavior on angle of impact and velocity of erodent and the dependence of coating adhesion strength on different operating power levels on different substrates. This technique helps in saving time and resources for experimental trials. The details of this methodology are described by Rajasekaran and Pai **[11]**.

3.1 Neural network model: Development and Implementation (for coating erosion wear rate)

An ANN is a computational system that simulates the microstructure (neurons) of biological nervous system. The most basic components of ANN are modeled after the structure of brain. Inspired by these biological neurons, ANN is composed of simple elements operating in parallel. It is the simple clustering of the primitive artificial neurons. This clustering occurs by creating layers, which are then connected to one another. The multilayered neural network which has been utilized in the most of the research works for material science, reviewed by Zhang and Friedrich **[12]**. A software package NEURALNET for neural computing developed by Rao and Rao **[13]** using back propagation algorithm is used as the prediction tool for coating erosion wear rate at different impact angles and impact velocity.

The database is built considering experiments at the limit ranges of each parameter. Experimental result sets are used to train the ANN in order to understand the input-output correlations. The database is then divided into three categories, namely: a validation category, which is required to define the ANN architecture and adjust the number of neurons for each layer. A training category, which is exclusively used to adjust the network weights and a test category, which corresponds to the set that validates the results of the training protocol. The input variables are normalized so as to lie in the same range group of 0-1. To train the neural network used for this work, about 25 data sets at different angles and different velocities are taken. It is ensured that these extensive data sets represent all possible input variations within the experimental domain. So a network that is trained with this data is expected to be capable of simulating the plasma spray process. Different ANN structures (I-H-O) with varying number of neurons in the hidden layer are tested at constant cycles, learning rate, error tolerance, momentum parameter and noise factor and slope parameter. Based on least error criterion, one structure, shown in table 4, is selected for training of the input-output data. The learning rate is varied in the range of 0.001-0.100 during the training of the input-output data. The network optimization process (training and testing) is conducted for 1000,000 cycles for which stabilization of the error is obtained. Here the hidden layer number is 1 and neuron numbers in the hidden layer is varied and in the optimized structure of the network, this number is 6. The number of cycles selected during training is high enough so that the ANN models could be rigorously trained.

Input Parameters for Training	Values
Error tolerance	0.002
Learning parameter(ß)	0.001
Momentum parameter(α)	0.001
Noise factor (NF)	0.002
Maximum cycles for simulations	1000000
Slope parameter (£)	0.6
Number of hidden layer neuron	6
Number of input layer neuron (I)	2
Number of output layer neuron (O)	1

Table 4. Input parameters selected for training (Coating erosion).

The impact angles and impact velocity have already been identified (from the outcome of Taguchi analysis) as the parameter significantly affecting the coating erosion wear rate. Each of these parameters is characterized by one neuron and consequently the input layer in the ANN structure has two neurons and the output layer in the ANN structure has one neuron. The optimized three-layer neural network having an input layer (I) with two input nodes, a hidden layer (H) with six neurons and an output layer (O) with one output node employed for this work is shown in Fig. 2.

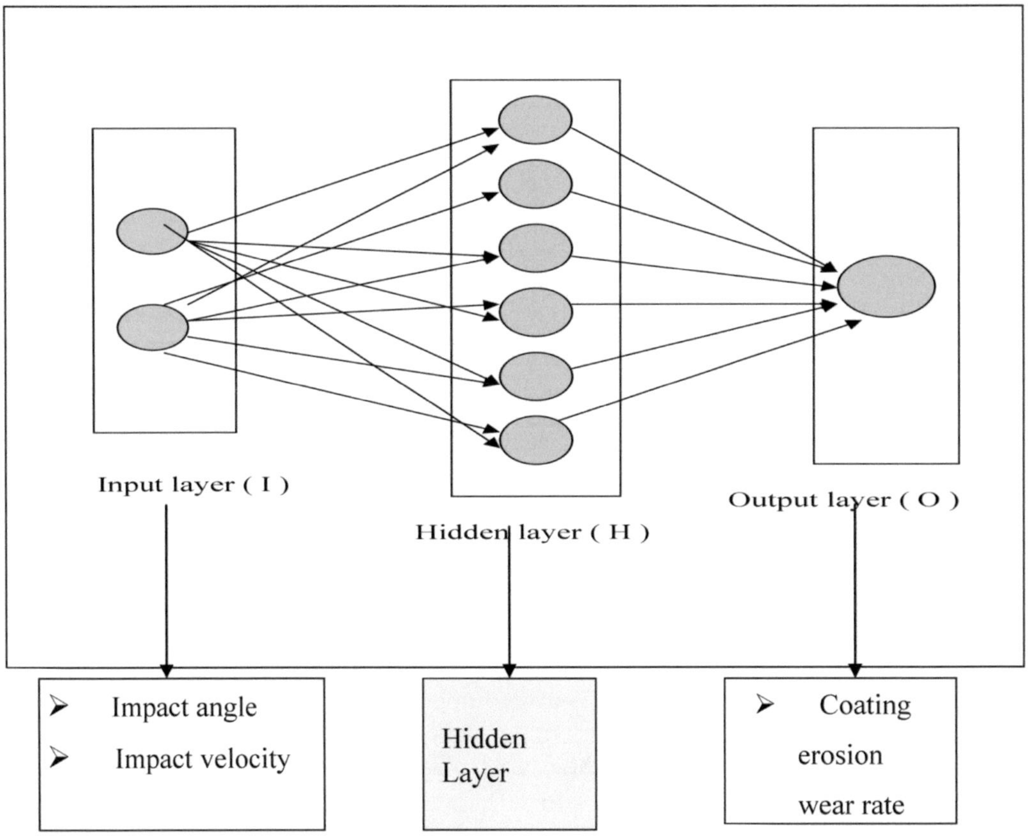

Fig. 2. The three Layer Neural network.

3.2 ANN prediction of erosion wear rate

The prediction neural network is tested with four data sets from the original process data. Each data set contained inputs such as impact angle and impact velocity and an output value i.e. erosion wear rate is returned by the network. As further evidence of the effectiveness of the model, an arbitrary set of inputs is used in the prediction network. Results are compared to experimental sets that may or may not be considered in the training or in the test procedures. Fig. 3 represents the comparison of predicted output values for erosion wear rate with those obtained experimentally at different impact angles of the erodent at different impact velocities i.e. 32m/sec, 45m/sec and 58 m/sec respectively.

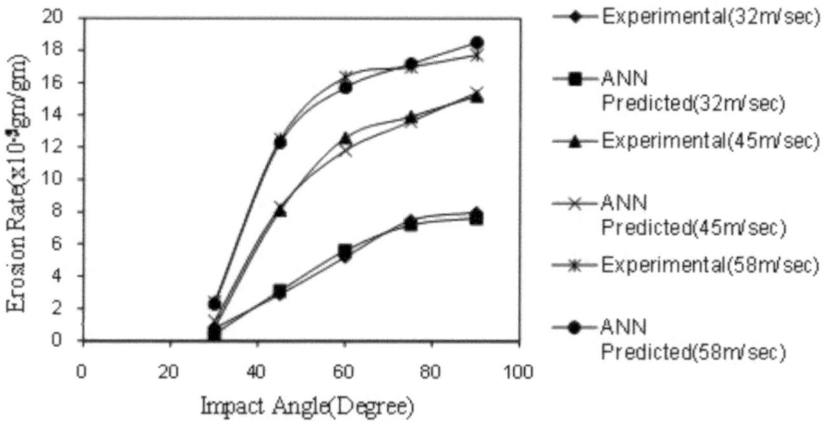

Fig. 3. Comparison plot for predicted and experimental values of coating erosion wear rate at different impact angles of the erodent at impact velocity 32m/sec, 45m/sec and 58m/sec (time of exposure 6 min, SOD 150mm, size of the erodent 400μm for the sample coated at 18 kW power).

Beside comparison of predicted and experimental values of erosion wear rate Fig.3 illustrates the effect of impact angle (α) on the erosion rate of coatings subjected to solid particle erosion. The erosion results for coatings of materials deposited at 18 kW operating power of the plasma torch at impact angles of 30^0,45^0, 60^0,75^0 and 90^0 for 32m/sec, 45m/sec and 58m/sec respectively at SOD of 150mm for size of the erodent 400μm are shown. Mass loss, then erosion rate (mass loss of coating (gm) per unit wt of erodent (gm) is measured after the samples are exposed to the erodent stream for 6 minutes. It is seen from the graph that irrespective of the feed material, the erosion mass loss is higher at larger angle of impact and the maximum erosion takes place at $\alpha = 90^0$. Such trend in generally observed for brittle materials.

It is interesting to note that the predictive results show good agreement with experimental sets realized after having generalizing the ANN structures. The optimized ANN structure further permits to study quantitatively, the effect of the selected impact angles. The range of the chosen parameter can be larger than the actual experimental limits, thus offering the possibility to use the generalization property of ANN in a large parameter space. In the present investigation, this possibility was explored by selecting the impact angle in a range from 10^0 to 90^0 for velocities 32m/sec, 45m/sec, 58m/sec and a set of prediction for erosion wear rate is evolved. Fig.4 illustrates the predicted evolution of erosion wear rate of alumina titania coatings on mild steel substrates with the impact angle for velocities 32m/sec, 45m/sec, 58m/sec. From the predicted graph in fig.4 with increasing impact angle erosion rate increases for different impact velocity, and it is maximum at 58m/sec.

In the present investigation, by selecting the impact velocity in a range from 20 to 70 m/sec at impact angles 30^0, 60^0 and 90^0 and a set of prediction for erosion wear rate is evolved. Fig.5 illustrates the predicted evolution of erosion wear rate of alumina titania coatings on mild steel substrates with the impact velocity at impact angles 30^0, 60^0 and 90^0.

From the predicted graph in fig.5 with increasing velocity erosion rate increases for different angles. It is obvious that, with increasing velocity the particles will have high kinetic energy, which transformed at impact and hence remove more particles from the impacted surface and it is maximum at 90⁰ angle. Beside that at low velocity and at low angle there may be one mechanism, so that the slope does not change much, but at high velocity and high angle there may be two mechanisms, so that may be the reason of large slope change.

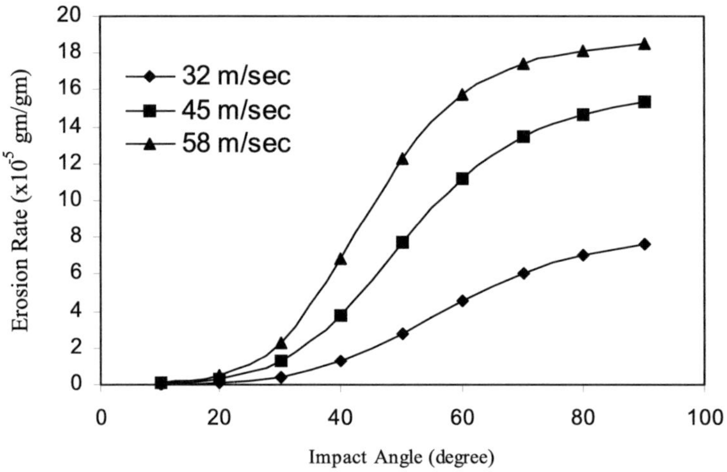

Fig. 4. Predicted erosion wear rate of the coating at different impact angles of the erodent for different impact velocities (for 6 minute time of exposure, SOD150mm, size of the erodent 400μm for the sample coated at 18 kW power level).

3.3 Neural network model: Development and implementation (for coating adhesion strength)

A software package NEURALNET for neural computing developed by Rao and Rao [13] using back propagation algorithm is used as the prediction of coating adhesion strength at different operating power levels for different substrates. To train the neural network used for this work, about 8 data sets at different operating power levels for different substrates are taken. Based on least error criterion, one structure, shown in table 5, is selected for training of the input-output data.

Input Parameters for Training	Values
Error tolerance	0.001
Learning parameter(ß)	0.002
Momentum parameter(α)	0.002
Noise factor (NF)	0.001
Maximum cycles for simulations	1000,000
Slope parameter (£)	0.6
Number of hidden layer neuron	6
Number of input layer neuron (I)	2
Number of output layer neuron (O)	1

Table 5. Input parameters selected for training (for coating adhesion strength).

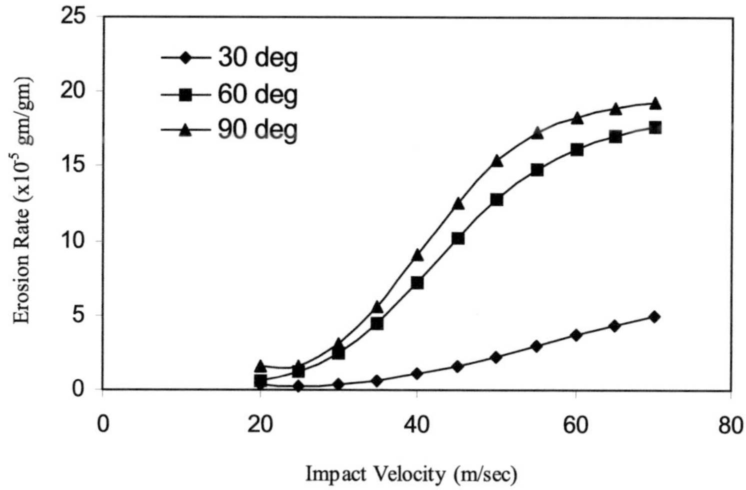

Fig. 5. Predicted erosion wear rate at different impact velocities impacted at different angles (for exposure time 6 min, SOD150mm, size of the erodent 400μm, for the sample coated at 18 kW power level).

The operating power levels and substrate materials are taken as the parameter significantly affecting the coating adhesion strength. Each of these parameters is characterized by one neuron and consequently the input layer in the ANN structure has two neurons and the output layer in the ANN structure has one neuron. The optimized three-layer neural network having an input layer (I) with two input nodes, a hidden layer (H) with six neurons and an output layer (O) with one output node employed for this work is as shown in Fig. 2.

3.4 ANN prediction of coating adhesion strength

The prediction neural network was tested with three data sets from the original process data. Each data set contained inputs such as torch input power, substrate material and an output value i.e. coating adhesion strength was returned by the network. As further evidence of the effectiveness of the model, an arbitrary set of inputs is used in the prediction network. Results were compared to experimental sets that may or may not be considered in the training or in the test procedures. Fig.6 presents the comparison of predicted output values for coating adhesion strength with those obtained experimentally with different torch input power on different substrates.

It is interesting to note that the predictive results show good agreement with experimental sets realized after having generalizing the ANN structures. The optimized ANN structure further permits to study quantitatively the effect of the considered input power. The range of the chosen parameter can be larger than the actual experimental limits, thus offering the possibility to use the generalization property of ANN in a large parameter space. In the present investigation, this possibility was explored by selecting the plasma torch input power in the range from 7 kW to 25 kW, and a set of prediction for coating adhesion strength is evolved. Fig.7 illustrates the predicted evolution of coating adhesion strength of alumina titania coatings on copper and mild steel substrates with torch input power.

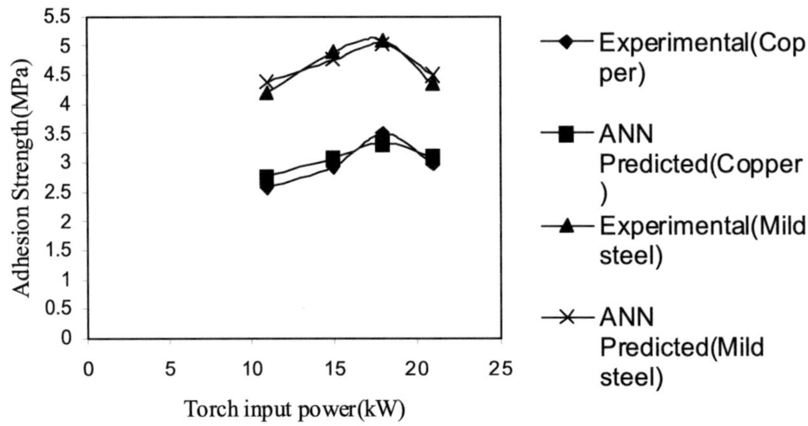

Fig. 6. Comparison plot for predicted and experimental values of coating adhesion strength with different torch input power on different substrates.

From the figure it can be visualized that, the interface bond strength increases with the input power of the torch up to a certain power level and then shows a decreasing trend in coating adhesion, irrespective of the substrate material. This might be due to the fact that, when the operating power level is increased, larger fraction of particles attain molten state as well as the velocity of the particles also increase. Therefore there is better splat formation and mechanical inter-locking of molten particles on the substrate surface leading to increase in adhesion strength [14]. But, at a much higher power level, the amount of fragmentation and vaporization of the particles increase. There is also a greater chance to fly off of smaller particles during in-flight traverse during plasma spraying and results in poor adhesion strength of the coatings. Coating adhesion strength is more in case of mild steel substrate than that of copper substrate may be due to the dependence of thermal conductivity for melted particle, dissipation of heat at metal interface and also may be due to thermal expansion coefficient mismatch at the ceramic metal interface [15].

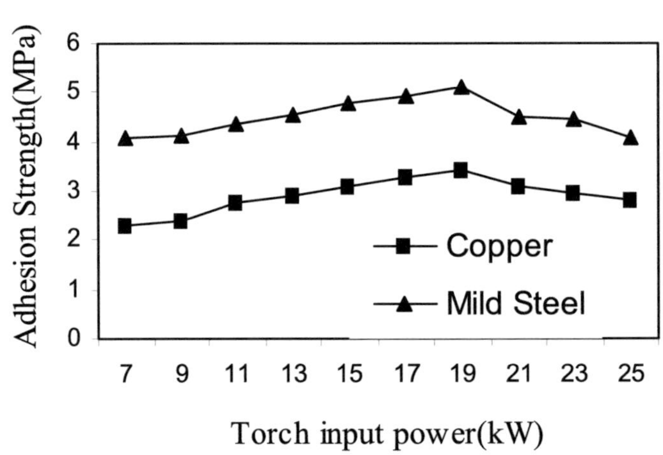

Fig. 7. Predicted values of coating adhesion strength of alumina titania coatings on copper and mild steel substrates at different torch input power.

4. Remarks

Functional coatings have to fulfill various requirements. The less erosion wear rate is one the main requirements of the coatings developed by plasma spraying. Solid particle erosion is considered as a non-linear process with respect to its variables: either materials or operating conditions. In order to achieve certain values of erosion rate accurately and repeatedly, the influence parameters of the process have to be controlled accordingly. Since the number of such parameters in plasma spraying is too large and the parameter-property correlations are not always known, statistical methods can be employed for precise identification of significant control parameters for optimization. Neural computation can be used as a tool to process very large data related to a spraying process like coating erosion wear rate and coating adhesion strength and to predict any desired coating characteristic the simulation can be extended to a parameter space larger than the domain of experimentation.

5. Conclusions

The conclusions drawn from the present work are as follows:

- Commercial grade alumina & titania mixed powders in the size range 40 to 100μm can be coated on metal substrates employing thermal plasma spray technique. Coatings made with alumina titania possess desirable coating characteristics comparable to those of other conventional plasma sprayed ceramic coatings.
- Adhesion strength of the coating varies with operating power. Maximum adhesion strength of 5.1 MPa on mild steel substrate and of 3.5Mpa on copper substrate is recorded at 18 kW. It is noted that invariably in all cases the interface bond strength increases with the input power of the torch up to a certain optimum power level and then shows a decreasing trend. Coating adhesion is higher in case of mild steel substrate than of copper substrate.
- Operating power level of the plasma torch influences the coating adhesion strength, deposition efficiency, coating thickness and coating hardness to a great extent. The coating morphology is also largely affected by the torch input power.
- It is observed that, the erosion wear rate is dependent on erodent dose, angle of attack, velocity of erodent, stand off distance and size of the erodent. Cumulative coating mass loss varies with time of erosion. Maximum amount /rate of erosion occur at 90^0 impact angle. The trend of erosion of the coatings seems to follow the mechanism predicted for brittle materials. Coating deposited at 18 kW power level shows a higher erosion rate than that of the sample deposited at 11kW power level.
- Erosion wear behavior is one of the main requirements of the coatings developed by plasma spraying for recommending specific application. In order to achieve tailored erosion wear rate accurately and repeatedly, the influence of the process parameters are to be controlled accordingly. The coating sustains erosion by solid particle impingement substantially and therefore alumina titania can be considered as a potential coating material suitable for various tribological applications.
- Impact velocity, impact angle, stand off distance and size of the erodent significantly affect the erosion wear rate of coating. Identification of these factors and their significance on the coating erosion wear rate is possible by statistical techniques like Taguchi experimental design. Artificial neural networks can be gainfully employed to simulate property-parameter correlations in a space larger than the experimental

domain. Neural computation can be gainfully employed as a tool to analyze, optimize and predict the erosion behavior, adhesion strength of the coatings purpose. It is evident that with an appropriate choice of processing conditions a sound and adherent ceramic coating is achievable using alumina and titania .

6. References

[1] Taylor R. - "Thermal Plasma Processing of Materials"— *Power Beams and Materials Processing PBAMP* 2002, Ed. A. K. Das et al., Allied Publishers Pvt. Ltd., Mumbai, India: 2002. pp.13-20 .

[2] Bandopadhyaya P.P. - *Processing and Characterization of Plasma sprayed Ceramic coatings on Steel Substrate*—Ph.D.Thesis, IIT, Kharagpur, India (2000).

[3] Pawlowski L. *The science and engineering of thermal spray coatings*. New York, USA: Wiley, 1995. p. 432.

[4] Normand B. , Fervel V., C. Coddet, Nikitine V., "Tribological properties of plasma sprayed alumina-titania coatings: role and control of the microstructure." *Surf. Coat.Technol.* Volume123, (2000) :p.278.

[5] Fervel V., Normand B., Coddet C.," Tribological behavior of plasma sprayed Al_2O_3-based cermet coatings." *Wear.* Volume230, (1999): p.70.

[6] Niemi K. , Vuoristo P. , Mantyla T., Lugscheider E., Knuuttila J., Jungklaus , in: *Proceedings of the 8th National Thermal Spray Conference*, Houston, TX: 1995, p. 645.

[7] Ramchandran K. and Selvarajan P. A., " In-flight particle behaviour and its effect on co-spraying of alumina-titania." *Thin Solid Film.* Volume 315, (1998): p.49.

[8] Steffens H.D., Haumann D., Gramlich M., Wilden J., Wewel M., Hohle M., Nestler M.C. , in: *Proceedings of the 8th National Thermal Spray Conference*, Houston, TX: 1995, p. 677.

[9] Ramachandran K. , Selvarajan V. , Ananthapadmanabhan P. V. , Sreekumar K.P., "Microstructure , adhesion, microhardness , abrasive wear resistance and electrical resistivity of the plasma sprayed alumina and alumina-titania coatings." *Thin Solid Films.* Volume 315, (1998): p. 144-152.

[10] Sahin Y.,"The Prediction of Wear Resistance Model for The Metal Matrix Composites." *Wear.* Volume 258, (2005): p. 1717-1722.

[11] Rajasekaran S., Vijayalakshmi Pai G. A., (2003),--Neural Networks, *Fuzzy Logic And Genetic Algorithms – Synthesis and Applications* - Prentice Hall of India Pvt. Ltd., New Delhi.

[12] Zhang Z., Friedrich K., *Artificial neural network applied to polymer composites: a review*, Comp. Sci. Technol. V-63(14), (2003), p. 2029-2044.

[13] Rao V. and Rao H., (2000),'C++ *Neural Networks and Fuzzy Systems*',BPB Publications.

[14] Halling, *Principles of Tribology*, The Mcmillan Press Ltd, NY, USA, 1975.

[15] Guilmad Y., Denape J. and Patit J . A.; " Friction and wear thresholds of alumina-chromium steel pairs sliding at high speeds underdry and wet condition; *Trib.Int.* Volume 26, (1993): p. 29-39.

Part 2

Plasma Spray in Biomaterials Applications

5

Plasma Sprayed Bioceramic Coatings on Ti-Based Substrates: Methods for Investigation of Their Crystallographic Structures and Mechanical Properties

Ivanka Iordanova[1], Vladislav Antonov[2],
Christoph M. Sprecher[3], Hristo K. Skulev[4] and Boyko Gueorguiev[3]
[1]University St. Kliment Ohridski, Sofia,
[2]Academy of Sciences, Sofia,
[3]AO Research Institute Davos, Davos,
[4]Technical University of Varna, Varna,
[1,2,4]Bulgaria
[3]Switzerland

1. Introduction

Recently, life expectancy has been rising due to improvements in public health, nutrition, medicine and health-related quality-of-life requirements. However, aging leads to osteoporosis, i.e. loss of bone substance, associated with an increased risk of fracture. That is why in the field of orthopaedics and traumatology the number of patients receiving metal, and especially titanium-based, implants to correct skeletal defects and diseases is increasing worldwide. Critical factors for long-term stability of fixation include biocompatibility, material selection and implant design. Their investigation and optimisation, especially in case of implants coated with bioceramics, require further improvement of the existing and development of new methods for applied research in the structure and operational properties.

2. Ti-based implants with bioceramic coatings

Titanium (Ti) and some of its alloys, such as Ti-6Al-4V (TAV) and Ti-6Al-7Nb (TAN), are widely used for repair or replacement of parts of the human musculoskeletal system because of their biocompatibility, corrosion resistance in biological environments, low specific weight and excellent mechanical properties (Duan & Wang, 2006). However, when used in medical implants these biomaterials are exposed to bodily fluids under intensive mechanical loading. In such cases, it is necessary to meet the requirements for chemical and structural stability combined with reliable strength, wear resistance and good mechanical fixation. Despite high biocompatibility, long term observations show a certain amount of implant loosening, attributed to the lack of sufficient bioactivity of the surface over time (Suchanek & Yoshimura, 1998).

It has been found that covering the implant surface with bioceramic coatings, such as calcium orthophosphates or titania (TiO_2), can speed up the process of bone integration, thus enhancing the long-term fixation and implant stability (LeGeros, 2002; Suchanek & Yoshimura, 1998). Therefore, the bioceramic coatings and calcium orthophosphates in particular, applied on Ti-based substrates, having a useful combination of biocompatibility, mechanical properties and relative stability in bodily environment, are subjects of an increasing interest for biomedical applications in modern orthopaedics and dentistry (Gaona et al., 2007; Salman et al., 2008; Soares et al., 2008; Thomas, 1994).

Calcium orthophosphates are compounds of calcium (Ca) and phosphorous (P) in differing of calcium-phosphate ratios. Calcium hydroxylapatite (HA) $Ca_5(PO_4)_3(OH)$ is the most chemically stable and least soluble in aqueous media (Nelea et al., 2007). The composition and structure of the calcium hydroxylapatite are very similar to those of the bone mineral content that comprises about 70 % of the volume. This is the main reason for the observed high biointegration of HA coatings.

In contrast, TiO_2 and its deposits are generally considered as bioinert, i.e. not initiating a response or interaction in contact with biological tissues. However, it was recently indicated that the titania surface bioactivity can be activated by different chemical treatments (Gaona et al., 2007; Zhao et al., 2007). This could provoke reconsideration of the titania coatings role, making them superior to other biomedical coatings, as so far TiO_2 is mostly recognized with its excellent corrosion resistance and high adhesion strength to different substrates (Liu et al., 2006).

The in vivo performance of an implant, whose surface is covered with a bioceramic coating, enables a more stable fixation, due to better bonding to bone tissue, and decreased release of metal ions into the human body (Heimann, 2002; Yu-Peng et al., 2004). The goal of coating implants is to take advantage of mechanical strength, ductility, wear resistance and ease of fabrication of their metallic body in combination with good osteointegration and chemical stability associated with the applied surface layer (Chou & Chang, 2002).

3. Plasma-spraying of bioceramic coatings

Bioceramic coatings can be applied on Ti-based substrates by a number of methods, but undoubtedly plasma-spraying with its three environmental variations air plasma-spraying (APS), controlled atmosphere plasma-spraying (CAPS) and vacuum plasma-spraying (VPS) has proved to be one of the leading technologies in the medical practice for implant production.

Utilizing this technology, coatings are usually formed from powder particles, injected into a high temperature field, created in a plasma torch (plasmotron), where they are accelerated, molten and propelled towards the coated surface (Dhiman & Chandra, 2007; Gueorguiev et al., 2009b). The plasma gas (usually argon, nitrogen, or their mixture with hydrogen) is introduced into the plasma torch, where an electric arc transforms it to a plasma state with high enthalpy and temperature in the range 10000 - 30000°C (Brossa & Lang, 1992; Gill & Tucker, 1986). Although the injected powder particles are exposed to this high temperature for a very short time ($10^{-3} - 10^{-4}$s), most of them are sufficiently heated to become molten (Brossa & Lang, 1992). Reaching the coated surface, each particle solidifies independently in strongly non-equilibrium conditions and transforms into a disk-shaped solid splat (lamella)

(Brossa & Lang, 1992). However, the initial powder particles usually differ in shape, dimensions, temperature and velocity that can result in evaporation, partial melting or even absence of melting of some particles (Brossa & Lang, 1992; Nicoll et al., 1985). The evaporated particles do not take part in the coating formation. However, the partially molten and some completely unmolten particles could get incorporated in the coating, causing formation of inhomogeneous and porous structure. The partially molten particles solidify on the surface as spherical grains that can get depleted or become sources of internal stress fields in the coating, influencing substantially its properties and performance (Brossa & Lang, 1992; Iordanova et al., 2001).

Residual stresses, preferred crystallographic orientations (crystallographic textures) and pores are often observed in plasma-sprayed coatings (Gergov et al., 1990; Iordanova & Forcey, 1991). In some cases decomposition of HA during plasma-spraying is possible according to the following reaction:

$$Ca_5(PO_4)_3(OH) \rightarrow CaO + Ca_3P_2O_8 \text{ (or } CaO + Ca_4P_2O_9) \quad (1)$$

It has been found that the decomposition process is more pronounced when the HA powder particles are smaller and the spraying voltages are higher (Morris & Ochi, 1998).

The main advantages of plasma-spraying technology to coat Ti-based implants with bioceramic coatings are the high speed of coating deposition onto large surfaces, economy and high efficiency, combined with superior mechanical coating properties and chemical stability in a bodily environment enabling stable mechanical fixation (Gaona et al., 2007; Khor et al., 2004; McPherson et al., 1995; Morris & Ochi, 1998). The main disadvantage concerns the relatively poor adhesion between implant surface and applied deposit, observed in a number of cases. That is why, in order to prevent depletion, a transition layer is often plasma-sprayed prior to the main coating. For example, titania, being used as a single layer coating, can alternatively be plasma-sprayed as a transition layer between the top HA coating and the substrate, or as an incorporated reinforcing additive in composite coatings together with HA (Salman et al., 2008; Yu-Peng et al., 2004). Thus, in most of the cases, the adhesive and cohesive strength of the coating can be further increased and the interface strengthened (Chou & Chang, 2002; Yu-Peng et al., 2004). It is assumed that double-layer implant systems, consisting of a Ti-based body, a TiO_2 transition layer and an osteoconductive top HA coating will be able to withstand not only high local compression stresses, but also tensile and shear stresses, occurring during the micro-movements of the patient in the first stages of the healing phase after implantation (Heimann, 2002; Heimann et al., 2004). The presence of TiO_2 transition layer could also reduce the mismatch of the thermal expansion coefficients between the substrate and the top HA coating, which appears to be the main reason for failure at the interface between dissimilar materials (Salman et al., 2008).

Despite the positive results obtained using plasma-sprayed bioceramic coatings, there are still some concerns, especially regarding the durability and successful long-term performance of the implants (Cofino et al., 2004; Morris & Ochi, 1998; Sun et al., 2002). Moreover, in spite of its importance, the structure-formation process of plasma-sprayed bioceramic coatings has not been studied in detail so far. That is why it is necessary to investigate the structure formation of the deposits as a function of the technological conditions and their influence on the final implant operational properties. Furthermore, the mechanism for modification of HA coating properties by deposition of TiO_2 transition layers or additives is not widely reported and the potential of plasma-sprayed titania coatings for bioactive applications has not been extensively explored yet (Salman et al., 2008; Yu-Peng et al., 2004).

The morphology and parameters of phase composition, interfaces, porosity, crystallographic textures, residual stresses, surface roughness, bond strength, availability of oxide inclusions and spherical grains are all critical parameters of the plasma-sprayed bioceramic coating structure, which must be considered, controlled and optimised. However, the knowledge about these parameters and their influence on the operational properties is still insufficient.

The application of advanced plasma-sprayed bioceramic coatings with superior properties in the contemporary medicine and dentistry requires a deep understanding of the structure formation processes, their dependence on the technological conditions and influence on the operational properties.

4. Relation between technological conditions, micro-structure formation and operational properties of bioceramic plasma-sprayed coatings for biomedical purposes

Due to the complexity of the plasma-spraying process, the technological conditions play a crucial role and can significantly influence the coating structure and practical performance (Azarmi et al., 2008; Frayssinet et al., 2006; Sampath et al., 2003).

The technological regime during plasma-spraying is mainly defined by the following parameters (Azarmi et al., 2008; Harsha et al., 2008; Sarikaya, 2005; Skulev et al., 2005):

- Heat power of the plasma jet;
- Type, pressure and flow rate of the plasma and transporting gas;
- Type, mean size and size distribution, flow rate and shape of the injected powder particles;
- Location, angle and type of powder injection into the plasma jet;
- Roughness, purity and temperature of the coated surface;
- Spraying-off distance between the spraying gun and substrate surface;
- Speed relation between the spraying gun and coated surface;
- Type of the ambient atmosphere in the plasmotron.

Not all of the parameters listed above are independent. That can complicate the selection and control of the technological conditions.

For instance, the type, pressure and flow rate of the plasma gas have an effect on the heat power of the plasma jet. The latter could be kept constant by a certain variation of the plasma gas pressure and flow rate.

The location, angle and type of the powder injection in order to ensure a complete melting of the initial particles could be pre-defined to a certain degree by the plasma jet heat power, powder type, particle size and plasma gas pressure.

The powder type, particles shape, mean size, size distribution and flow rate, together with the location and the type of powder injection determine the transporting gas flow rate.

The plasma jet heat power, spraying-off distance and relative speed between the spraying gun and substrate influence significantly the temperature of the coated surface.

The type and flow rates of the plasma and transporting gas have a significant effect on the ambient atmosphere.

A scheme, showing the interaction between the main technological parameters during plasma-spraying, is given in Figure 1.

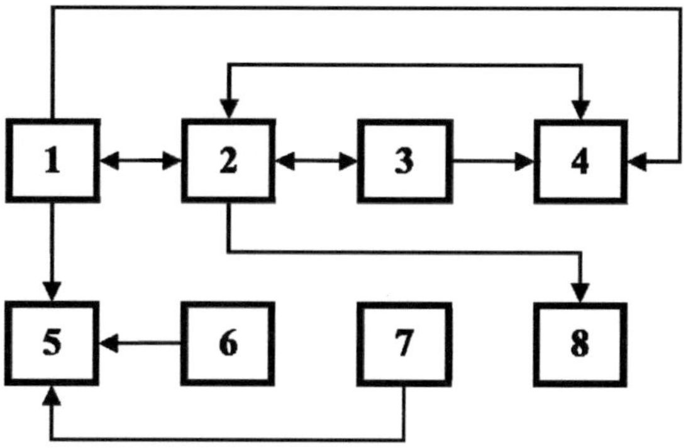

Fig. 1. Interaction scheme between the main parameters (1-8) determining the technological conditions during plasma-spraying: 1-heat power of the plasma jet; 2-type, pressure and flow rate of the plasma and transporting gas; 3-type, mean size and size distribution, flow rate and shape of the injected powder particles; 4-location, angle and type of powder injection into the plasma jet; 5-roughness, purity and temperature of the coated surface; 6-spraying-off distance between the spraying gun and substrate surface; 7-speed relation between the spraying gun and coated surface; 8-type of the ambient atmosphere in the plasmotron.

Initially the plasma-spraying processes and their influence on the coating properties have been investigated mainly empirically (Clyne & Gill, 1996). However, the contemporary requirements and high standards for operational properties of the bioceramic coatings lead to the necessity of a more complex approach to obtain fundamental knowledge about the structure formation processes and explore the dependence of their mechanisms and kinetics on the technological conditions. On the other hand, this complex approach is necessary to investigate the influence of coating composition and micro-structure parameters on their operational properties. This in turn has brought the necessity of more precise evaluation and control of the micro-structure parameters and technological conditions during plasma-spraying. As a result, further improvement of the existing and development of new experimental and theoretical methods and models for prognosis, investigation and control of the structure formation processes, technological parameters and operational properties has been provoked (Azarmi et al., 2008; Chen et al., 1993; Fan et al., 2005; Iordanova & Forcey, 1997; Kuroda & Clyne, 1991; Liu et al., 2005; Mawdsley et al., 2001; Moreau et al., 2005; Parizi et al., 2007; Sampath et al., 2003; Sarikaya, 2005; Toma et al., 2003). The main aim of this general trend for research and investigation, observed at present, is optimisation of the plasma-spraying technology to achieve superior parameters of the deposited bioceramic coatings enabling their successful and reliable long-term utilisation in a wide range of biomedical applications.

For this purpose, new techniques for visualisation and diagnostics of the injected powder particles during their flight to the coated surface have been developed (Moreau et al., 2005; Srinivasan et al., 2006; Streibl et al., 2006). New methods for non-destructive control, including statistical evaluation of the collected data, together with prognosis of the coating parameters as a function of the technological conditions have been offered as well (Zhu & Ding, 2000). Some work has already been done to characterise the following main dependencies that play an important role for successful performance of the discussed biomedical coatings (Azarmi et al., 2008; Boulos et al., 1993; Celik et al., 1999; Chen et al., 1993; Choi et al., 1998; Dyshlovenko et al., 2005; Fan et al., 2005; Gueorguiev et al., 2008a, 2008c; Iordanova & Forcey, 1997; Keller et al., 2003; Khor et al., 2004; Kreye et al., 2007; Kuroda & Clyne, 1991; Leigh & Berndt, 1997; Li & Khor, 2002; Liu et al., 2005; Lugscheiber et al., 1993; Mawdsley et al., 2001; Montavon et al., 1997; Moreau et al., 2005; Neufuss et al., 2001; Ohmori & Li, 1991; Parizi et al., 2007; Sampath et al., 2003; Sarikaya, 2005; Srinivasan et al., 2006; Staia et al., 2001; Streibl et al., 2006; Toma et al., 2003; Tong et al., 1996; Wallace & Ilavsky, 1998; Zhu & Ding, 2000):

- Relation between technological parameters, thermal and kinetic energy of the powder particles, their trajectories and stage of melting;
- Influence of the chemical composition and size of the powder particles and substrate parameters (composition, roughness, purity, temperature) on the morphology, phase composition, residual stresses and crystallographic texture of the applied coatings;
- Influence of the parameters, discussed above, on such important operational properties of the coatings as density, hardness, toughness, frictional characteristics, wear resistance, elasticity, thermal conductivity, structural and chemical stability.

Despite the lot of work that has been done so far, a generalised model allowing planning of technological conditions for production of coatings with pre-defined properties has not been developed yet. This is mainly due to the complexity of the plasma-spraying and structure formation processes. Their characterisation and understanding require development of new methods for further fundamental and applicable interdisciplinary analyses by involvement of highly qualified specialists from different fields.

5. Methods for analysis of crystallographic structures and mechanical properties of plasma-sprayed bioceramic coatings

The main methods applied and the latest results obtained by the authors of this chapter during the investigation of initial powders and plasma-sprayed hydroxylapatite and titania bioceramic coatings are discussed below.

5.1 Optical metallography and scanning electron microscopy

Optical metallography and scanning electron microscopy (SEM) have been applied to evaluate particle shape and size distribution parameters of initially injected powders as well as to analyse coating morphology, spherical grains, chemical composition and distribution of elements in bioceramic plasma-sprayed coatings.

5.1.1 Particle shape and size distribution parameters of the initial powders

Mean size and size distribution of the initial powder particles are important parameters influencing the morphology and the macro-properties of the plasma-sprayed coatings.

Images, taken with transmission optical microscope of the initial bioceramic powders XPT-D-703 (HA) and AMDRY-6505 (TiO_2), produced by Sulzer Metco Europe GmbH (Switzerland), are shown in Figure 2.

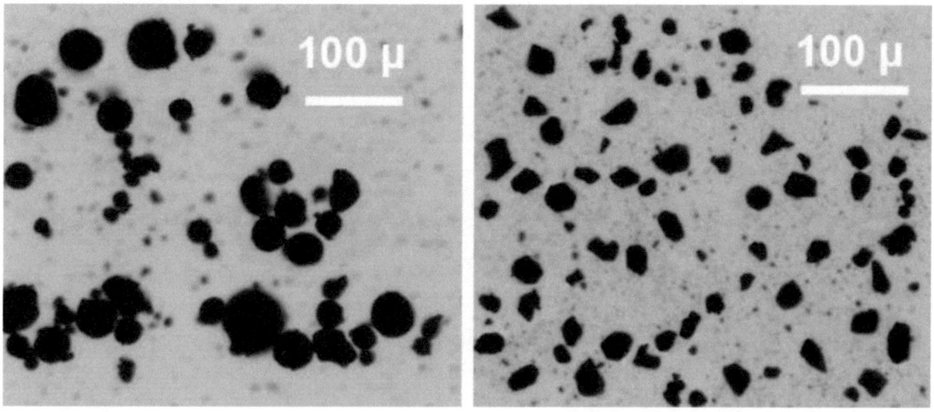

Fig. 2. Particles of powders XPT-D-703 (HA, left) and AMDRY-6505 (TiO_2, right), observed in transmission optical microscope (Gueorguiev et al., 2008a, 2008b).

It is obvious that the majority of the HA particles are spherical in contrast to the sharp-edged TiO_2 ones. Whereas the size of spherical particles is characterised with the diameter, the more asymmetric particle size has been evaluated via the following two parameters (Gueorguiev et al., 2008a, 2008b; Iordanova et al., 2001):

- Mean value between the measured maximum and minimum particle size;
- Average diameter d_{opt} which is a function of the area s_{opt} of the particle projection, evaluated with image analysis software (for example, Carl Zeiss AxioVision). The following equation is in power:

$$d_{opt} = 2\sqrt{s_{opt} / \pi} \tag{2}$$

Particle size distribution curves, based on several hundred measurements of randomly chosen particles, can be plotted and fitted with the following type of Gaussian function:

$$f = A \exp\left[-(x-c)^2 / b^2\right] \tag{3}$$

Characteristics of powder particle dimensions of the initial powders XPT-D-703 and AMDRY-6505, used for deposition of bioceramic plasma-sprayed coatings, are given in Table 1.

Initial powder	Particle size (d_{opt}), μm		
	Minimal	Maximal	Mean
XPT-D-703 (HA)	3.76	89.15	39.70±11.35
AMDRY-6505 (TiO_2)	5.65	33.52	18.0±5.15

Table 1. Particle size parameters of two initial powders for plasma-spraying (Gueorguiev et al., 2008b).

5.1.2 Coating morphology, chemical composition and distribution of elements

Plasma-sprayed bioceramic coatings have been investigated at a cryo-fractured cross-section using scanning electron microscope (Hitachi S-4100 FESEM) equipped with a secondary electron (SE) detector for topographical imaging, a back scattered electron (BSE) one for density imaging and an energy dispersive X-ray (EDX) detector to analyse chemical composition (Burgess et al., 1999; Gueorguiev et al., 2009b; Iordanova et al., 1994, 1995). EDX has been applied for analysis of interfaces in double-layer plasma-sprayed bioceramic coatings, shown in Figure 3, together with the distribution of the main elements Ca, P and Ti at the interface between the transition TiO_2 and top HA layers.

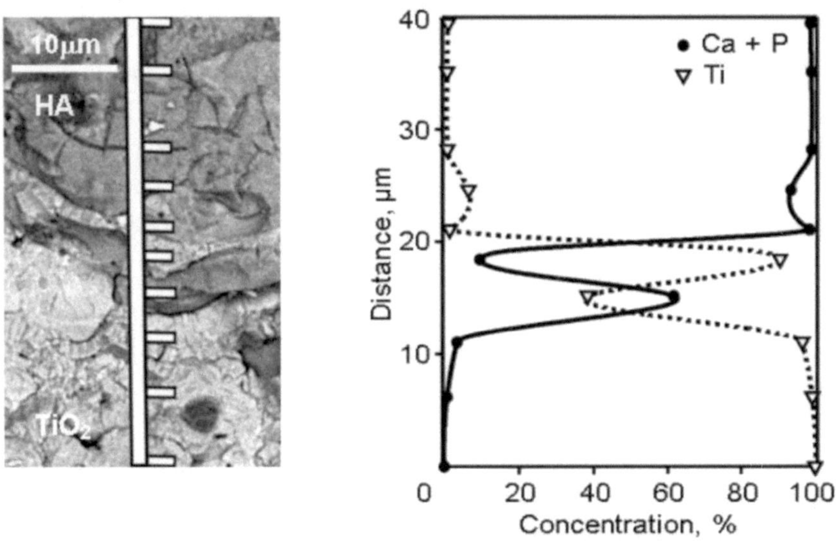

Fig. 3. SEM image (BSE mode) of the interface between the transition TiO_2 and top HA layers of a plasma-sprayed bioceramic coating (left) together with the corresponding EDX profile of Ca, P and Ti concentrations (right) (Gueorguiev et al., 2009b).

The observed element distribution shows that a diffusion zone, around 15 μm wide, with non-smooth change of the element concentration has been formed at the TiO_2-HA interface.

5.1.3 Stereological metallographic analysis of spherical grains

If during plasma-spraying the injected powder particles reach the coated surface in a partially molten state they can form grains with a spherical shape in the coating. The latter, being inhomogeneous inclusions in the matrix with a predominantly lamellar structure, can significantly influence the macro-properties and practical performance of the bioceramic coatings. For that reason the parameters of such spherical grains have to be analysed and controlled. Stereological metallographic methods developed for characterisation of plasma-sprayed and in particular bioceramic coatings are an excellent tool for this analysis (Gueorguiev et al., 2008b; Iordanova et al., 2001). Principally, these methods allow estimation of volume parameters of microstructure elements, based on their two-dimensional cross-section plane parameters that can be experimentally measured applying standard metallographic methods performed via optical or scanning electron microscopy.

All quantitative stereological analyses are based on the principle that the share of a phase (or any other structure element) in a volume is equal to the surface or linear share of its cross-sections with a plane or a line intersecting this volume, respectively. Only structure elements that can be represented as convex bodies are suitable for stereological analyses, as a concave ones can be crossed more than once by the intersecting plane or line. In addition, stereological methods are applicable when the distribution of the analyzed structure element in the volume is close to random.

A method, based on the Scheil-Schwartz-Saltikow technique, was developed by the authors of this chapter to evaluate volume parameters of spherical grains of interest in bioceramic coatings (Gueorguiev et al., 2008b). The analysis on double-layer coatings showed that the share of spherical grains in the top HA and in the transition TiO_2 layer was less than 1% and about 1.5% of the layer's volume, respectively.

5.2 Phase composition and crystallographic parameters characterised with X-ray powder diffraction

X-ray powder diffraction (XRD) experiments are performed in two configurations, namely symmetric Bragg-Brentano (B-B) and Grazing Incidence Asymmetric Bragg Diffraction (GIABD) modes (Gueorguiev et al., 2008c). Their schematics are represented in Figure 4.

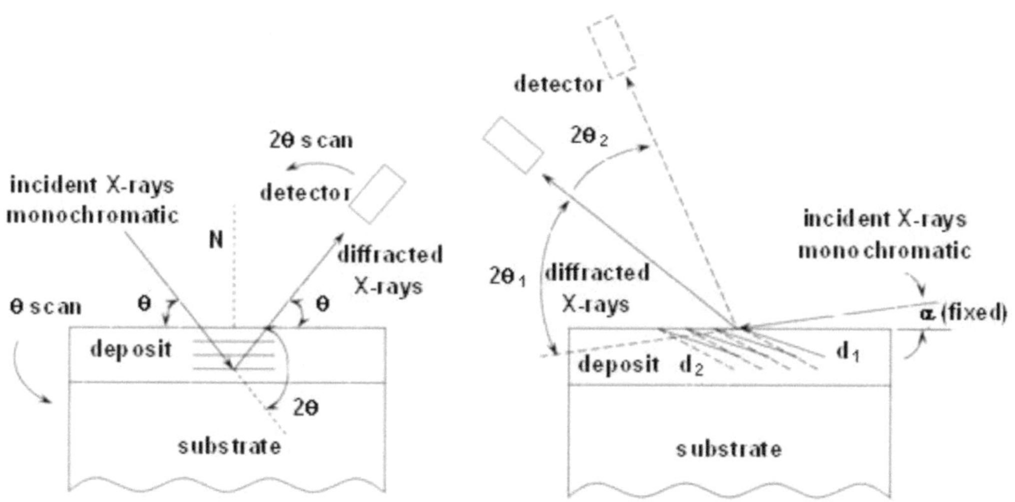

Fig. 4. Schematics of X-ray diffraction pattern registration in Bragg-Brentano (B-B, left) and Grazing Incidence Asymmetric Bragg Diffraction (GIABD, right) modes.

The registration in B-B mode is realized via standard synchronized θ-2θ scans of the sample and detector, where θ is the Bragg angle, and the registered patterns are resultant from crystallographic planes that are parallel to the surface (Figure 4, left).

In GIABD mode the sample is fixed in respect to the incident beam at a preset angle (α) and only the detector performs 2θ scans (Figure 4, right). Thus, the diffraction pattern is resultant from the interaction of the incident beam with the crystallographic planes in Bragg orientation, that are declined to the investigated surface at an angle $\psi = \theta - \alpha$ (Iordanova et al., 2002). In order to increase the ratio between the peak intensities and the background, a plane monochromator is usually used on the diffracted beam.

In both B-B and GIABD modes the XRD patterns have been registered stepwise with step size and counting times dependent on the parameter analyzed. The diffraction peaks have been background corrected and then least square fitted for estimation of their angular position (i.e. the peak centroid 2θ), intensity and broadening. Three analytical functions, namely Gaussian, Lorentzian or Pseudo-Voight have been used for this purpose. The analysed crystallographic parameters are described below.

5.2.1 Phase composition

In order to characterize the phase composition of the initial powders or plasma-sprayed ceramic coatings, XRD patterns have been registered in B-B mode with a relatively large angular interval (for example, 2θ between 15° and 105°) using Cu Kα characteristic radiation. Typical XRD patterns are presented in Figure 5.

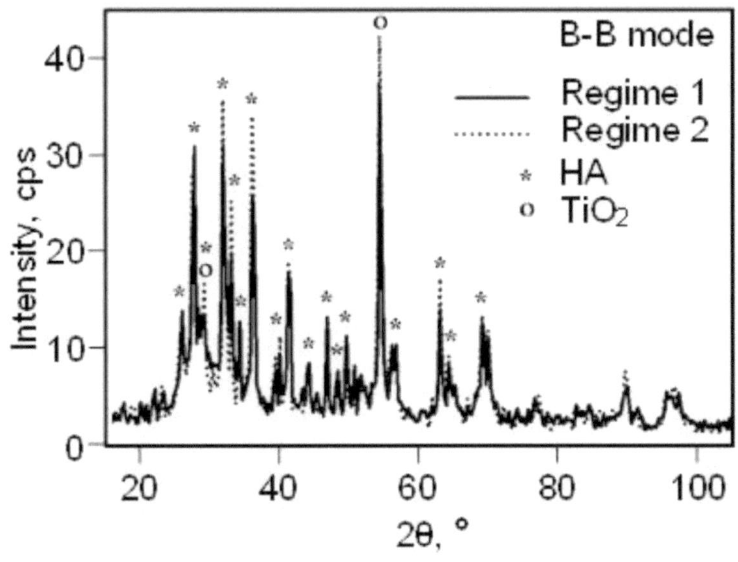

Fig. 5. XRD patterns registered in B-B mode from two double-layer bioceramic coatings with transition TiO_2 and top HA layers, deposited on TAV substrates in two different plasma-spraying regimes (Gueorguiev et al., 2008c).

The patterns are typical for materials with polycrystalline structure containing a number of diffraction peaks. However, the slightly increased hallo-like background within the angular interval 2θ between 23° and 35° points to the availability of a negligible amount of amorphous phase.

The qualitative phase analysis based on comparison between the angular positions 2θ of the registered diffraction peaks and the respective standard ASTM (JCPDS) values showed that the top and the transition layers correspond to crystalline HA with hexagonal unit cell, space group 176, $P6_3/m$ and β-TiO_2 with monoclinic symmetry, respectively (Gueorguiev et al., 2008c, 2009b). Similar patterns have been registered for the respective initial powders as well. The comparison between the XRD patterns of the analyzed layers and the respective injected powders showed a qualitative coincidence, suggesting no decomposition of the initial powders during the plasma-spraying in both regimes.

5.2.2 Line broadening analysis

Broadening of diffraction peaks is analyzed via their full width at half maximum (FWHM). In general, this parameter is a function of lattice imperfections and non-homogeneous elastic deformation in the investigated material. Concerning plasma-sprayed coatings it has been found that in most cases the FWHM is higher for thinner coatings and coatings, deposited at lower substrate temperature, compared to thicker coatings and coatings on substrates, preheated to higher temperature. In our analyses a decrease of the FWHM during the growth of plasma-sprayed HA coatings was observed, which is considered to result from the steadily increasing substrate temperature during plasma-spraying.

Basic parameter related to the crystallographic structure perfection is the physical broadening β_t that can be evaluated from the experimental FWHM. According to the Williamson-Hall model (Hemmati et al., 2006; Herrmann et al., 2002) the physical broadening of the XRD peaks is related to the crystallographic parameters of the investigated material according to equations (4) and (5) for Lorentzian and Gaussian fitting functions of the peak, respectively.

$$\beta_t = \frac{K\lambda}{D\cos\theta} + 2\varepsilon\tan\theta + \beta_0 \tag{4}$$

$$\beta_t = \sqrt{\left(\frac{K\lambda}{D\cos\theta}\right)^2 + (2\varepsilon\tan\theta)^2 + \beta_0^2} \tag{5}$$

K is the Sherrer constant, D is the dimension of the regions of coherent X-ray scattering, related to the crystalline size, ε is the lattice residual elastic micro-deformation, resultant from the available crystallographic defects and β_0 is an instrumental factor, influencing the line broadening.

5.2.3 Crystallographic texture analysis

It is well known that in plasma-sprayed coatings, being deposited in non-equilibrium conditions, crystallographic texture with axial symmetry, same as that of the spraying process, could be formed. According to the theoretical model, suggested by Iordanova and Forcey (Iordanova & Forcey, 1991; Gueorguiev et al., 2008a), if the energetic potential during plasma-spraying and the temperature at the substrate surface are relatively low, the axial (fiber, out-of-plane) texture is usually characterized with a preferable orientation of the most densely packed crystallographic planes of the coating crystallites parallel to the substrate surface.

Fiber crystallographic textures can be qualitatively characterized with one-dimensional orientation distribution functions (1D ODFs). For this purpose, a particular diffraction peak of the bioceramic coating with Bragg angle θ is registered in GIABD mode at several preset incident angles (α). Analogous registrations are performed for the same peak of a random sample without texture (for example, powder of the same material fixed with special glue in the sample holder). Then the intensities of the coating peak at different incident angles are normalized to the corresponding intensities of the peak, registered for the random standard. The plotted normalised intensities versus angle ψ (where ψ=θ-α) represent the 1D ODF that provides a qualitative information about the formed fiber texture.

The availability of preferred crystallographic orientation in plasma-sprayed films and coatings can be characterized quantitatively with pole density $P_{\{hkl\}}$, which is proportional to the probability that a particular family of crystallographic planes {hkl} is parallel to the sample surface. The pole density can be calculated according to the following equation:

$$P_{\{hkl\}} = \frac{I^{film}_{\{hkl\}} / I^{st}_{\{hkl\}}}{(1/n)\sum(I^{film}_{\{hkl\}} / I^{st}_{\{hkl\}})} \tag{6}$$

where $I^{film}_{\{hkl\}}$ and $I^{st}_{\{hkl\}}$ are the intensities of the {hkl} diffraction peaks registered in B-B mode for the analyzed deposits and the texture-less powder standard, respectively, and n is the number of the analyzed peaks (only the lowest diffraction orders are considered).

Two histograms representing the relative shares of the HA crystallites, oriented with their {hkl} crystallographic planes parallel to the coating surfaces, are presented in Figure 6. The two analysed coatings differ by the thickness of the top HA layer (thin - 20μm, thick - 40 μm). It can be concluded that in none of these HA coatings a pronounced crystallographic texture has been formed.

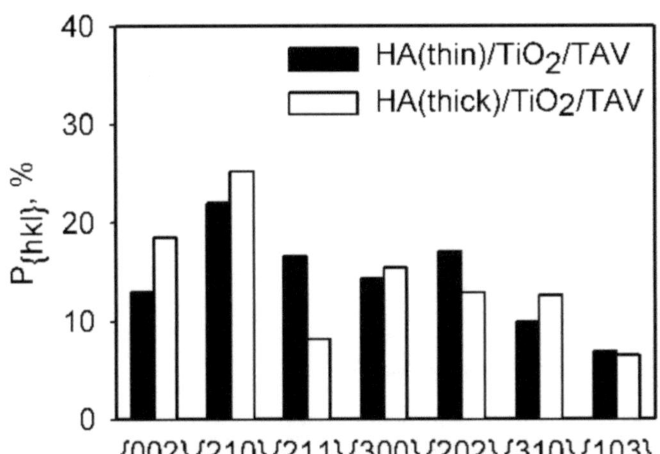

Fig. 6. Pole density histograms, representing the share of HA crystallites, oriented with their {hkl} planes parallel to the substrate surface in two double-layer plasma-sprayed coatings (thin and thick) with transition TiO_2 and top HA layers (Gueorguiev et al., 2009b).

5.2.4 Lattice parameters

Lattice unit-cell parameters of crystalline materials are among their most important crystallographic characteristics. They can give additional information about the phase composition and crystallographic perfection of the available phases. Lattice parameters can be evaluated from the XRD patterns via the interplanar distances $d_{\{hkl\}}$ according to the equation based on the Bragg law:

$$d_{\{hkl\}} = \lambda/2\sin\theta_{\{hkl\}} \quad (7)$$

where λ is the wavelength of the used characteristic radiation and $\theta_{\{hkl\}}$ is the experimentally defined centroid of the {hkl} XRD peak.

Evaluation of lattice parameters for crystallites belonging to the {hkl} texture component requires registration of the corresponding {hkl} diffraction peak in B-B mode. Such an approach has been applied for calculation of the lattice parameter (c) for two plasma-sprayed HA coatings with hexagonal unit cell as follows (Gueorguiev et al., 2008c). Firstly, the interplanar $d_{\{002\}}$ distance has been evaluated from the {002} diffraction peak of the HA coating, registered in B-B mode as described above. Then the following relation, valid for the hexagonal HA lattice, has been used:

$$\frac{1}{d^2} = \frac{4}{3}\frac{(h^2 + hk + k^2)}{a^2} + \frac{l^2}{c^2} \quad (8)$$

For {002} diffraction peaks it transforms to:

$$\frac{1}{d_{002}^2} = \frac{4}{c_{002}^2} \quad (9)$$

Finally, the $c_{\{002\}}$ parameter has been evaluated via the relation:

$$c_{\{002\}} = 2d_{\{002\}} \quad (10)$$

The evaluated lattice parameters $c_{\{002\}}$ for the plasma-sprayed HA coatings are presented in Figure 7 together with the standard ASTM value. It is obvious that the experimental parameters are higher than the standard due to fixed compressive residual macro-stresses, acting parallel to the HA coating surface (Gueorguiev et al., 2008c).

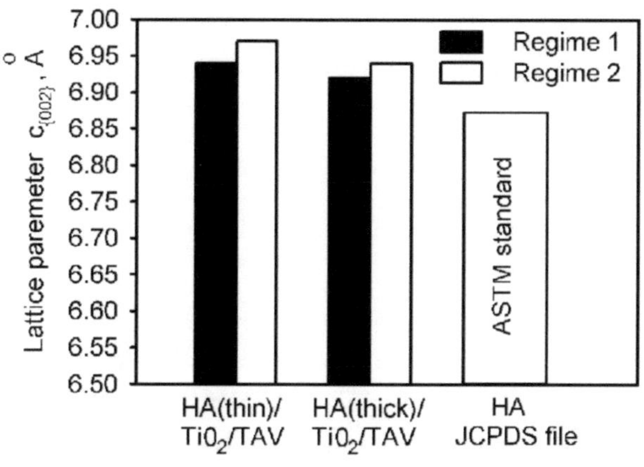

Fig. 7. Lattice parameters $c_{\{002\}}$ for the HA layer in two double-layer plasma-sprayed coatings with transition TiO_2 and top HA layers, differing in spraying regime and thickness, together with the respective standard JCPDS value (Gueorguiev et al., 2008c).

5.2.5 Residual macro-stresses

Residual macro-stresses are an important parameter of bioceramic coatings creating irreversible elastic deformation of the crystal lattice and significantly affecting properties and coating performance. Their evaluation via XRD methods is most often based on the Poisson's model, according to which the fixed residual stresses cause different elastic deformations in crystallites having different crystallographic orientations. The deformations can be evaluated via change of interplanar spacing in respect to the one existing prior to formation of residual stresses. The most suitable method for evaluation is called '$\sin^2\psi$' (Prevey, 1986), where a chosen {hkl} peak has to be firstly registered in GIABD mode at different incident angles (α), followed by evaluation of the interplanar spacings $d_{\{hkl\}}$. Then the dependence $d_{\{hkl\}}$ versus $\sin^2\psi$, where $\psi=\theta-\alpha$ needs to be plotted and fitted linearly as shown in Figure 8.

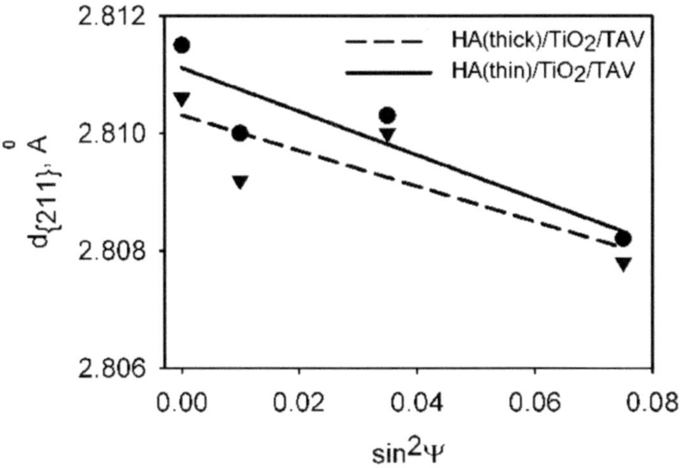

Fig. 8. Experimental data and least square linear fits of the dependencies of the interplanar HA spacing $d_{\{211\}}$ versus $\sin^2\psi$ for two double-layer plasma sprayed coatings with transition TiO_2 and top HA layers differing in thickness (Gueorguiev et al., 2009b).

According to the plane stress-strain model (Prevey, 1986) the values of the residual macro-stresses, acting parallel to the surface can be evaluated from the following equation (Prevey, 1986; Clyne & Gill, 1996):

$$\sigma = \frac{E}{1+\nu} \cdot \frac{1}{d_0} \cdot \left(\frac{\partial d_\psi}{\partial \sin^2 \psi} \right) \tag{11}$$

where E and ν are the Young's modulus and Poisson's ratio of the analysed material (HA in this particular case); d_0 and $\partial d_\psi / \partial \sin^2\psi$ are respectively the experimentally obtained y-intercept and gradient of the linear plots of the dependence $d_{\{hkl\}}$ versus $\sin^2\psi$.

As the XRD methods are based on diffraction and Young's modulus and Poisson's ratio are anisotropic parameters, also called X-ray elastic parameters as well, it is necessary to know their values in the crystallographic direction of the estimated interplanar distance. In addition, if E and ν have not been experimentally evaluated for the particularly investigated material, they would need some corrections. For example, it is known that air plasma-sprayed coatings usually contain between 15 and 30 volume percent pores that are expected to decrease the E/v ratio about five times in respect that one obtained for monocrystals or bulk materials with the same composition (Kuroda & Clyne, 1991).

According to equation (11) for residual stress evaluation, the negative gradients observed in both fitted linear plots in Figure 8 indicate existence of compressive (negative) residual macro-stresses, acting parallel to the surface. Contrary to that, positive gradients indicate tensile (positive) residual stresses.

According to the model suggested by Kuroda and Clyne (Kuroda & Clyne, 1991), main components of fixed macro-stresses in plasma-sprayed coatings are quenching and thermal stresses. They can be quantitatively estimated from the equations offered by the same authors. However, some additional factors, such as possible plastic deformation, polymorphous phase transformations, relaxation processes (for example, cracking and bending), that could accompany the coating formation, must also be considered.

5.3 Porosity characterization

Due to micro-pores contained in the initial powder material or incomplete bonding between the lamellae in plasma-sprayed bioceramic coatings, a certain degree of porosity exists (Li & Ohmori, 2002; Li et al., 2004).

The relative porosity p is defined as follows [84]:

$$p[\%] = \left(1 - \frac{\rho_1}{\rho_0}\right) \cdot 100 \tag{12}$$

where ρ_1 and ρ_0 are coating and initial powder material densities, respectively (Tsui et al., 1998).

A suitable and relatively simple method for porosity estimation is a hydrostatic weighing method, involving measurements of the weight in air and in a liquid with known density (usually distilled water) (Gergov et al., 1990; Iordanova et al., 1995). For analysis of a plasma-sprayed coating, it is firstly necessary to separate a part of it from the substrate. Prior to immersing in the liquid, the analysed sample needs to be covered with vaseline in order to isolate its surface pores from the liquid. Then the porosity can be evaluated according to the following formula:

$$p[\%] = \left(1 - \frac{\dfrac{W_z \cdot \rho_w}{\rho_z}}{W - W' - \dfrac{W_v \cdot \rho_w}{\rho_v}}\right) \cdot 100 \tag{13}$$

where W_z is the weight of the coating in air, ρ_z is the density of the initial powder, W is the weight of the coating with vaseline and the elastic thread from which it is hung during weighing, W' is the weight of the sample on the thread immersed in water, W_v is the weight of the applied vaseline film, ρ_v is the density of vaseline (handbook value usually taken as 0.88 g/cm^3) and ρ_w is the distilled water density.

The density of the initial powder ρ_z can be estimated by subsequent weighing in air and in distilled water from the following relation:

$$\rho_z = \frac{G_3 - G_2}{V_k - \dfrac{G_4 - G_3}{\rho_w}} \tag{14}$$

where G_3 is the weight of powder with the test tube, G_2 is the weight of the empty test tube, V_k is the volume of the powder and water in the test tube, G_4 is the weight of the test tube with powder and distilled water, and ρ_w is the distilled water density.

The porosity of plasma-sprayed ceramic coatings for medical needs must be controlled to be relatively low in order to achieve better adhesion, suitable mechanical properties and lower brittleness of the deposits. On the other hand, higher porosity is expected to enhance bone ingrowth during the healing process in the human body.

5.4 Tensile bond strength between substrate and plasma-sprayed coating

Adhesion (tensile bond strength) between substrate and plasma-sprayed coating is usually measured via pull-out mechanical tests according to ASTM C633-79 (Ding et al., 2001; Gueorguiev et al., 2008c). For this purpose, prior to testing the coated surface is glued to a grit-blasted stainless-steel cylindrical counter body with special adhesives (for example, HTK Ultra Bond, Hanseatisches Technologie Kontor GmbH). Then a tensile force is applied to deplete the counter body. A schematic of a pull-out test, together with a picture of a test setup, are shown in Figure 9.

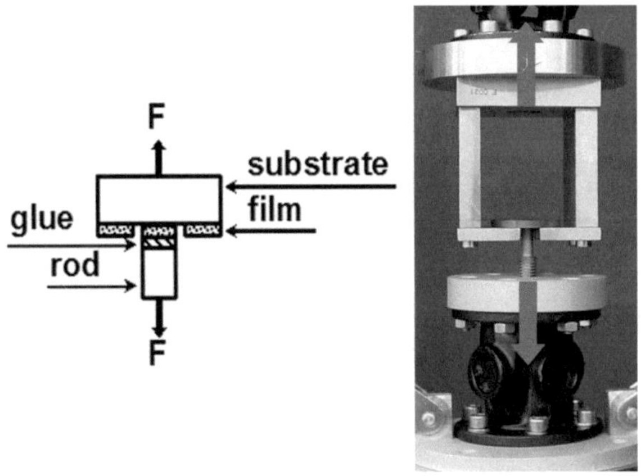

Fig. 9. Schematic of a pull-out test (left) and example picture of a test setup (right) (Gueorguiev et al., 2008c).

Several coatings of each kind are usually tested and the average of the maximum applied tensile mechanical stress γ, calculated according to the formula below, is defined as tensile bond strength of the deposit:

$$\gamma = \frac{F_{max}}{S} \tag{15}$$

where F_{max} is the depletion force and S is the area of the glued cylindrical counter body.

In general, similar tests can be applied for evaluation of the cohesion between different layers in multi-layer coatings.

5.5 Numerical methods for investigation of structural and mechanical parameters based on the density functional theory (DFT)

In order to estimate precisely residual stresses in HA bioceramic deposits, their Young's modules $E_{<uvw>}$ and Poisson's ratios $\nu_{<uvw>}$ in different crystallographic directions <uvw> have to be evaluated. A suitable way way can be the development of a numerical method, based on the density functional theory (DFT). This is a quantum mechanical theory to investigate electronic structure of many-body systems and condensed phases, where the electronic structure can be determined using functionals, i.e. functions of another function, the latter being in this case the electron density distribution. DFT is one of the most reliable and prospective methods for evaluation of structure and elastic properties of such systems. The calculations can be performed using appropriate software packages (for example, Quantum Espresso program code) to simulate the unit cell and estimate the values of the elastic constant matrix C_{ij} and elastic compliance matrix s_{ij} of the respective monocrystals (Giannozzi et al., 2009).

Then the $E_{<uvw>}$ and $\nu_{<uvw>}$ values can be calculated by means of pseudo-potentials and known crystallophysics relationships. The main algorithms are published in (Gonze, 1996; Monkhorst & Pack, 1976; Schlegel, 1982). Verification of the results is performed by comparison of the simulated unit cell parameters to those obtained from respective standard ASTM files. In a previous paper (Gueorguiev et al., 2009a) DFT was applied to simulate the crystallographic structure and estimate elastic constant and elastic compliance matrixes of HA monocrystal with hexagonal symmetry. The obtained HA unit cell is shown in Figure 10. Its symmetry, corresponding to $P6_3/m$ (space group 176) together with the values of the parameters a=9.418 Å and c=6.875 Å are in very good agreement with those from the standard ASTM file JCPDS 34-0010 for $P6_3/m$ HA (space group 176), which is a proof of correctly performed numerical simulations.

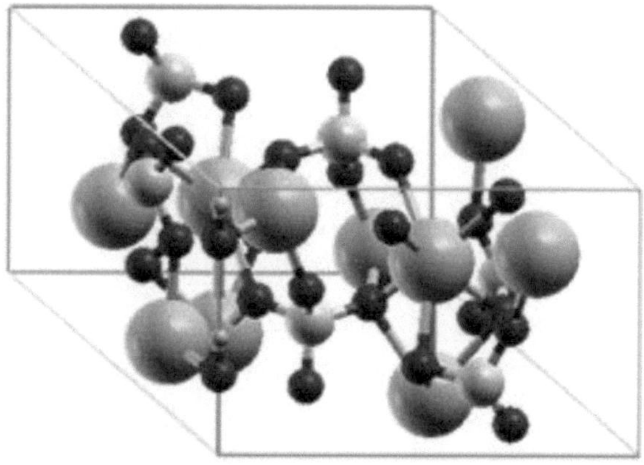

Fig. 10. DFT simulated unit cell of HA, space group 176, $P6_3/m$, with Ca, O, P and H atoms presented with big gray, black, middle-sized gray and small gray balls, respectively.

The values of the Young's modulus $E_{<uvw>}$ and Poisson's ratio $\nu_{<uvw>}$ have been derived from the values of the elastic constant matrix C_{ij} and elastic compliance matrix s_{ij} and the parameters of the unit cell a and c via the following relations (Sirotin & Shaskolskaya, 1983):

$$E_{<hkl>}^{-1}=\frac{\left\{\left(h^2+k^2-hk\right)^2 a^4 s_{11}+l^4c^4 s_{33}+\left(h^2+k^2-hk\right)l^2a^2c^2\left(s_{44}+2s_{13}\right)\right\}}{\left[\left(h^2+k^2-hk\right)a^2+l^2c^2\right]^2}+ \\ +\frac{a^3c\left[3\sqrt{3}hkl(h-k)s_{14}+l(2h-k)\left(2k^2-h^2+hk\right)s_{25}\right]}{\left[\left(h^2+k^2-hk\right)a^2+l^2c^2\right]^2} \quad (16)$$

$$\nu_{<hkl>}=\frac{3B-E_{<hkl>}}{6B} \quad (17)$$

where

$$B=\frac{2c_{11}+C_{33}+2C_{12}+4C_{13}}{9} \quad (18)$$

The calculated DFT values of the Young's modulus $E_{<uvw>}$ and Poisson's ratio $\nu_{<uvw>}$ in different crystallographic directions <uvw> of the $P6_3/m$ HA unit cell are given in Table 2.

Cristallographic direction	Young's modulus, GPa	Poisson's ratio
<102>	96	0.313
<103>	94	0.315
<311>	96	0.313
<002>	93	0.319
<211>	98	0.309
<202>	98	0.309
<210>	96	0.313

Table 2. Young's modulus and Poisson's ratio in different crystallographic directions of a $P6_3/m$ HA unit cell.

The values of $E_{<211>}$ = 98 GPa and $\nu_{<211>}$ = 0.309 were further used for a more precise estimation of the residual stresses in plasma-sprayed double-layer coatings with TiO_2 transition and top HA layers (Gueorguiev et al., 2009a).

3. Conclusion

Despite the recent advances in research on structural parameters of plasma-sprayed bioceramic coatings and their influence on the operational properties, fundamental knowledge in this field is still insufficient. Non-destructive methods, focused on investigation and control of crystallographic structure and mechanical properties of such coatings are described and adopted. In addition, the complex approach for fundamental and applicable interdisciplinary analyses contributes to better understanding of the structure formation processes from a physical point of view and has a positive impact on the practical applications of plasma-sprayed bioceramic coatings.

4. Acknowledgment

Prof. R. Geoff Richards, director of the AO Research Institute Davos for fruitful discussions and expert opinions.

National Science Fund grants VU-F-205/2006, DO-02-136/2008 and DO-02-167/2008 for calculation time on PHYSON computer cluster.

5. References

Azarmi, F.; Coyle, T. & Mostaghimi, J. (2008). Optimization of Atmospheric Plasma Spray Process Parameters using a Design of Experiment for Alloy 625 coatings. *Journal of Thermal Spray Technology*, Vol.17, No.1, (March 2008), pp. 144-155, ISSN 1059-9630

Brossa, F. & Lang, E. (1992). Plasma Spraying - a Versatile Coating Technique, In: *Mechanical and Materials Science: Advanced Techniques for Surface Engineering*, Gissler, W. & Jehn, H. (Eds.), pp. 199-252, ISBN 0-7923-2006-9, Kluwer Academic Publishers, London, UK

Boulos, M.; Fauchais, P.; Vandelle, A. & Pfender, E. (1993). Fundamentals of Plasma Particle Momentum and Heat Transfer, In: *Plasma Spraying: Theory and Applications*, Suryanarayanan, R. (Ed.), pp. 3-60, ISBN 981-02-1363-8, World Scientific, New Jersey, USA

Burgess, A.; Story, B.; La, D.; Wagner, W & LeGeros, J. (1999). Highly Crystalline MP-1 Hydroxylapatite Coating Part I: In Vitro Characterization and Comparison to Other Plasma-Sprayed Hydroxylapatite Coatings. *Clinical Oral Implants Research*, Vol.10, No.4, (August 1999), pp. 245-256, ISSN 1600-0501

Celik, E.; Demirkiran, A. & Avci, E. (1999). Effect of Grit Blasting of Substrate on the Corrosion Behaviour of Plasma-Sprayed Al_2O_3 Coatings. *Surface and Coatings Technology*, Vol.116-119, No.1, (September 1999), pp. 1061-1064, ISSN 0257-8972

Chen, S.; Siitonen, P. & Kettunen, P. (1993). Experimental Design and Optimization of Plasma Sprayed Coatings, In: *Plasma Spraying: Theory and Applications*, Suryanarayanan, R. (Ed.), pp. 95-120, ISBN 981-02-1363-8, World Scientific, New Jersey, USA

Choi, H.; Kang, B.; Choi, W.; Choi, D., Choi, S.; Kim, J.; Park, Y & Kim, G. (1998). Effect of the Thickness of Plasma-Sprayed Coating on Bond Strength and Thermal Fatigue Characteristics. *Journal of Materials Science*, Vol.33, No.24, (December 1998), pp. 5895-5899, ISSN 0022-2461

Chou, B. & Chang, E. (2002). Plasma-Sprayed Hydroxyapatite Coating on Titanium Alloy with ZrO_2 Second Phase and ZrO_2 Intermediate Layer. *Surface and Coatings Technology*, Vol.153, No.1, (April 2002), pp. 84-92, ISSN 0257-8972

Clyne, T. & Gill, S. (1996). Residual Stresses in Thermal Spray Coatings and Their Effect on Interfacial Adhesion: A Review of Recent Work. *Journal of Thermal Spray Technology*, Vol.5, No.4, (December 1996), pp. 401-418, ISSN 1059-9630

Cofino, B.; Fogarassy, P.; Millet, P. & Lodini, A. (2004). Thermal Residual Stresses near the Interface between Plasma Sprayed Hydroxyapatite and Titanium Subtrate. *Journal of Biomedical Materials Research Part A*, Vol.70A, No.1, (July 2004), pp. 20-27, ISSN 1549-3296

Dhiman, R. & Chandra, S. (2007). Predicting Splat Morphology in a Thermal Spray Process, In: *Thermal Spray 2007: Global Coating Solutions,* Marple, B.; Hyland, M.; Lau, Y.; Lima, R. & Montavon, G. (Eds.), pp. 207-212, ASM International, Materials Park, Ohio, USA

Ding, S.; Su, Y.; Ju, C. & Lin, J. (2001). Structure and Immersion Behavior of Plasma-Sprayed Apatite-Matrix Coatings. *Biomaterials,* Vol.22, No.8, (April 2001), pp. 833-845, ISSN 0142-9612

Duan, K. & Wang, R. (2006). Surface Modifications of Bone Implants through Wet Chemistry. *Journal of Materials Chemistry,* Vol.6, No.24, (April 2006), pp. 2309-2322, ISSN 0959-9428

Dyshlovenko, S.; Pawlowski, L. & Roussel, P. (2005). Experimental Investigation of Influence of Plasma Spraying Operational Parameters on Properties of Hydroxyapatite, *Proceedings of the International Thermal Spray Conference,* pp. 726-731, Basel, Switzerland, May 2-4, 2005

Fan, Q.; Wang, L.; Wang, F. & Wang, Q. (2005). Modeling of Temperature and Residual Stress Fields Resulting from Impacting Process of a Molten Ni Particle onto a Flat Substrate, *Proceedings of the International Thermal Spray Conference,* pp. 275-279, Basel, Switzerland, May 2-4, 2005

Frayssinet, P.; Hardy, D. & Rouquet, N. (2006). The Role of Hydroxylapatite Coating Characteristics in Bone Integration after Two Decades of Follow-up in Human Beings, *Proceedings of the International Thermal Spray Conference,* pp. 35-40, Seattle, Washington, USA, May 15-18, 2006

Gaona, M.; Lima, R. & Marple, B. (2007). Nanostructured Titania/Hydroxyapatite Composite Coatings Deposited by High Velocity Oxy-Fuel (HVOF) Spraying. *Materials Science and Engineering: A,* Vol.458, No.1-2, (June 2007), pp. 141-149, ISSN 0921-5093

Gergov,B.; Iordanova, I. & Velinov, T. (1990). A Complex Investigation of Structure and Properties of Thermally Sprayed Ni and Cu-Based Coatings. *Revue de Physique Appliquee,* Vol.20, No.12, (December 1990), pp. 1197-1204, ISSN 0302-0738

Giannozzi, P.; Baroni, S.; Bonini, N.; Calandra, M.; Car, R.; Cavazzoni, C.; Ceresoli, D.; Chiarotti, G. L.; Cococcioni, M.; Dabo, I.; Dal Corso, A.; Fabris, S.; Fratesi, G.; de Gironcoli, S.; Gebauer, R.; Gerstmann, U.; Gougoussis, C.; Kokalj, A.; Lazzeri, M.; Martin-Samos, L.; Marzari, N.; Mauri, F.; Mazzarello, R.; Paolini, S.; Pasquarello, A.; Paulatto, L.; Sbraccia, C.; Scandolo, S.; Sclauzero, G.; Seitsonen, A. P.; Smogunov A.; Umari, P. & Wentzcovitch, R. M. (2009). Quantum Espresso: a Modular and Open-Source Software Project for Quantum Simulations of Materials. *Journal of Physics: Condensed Matter,* Vol.21, No.39, (September 2009), pp. 395502-395521, ISSN 0953-8984

Gill, B. & Tucker, R. (1986). Plasma Spray Coating Processes. *Journal of Materials Science and Technology,* Vol.2, No.1, (January 1986), pp. 207-213, ISSN 1005-0302

Gonze, X. (1996). Towards a Potential-Based Conjugate Gradient Algorithm for Order-N Self-Consistent Total Energy Calculations. *Physical Review B: Condensed Matter and Materials Physics,* Vol.54, No.7, (August 1996), pp. 4383-4386, ISSN 0163-1829

Gueorguiev, B.; Iordanova, I. & Sprecher, C. (2008). Crystallography of Hydroxylapatite Films Applied by Flame Spraying on TAV Substrates, *Proceedings of the 15th Workshop on Plasmatechnik,* pp. 25-32, Ilmenau, Germany, June 26-27, 2008

Gueorguiev, B.; Iordanova, I. & Sprecher, C. (2008). Evaluation of Spherical Grains in Flame-Sprayed Coatings for Medical Purposes by Stereological Methods. *Bulgarian Journal of Physics,* Vol.35, No.2, (December 2008), pp. 119-128, ISSN 1310-0157

Gueorguiev, B.; Iordanova, I.; Sprecher, C. & Skulev, H. (2008). Surface Engineered Titanium Alloys by Application of Bioceramic Coatings for Medical Purposes. *Galvanotechnik,* Vol.99, No.12, (December 2008), pp. 3070-3076, ISSN 0016-4232

Gueorguiev, B.; Sprecher, C.; Antonov, V.; Iordanova, I. & Skulev, H. (2009). Investigation of TiO_2 and HA plasma-Sprayed Biomedical Coatings by Structural and DFT Methods, *Proceedings of the 16th Workshop on Plasmatechnik*, pp. 20-26, Ilmenau, Germany, June 25-26, 2009

Gueorguiev, B.; Iordanova, I.; Sprecher, C.; Skulev, H.; Wahl, D. & Hristov, A. (2009). Plasma-Spraying Methods for Applications in the Production of Quality Biomaterials for Modern Medicine and Dentistry. *Journal of Optoelectronics and Advanced Materials*, Vol.11, No.1, (September 2009), pp. 1331-1334, ISSN 1454-4164

Harsha, S.; Dwivedi, D. & Agarwal, A. (2008). Performance of Flame Sprayed Ni-WC Coating under Abrasive Wear Conditions. *Journal of Materials Engineering and Performance*, Vol.17, No.1, (February 2008), pp. 104-110, ISSN 1059-9495

Heimann, R. (2002). Materials Science of Crystalline Bioceramics: a Review of Basic Properties and Applications. *Chiang Mai University Journal of Natural Sciences*, Vol.1, No.1, (January 2002), pp. 23-46, ISSN 16851994

Heimann, R.; Schürmann, N. & Müller, R. (2004). In Vitro and In Vivo Performance of Ti6Al4V Implants with Plasma-Sprayed Osteoconductive Hydroxylapatite–Bioinert Titania Bond Coat "Duplex" Systems: an Experimental Study in Sheep. *Journal of Materials Science: Materials in Medicine*, Vol.15, No.9, (September 2004), pp. 1045-1052, ISSN 0957-4530

Hemmati, I.; Hosseini, H. & Kianvash, A. (2006). The Correlations between Processing Parameters and Magnetic Properties of an Iron–Resin Soft Magnetic Composite. *Journal of Magnetism and Magnetic Materials*, Vol.305, No.1, (October 2006), pp. 147-151, ISSN 0304-8853

Herrmann, M.; Engel, W. & Göbel, H. (2002). Micro Strain in HMX Investigated with Powder X-Ray Diffraction and Correlation with the Mechanical Sensitivity, *Proceedings of the 51th Annual Denver X-Ray Conference Advances in X-Ray Analysis. JCPDS-International Centre for Diffraction Data, Advances in X-Ray Analysis*, Vol.45, No.1, (2002), pp. 212-217, Colorado Springs, Colorado, USA, July 29- August 2, 2002

Iordanova, I. & Forcey, K. (1991). Investigation by Rutherford Back-Scattering Spectrometry of Tin Coatings Electrolytically Applied on Steel Strip. *Materials Science and Technology*, Vol.7, No.1, (January 1991), pp. 20-23, ISSN 0267-0836

Iordanova, I.; Forcey, K.; Valtcheva, J. & Gergov, B. (1994). An X-ray Study of Thermally-Sprayed Metal Coatings. *Materials Science Forum*, Vol.166-169, No.1, (June 1994), pp. 319-324, ISSN 0255-5476

Iordanova, I.; Forcey, K.; Gergov, B. & Bojinov, V. (1995). Characterization of Flame-Sprayed and Plasma-Sprayed Pure Metallic and Alloyed Coatings. *Surface and Coatings Technology*, Vol.72, No.1-2, (May 1995), pp. 23-29, ISSN 0257-8972

Iordanova, I. & Forcey, K. (1997). Texture and Residual Stresses in Thermally Sprayed Coatings. *Surface and Coatings Technology*, Vol.91, No.3, (May 1997), pp. 174-182, ISSN 0257-8972

Iordanova, I.; Surtchev, M. & Forcey, K. (2001). Metallographic and SEM Investigation of the Microstructure of Thermally Sprayed Coatings on Steel Substrates. *Surface and Coatings Technology*, Vol.139, No.2-3, (May 2001), pp. 118-126, ISSN 0257-8972

Iordanova, I.; Antonov, V. & Gurkovsky, S. (2002). Changes of Microstructure and Mechanical Properties of Cold-Rolled Low Carbon Steel Due to Its Surface Treatment by Nd:Glass Pulsed Laser. *Surface and Coatings Technology*, Vol.153, No.2-3, (April 2002), pp. 174-182, ISSN 0257-8972

Keller, N.; Bertrand, G.; Coddet, C. & Meunier, C. (2003). Influence of Plasma Spray Parameters on Microstructural Characteristics of TiO_2 Deposits, In: *Thermal Spray*

2003: Advancing the Science & Applying the Technology, Moreau, C. & Marple, B. (Eds.), pp. 1403-1408, ASM International, Materials Park, Ohio, USA

Khor, K.; Gu, Y.; Pan, D. & Cheang, P. (2004). Microstructure and Mechanical Properties of Plasma Sprayed HA/YSZ/Ti-6Al-4V Composite Coatings. *Biomaterials,* Vol.25, No.18, (August 2004), pp. 4009-4017, ISSN 0142-9612

Kreye, H.; Schwetzke, R. & Zimmermann, S. (2007). High Velocity Oxy-Fuel Flame Spraying Process and Coating Characteristics, In: *Thermal Spray 1996: Practical Solutions for Engineering Problems,* Berndt, C. (Ed.), pp. 451-456, ASM International, Materials Park, Ohio, USA

Kuroda, S. & Clyne, T. (1991). The Quenching Stress in Thermally Sprayed Coatings. *Thin Solid Films,* Vol.200, No.1, (May 1991), pp. 49-66, ISSN 0040-6090

LeGeros, W. (2002). Properties of Osteoconductive Biomaterials: Calcium Phosphates. *Clinical Orthopaedics and Related Research,* Vol.395, No.1, (February 2001), pp. 81-95, ISSN 1528-1132

Leigh, S. & Berndt, C. (1997). Evaluation of Off-Angle Thermal Spray. *Surface and Coatings Technology,* Vol.89, No.3, (March 1997), pp. 213-224, ISSN 0257-8972

Li, C. & Ohmori, A. (2002). Relationships between the Microstructure and Properties of Thermally Sprayed Deposits. *Journal of Thermal Spray Technology,* Vol.11, No.3, (September 2002), pp. 365-374, ISSN 1059-9630

Li, H.; Khor, K. & Cheang, P. (2004). Thermal Sprayed Hydroxyapatite Splats: Nanostructures, Pore Formation Mechanisms and TEM Characterization. *Biomaterials,* Vol.25, No.17, (August 2004), pp. 3463-3471, ISSN 0142-9612

Li, Y. & Khor, K. (2002). Microstructure and Composition Analysis in Plasma Sprayed Coatings of Al_2O_3/$ZrSiO_4$ Mixtures. *Surface and Coatings Technology,* Vol.150, No.2-3, (February 2002), pp. 125-132, ISSN 0257-8972

Liu, F.; Zeng, K.; Wang, H.; Zhao, X.; Ren,X. & Yu, Y. (2005). Numerical Investigation on the Heat Insulation Behaviour of Thermal Spray Coating by Unit Cell Model, *Proceedings of the International Thermal Spray Conference,* pp. 806-810, Maastricht, Netherlans, June 2-4, 2008

Liu, X.; Zhao, X. & Ding, C. (2006). Introduction of Bioactivity to Plasma Sprayed TiO_2 Coating with Nanostructured Surface by Post-Treatment, *Proceedings of the International Thermal Spray Conference,* pp. 53-57, Seattle, Washington, USA, May 15-18, 2006

Lugscheiber, E.; Oberländer, B. & Rouhaghdam, A. (1993). Optimising the APS-Process Parameters for New Ni-Hardfacing Alloys Using a Mathematical Model, In: *Plasma Spraying: Theory and Applications,* Suryanarayanan, R. (Ed.), pp. 141-162, ISBN 981-02-1363-8, World Scientific, New Jersey, USA

Mawdsley, J.; Su, Y.; Faber, K. & Bernecki, T. (2001). Optimization of Small-Particle Plasma-Sprayed Alumina Coatings Using Designed Experiments. *Materials Science and Engineering: A,* Vol.308, No.1-2, (June 2001), pp. 189-199, ISSN 0921-5093

McPherson, R.; Gane, N. & Bastow, J. (1995). Structural haracterization of Plasma-Sprayed Hydroxylapatite Coatings. *Journal of Materials Science: Materials in Medicine,* Vol.6, No.6, (June 1995), pp. 327-334, ISSN 0957-4530

Monkhorst, H. & Pack, J. (1976). Special Points for Brillouin-Zone Integrations. *Physical Review B: Condensed Matter and Materials Physics,* Vol.13, No.12, (June 1976), pp. 5188-5192, ISSN 0163-1829

Montavon, A.; Sampath, S.; Berndt, C.; Herman, H. & Coddet, C. (1997). Effects of the Spray Angle on Splat Morphology During Thermal Spraying. *Surface and Coatings Technology,* Vol.91, No.1-2, (May 1997), pp. 107-115, ISSN 0257-8972

Moreau, C.; Bisson, J.; Lima, R. & Marple, B. (2005). Diagnostics for Advanced Materials Processing by Plasma Spraying. *Pure and Applied Chemistry,* Vol.77, No.2, (February 2005), pp. 443-462, ISSN 0033-4545

Morris, H. & Ochi, K. (1998). Hydroxyapatite-Coated Implants: a Case for Their Use. *Journal of Oral and Maxillofacial Surgery,* Vol.56, No.11, (November 1998), pp. 1303-1311, ISSN 0278-2391

Nelea, V.; Mihailescu, I. & Jelinek, M. (2007). Biomaterials: New Issues and Breakthroughs for Biomedical Applications, In: *Pulsed Laser Deposition of Thin Films: Applications-Led Growth of Functional Materials,* Eason, R. (Ed.), pp. 421-459, John Wiley & Sons, Inc., London, UK

Neufuss, K.; Ilavsky, J.; Kolman, B.; Dubsky, J.; Rohan, P. & Chraska, P. (2001). Variation of Plasma Spray Deposits Microstructure and Properties Formed by Particles Passing through Different Areas of Plasma Jet. *Ceramics-Silikaty,* Vol.45, No.1, (March 2001), pp. 1-8, ISSN 0862-5468

Nicoll, A.; Gruner, H.; Prince, R. & Wuest, G. (1985). Thermal Spray Coatings for High Temperature Protection. *Surface Engineering,* Vol.1, No.1, (February 1985), pp. 59-71, ISSN 0267-0844

Ohmori, A. & Li, C. (1991). Quantitative Characterization of the Structure of Plasma-Sprayed Al_2O_3 Coating by Using Copper Electroplating. *Thin Solid Films,* Vol.201, No.2, (June 1991), pp. 241-252, ISSN 0040-6090

Parizi, H.; Rosenzweig, L.; Mostaghimi, J.; Chandra, S.; Coyle, T.; Salimi, L.; Pershin, L.; McDonald, A. & Moreau, C. (2007). Numerical Simulation of Droplet Impact on Patterned Surfaces, In: *Thermal Spray 2007: Global Coating Solutions,* Marple, B.; Hyland, M.; Lau, Y.; Lima, R. & Montavon, G. (Eds.), pp. 213-218, ASM International, Materials Park, Ohio, USA

Prevey, P. (1986). X-Ray Diffraction Residual Stress Techniques, In: *Metals Handbook Volume 10: Materials Characterization,* Mills, K.; Davis, J.; Destefani, J. & Dietrich, D. (Eds.), pp. 380-392, ISBN 981-02-1363-8, ASM International, Materials Park, Ohio, USA

Salman, S.; Cal, B.; Gunduz, O.; Agathopoulos, S. & Oktar, F. (2008). The Influence of Bond-Coating on Plasma Sprayed Alumina-Titania, Doped with Biologically Derived Hydroxyapatite, on Stainless Steel, In: *Virtual and Rapid Manufacturing,* Bartolo, P. (Ed.), pp. 289-292, ISBN 978-0-415-41602-3, Taylor & Francis Group, London, UK

Sampath, S.; Jiang, X.; Kulkarni, A.; Matejicek, J.; Gilmore, D. & Neiser, R. (2003). Development of Process Maps for Plasma Spray: Case Study for Molybdenum. *Materials Science and Engineering: A,* Vol.348, No.1-2, (May 2003), pp. 54-66, ISSN 0921-5093

Sarikaya, O. (2005). Effect of Some Parameters on Microstructure and Hardness of Alumina Coatings Prepared by the Air Plasma Spraying Process. *Surface and Coatings Technology,* Vol.190, No.2-3, (January 2005), pp. 388-393, ISSN 0257-8972

Schlegel, H. (1982). Optimization of Equilibrium Geometries and Transition Structures, Vol.3, No.2, (July 1982), pp. 214-218, ISSN 1096-987X

Sirotin, Y. & Shaskolskaya, M. (1983). *Fundamentals of Crystal Physics*, ISBN 978-082-8524-64-3, Imported Publications, New York, USA

Skulev, H.; Malinov, S.; Sha, W. & Basheer, P. (2005). Microstructural and Mechanical Properties of Nickel-Base Plasma Sprayed Coatings on Steel and Cast Iron Substrates. *Surface and Coatings Technology,* Vol.197, No.2-3, (July 2005), pp. 177-184, ISSN 0257-8972

Soares, P.; Mikowski, A.; Lepienski, C.; Santos, E.; Soares, G.; Filho, V. & Kuromoto, N. (2008). Hardness and Elastic Modulus of TiO_2 Anodic Films Measured by Instrumented Indentation. *Journal of Biomedical Materials Research Part B: Applied Biomaterials,* Vol.84B, No.2, (February 2008), pp. 524-530, ISSN 1552-4973

Srinivasan, V.; Sampath, S.; Vaidya, A.; Streibl, T. & Friis, M. (2006). On the Reproducibility of Air Plasma Spray Process and Control of Particle State. *Journal of Thermal Spray Technology,* Vol.15, No.4, (December 2006), pp. 739-743, ISSN 1059-9630

Staia, M.; Valente, T.; Bartuli, C.; Lewis, D.; Constable, C.; Roman, A.; Lesage, J.; Chicot, D. & Mesmacque, G. (2001). Part II: Tribological Performance of Cr_3C_2-25% NiCr Reactive Plasma Sprayed Coatings Deposited at Different Pressures. *Surface and Coatings Technology,* Vol.146-147, No.1, (September-October 2001), pp. 563-570, ISSN 0257-8972

Streibl, T.; Vaidya, A.; Friis, M.; Srinivasan, V. & Sampath, S. (2006). A Critical Assessment of Particle Temperature Distributions During Plasma Spraying: Experimental Results for YSZ. *Plasma Chemistry and Plasma Processing,* Vol.26, No.1, (February 2006), pp. 73-102, ISSN 0272-4324

Suchanek, W. & Yoshimura, M. (1998). Processing and Properties of Hydroxyapatite-Based Biomaterials for Use as Hard Tissue Replacement Implants. *Journal of Materials Research,* Vol.13, No.1, (January 1998), pp. 94-117, ISSN 0884-2914

Sun, L.; Berndt, C.; Khor, K.; Cheang, H. & Gross, K. (2002). Surface Characteristics and Dissolution Behavior of Plasma-Sprayed Hydroxyapatite Coating. *Journal of Biomedical Materials Research Part A,* Vol.62A, No.2, (November 2002), pp. 228-236, ISSN 1549-3296

Thomas, K. (1994). Hydroxyapatite Coatings. *Orthopedics,* Vol.17, No.3, (March 1994), pp. 267-278, ISSN 0147-7447

Toma, L.; Keller, N.; Bertrand, G.; Klein, D. & Coddet, C. (2003). Elaboration and Characterization of Environmental Properties of TiO_2 Plasma Sprayed Coatings. *International Journal of Photoenergy,* Vol.5, No.3, (March 2003), pp. 141-151, ISSN 1110-662X

Tong, W.; Chen, J.; Li, X.; Cao, Y.; Yang, Z.; Feng, J. & Zhang, X. (1996). Effect of Particle Size on Molten States of Starting Powder and Degradation of the Relevant Plasma-Sprayed Hydroxyapatite Coatings. *Biomaterials,* Vol.17, No.15, (August 1996), pp. 1507-1513, ISSN 0142-9612

Tsui, Y.; Doyle, C. & Clyne, T. (1998). Plasma Sprayed Hydroxyapatite Coatings on Titanium Substrates Part 1: Mechanical Properties and Residual Stress Levels. *Biomaterials,* Vol.19, No.22, (November 1998), pp. 2015-2029, ISSN 0142-9612

Wallace, J. & Ilavsky, J. (1998). Elastic Modulus Measurements in Plasma Sprayed Deposits. *Journal of Thermal Spray Technology,* Vol.7, No.4, (December 1998), pp. 521-526, ISSN 1059-9630

Yu-Peng, L.; Mu-Sen, L.; Shi-Tong, L.; Zhi-Gang, W. & Rui-Fu, Z. (2004). Plasma-Sprayed Hydroxyapatite+Titania Composite Bond Coat for Hydroxyapatite Coating on Titanium Substrate. *Biomaterials,* Vol.25, No.18, (August 2004), pp. 4393-4403, ISSN 0142-9612

Zhao, X.; Chen, Z.; Liu, X. & Ding, C. (2007). Preparation, Microstructure and Bioactivity of Plasma-Sprayed TiO_2 Coating, In: *Thermal Spray 2007: Global Coating Solutions,* Marple, B.; Hyland, M.; Lau, Y.; Lima, R. & Montavon, G. (Eds.), pp. 397-400, ASM International, Materials Park, Ohio, USA

Zhu, S. & Ding, C. (2000). Characterization of Plasma Sprayed Nano-Titania Coatings by Impedance Spectroscopy. *Journal of the European Ceramic Society,* Vol.20, No.2, (February 2000), pp. 127-131, ISSN 0955-2219

6

Effect of Hydrothermal Self-Healing and Intermediate Strengthening Layers on Adhesion Reinforcement of Plasma-Sprayed Hydroxyapatite Coatings

Chung-Wei Yang and Truan-Sheng Lui
Department of Materials Science and Engineering, National Formosa University,Yunlin,
Department of Materials Science and Engineering,
National Cheng Kung University, Tainan,
Taiwan

1. Introduction

Biomaterials employed in calcified hard tissue repair generally serve the purpose of load carrying in cases of fractures, defects and joint replacement. Metallic materials are more suitable for load-bearing applications compared with ceramics and polymeric materials due to their combination of high mechanical strength and fracture toughness. Among generally used metallic biomaterials such as 316L stainless steel and Co-Cr-Mo alloys (ASTM F75), grade II commercial pure titanium (ASTM F67) and Ti-6Al-4V alloys (ASTM F136ELI) exhibit the most suitable characteristics for biomedical applications because of their high biocompatibility, specific strength and corrosion resistance [Niinomi, 2001]. The apparent success of titanium and its alloys in implants has been attributed to the existence of a thin, stable passivation TiO_2 layer. Another advantage of titanium and its alloys for using in hard tissue replacements is their low Young's modulus because a low Young's modulus equivalent to that of human cortical bone is simultaneously required to inhibit stress shielding effect and bone absorption [Pilliar et al., 1979; Kuroda et al., 1998; Niinomi et al., 2002]. Nowadays, they are commonly clinical used in hard tissue implants such as artificial hip prosthesis, knee joints and dental roots. A biological fixation between these hard tissue implants and surrounding bones can be successfully achieved by the bone ingrowth with a mechanical interlocking [Engh et al., 1987; Callaghan, 1993]. However, limitations of metallic biomaterials are the release of toxic metallic ions and corrosion/wear products into surrounding tissues and fluids [Sunderman et al., 1989; Healy & Ducheyne, 1992; Niinomi et al., 1999; Akahori et al., 2004].

In the biomedical applications, another concept to design a bioactive surface fixation has been achieved by the bone apposition method. Bioactive ceramics have often been used as coatings to modify the surface and create a new surface for the bioinert metallic implants. With the same chemical and crystallographic structure as the major inorganic constitute of hard tissues, bioactive hydroxyapatite ($Ca_{10}(PO_4)_6(OH)_2$, HA) is a widely preferred calcium phosphate bioceramic, which is considered as suitable bone graft substitutes [Holmes et al.,

1986; Bucholz et al., 1989] in both dentistry and orthopaedics due to its favorable bioactive properties and osteoconductivity [Munting et al., 1990; Jansen et al., 1991; Yuan et al., 2001]. The advantages of HA including: (1) earlier stabilization, rapid fixation and stronger chemical bonding between the host bone and the implant [Jansen et al., 1991; Hench, 1991; Schreurs et al., 1996], and (2) increased uniform bone ingrowth and ongrow the bone-implant interface. In spite of the good biocompatibility and osteoconductivity of HA, the limitations for the usage of the dense HA sintering bulks for bone replacement are their low fracture toughness [van Audekercke & Martens, 1984] and bending strength under load-bearing situations [de Groot et al., 1990; Choi et al., 1998]. Therefore, HA is generally applied as coatings for the purpose of improving the bioactivity of the bioinert metallic implants including the stem and the acetabular cup. The combination of high strength metallic substrates with osteoconductive properties of bioceramic makes HA-coated titanium implants attractive for the load-bearing situations in orthopedic and dental surgery. In addition to promote earlier stabilization of the implant with surrounding bone, another reason for coating HA is to extend the functional life of the prosthesis and to improve the adhesion of the prosthesis to the bone. Studies demonstrated that HA-coated titanium implants show higher push-out strength compared to uncoated titanium implants [Geesink et al., 1988; Wolke et al., 1991; Cook et al., 1992; Wang et al., 1993a], and post-mortem studies reported direct bone contact with implants without a fibrous tissue interface in patients who have had successful HA-coated total hip arthroplasties [Bauer et al., 1991; Lintner et al., 1994]. Moreover, the bone bonding capacity of the HA coatings can help cementless fixation of orthopedic prostheses. It has been shown that the skeletal bonding is enhanced immediately after implantation [Jarcho, 1981; Geesink et al., 1987; Cook et al., 1988].

Advances in coating technology have brought about a new dimension in processing of biomaterials. It is clear that surface modification of metallic biomaterials gives rise to enhance biocompatibility. Many coating techniques have been used for HA coatings preparation onto metallic substrates, including plasma spraying [Wang et al., 1993b; Gross & Berndt, 1998], HVOF spraying [Sturgeon & Harvey, 1995; Lugscheider et al., 1996], chemical vapor deposition (CVD) [Liu et al., 2007], physical vapor deposition (PVD, including the RF-sputtering method) [Ozeki et al., 2006], sol-gel coating [Ben-Nissan & Choi, 2006], electrochemical deposition [Peng et al., 2006], electrophoresis method [Wei et al., 2005], and biomimetic coating methods [Kokubo et al., 1987]. The HA coatings obtained from these various techniques differ in chemistry and crystallinity, which will affect the biological responses and their performances. Therefore, in addition to consider biological advantages of fast bony adaptation, firm implant-bone attachment, reduced healing time of HA to surrounding bone, the phase composition, mechanical properties and operation feasibility of HA-coated implants should also be considered for using in long-term load-bearing applications of dental implants and orthopedic prostheses. Compared with these techniques, plasma spraying constitutes the state-of-the-art procedure to improve the biological integration implants and the main industrial process to deposit thick HA coatings. The attraction lies in its easy operation, relatively low substrate temperature, high HA coating efficiency and its ability to deposit tailored HA coatings on implants with complex shapes.

The plasma spraying process was patented in 1960s, and the technical utilization of plasma as a high-temperature source is realized in the plasma torch. The torch operates with a central cone-shaped tungsten cathode and a water-cooled cylindrical copper anode. A typical plasma spraying process is shown in Fig. 1 [Suryanarayanan, 1993]. The principle of plasma spraying is that inducing an arc by a high current density and a high electric potential between the anodic copper nozzle and tungsten cathode. The plasma gases flow is injected into the annular gap between the two electrodes, and an arc is initiated by a high-frequency discharge. Noble gases of helium (He) and argon (Ar) are usually used as the primary plasma-generating gas. Diatomic gases of hydrogen (H_2) and nitrogen (N_2) can be used as the secondary gas to increase the enthalpy of plasma torch. As the plasma gases pass around the arc created between the electrodes, they are heated and partially ionized emerging from the nozzle with high velocity and high temperature. For atmospheric plasma spraying (APS), the processing temperatures are typically in the range of 1×10^4 to 1.5×10^4 K depending on the type of the plasma gas used and the power input. Factors influence the degree of particles melting during plasma spraying includes variables which control the temperature of the plasma, such as current density, anode-cathode gap distance and gas mixture. The widely used plasma-generating gas is pure Ar (purity > 99.95 wt. %).

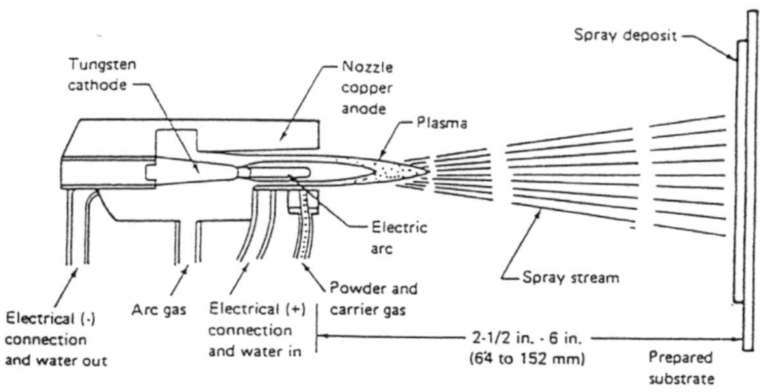

Fig. 1. Schematic illustration of plasma spraying process operation.

Since the thermal conductivity and the heat conduction potential for diatomic gases, such as H_2 and N_2, are much higher than Ar [Bourdin et al., 1983], a mixed gas composition with Ar and H_2/N_2 gives a quite hotter plasma torch than 100% Ar gas. Figure 2 displays the variation of heat content and temperature during ionization and dissociation stages of these plasma gases [Ingham & Shepard, 1965]. When well-crystallized HA powders are injected into the high temperature plasma torch, small granules will be evaporated in the torch, and larger particles are melted or partial-melted quickly by the high temperature plasma torch. Then these melted droplets are accelerated to about 200 m/s before impacting the substrate [Fauchias et al., 1992; Pfender, 1994]. The high impact velocity supplies high kinetic energy, which is expended in spreading the molten or semi-molten droplets and creating a lamellar microstructure. In addition, high cooling rate upon impact is estimated to be of the order of 10^6 to 10^8 K s^{-1}. Therefore, the large contact area with the substrate and the rapid solidification result in producing amorphous calcium phosphate (ACP) component within coatings, and it is more commonly found at the coating/substrate interface.

Because of the extremely high temperature, high enthalpy of the plasma torch and rapid solidification, a significant phase transformation or decomposition of HA is occurred during the plasma spray coating process. It results in large scale dehydroxylation and decomposition effects of crystalline HA phase into tri-calcium phosphate ($Ca_3(PO_4)_2$, TCP), tetra-calcium phosphate ($Ca_4P_2O_9$, TP), calcium oxide (CaO), oxyhydroxyapatite and ACP within the sprayed coatings. Plasma-sprayed HA coatings (HACs) with a higher content of impurity phases and ACP component will display higher dissolution rate than crystalline HA in aqueous solutions and body fluids [Ducheyne et al., 1993; Radin & Ducheyne, 1993]. It results in some problems with decreasing the structural homogeneity and the degradation of mechanical properties in the firm fixation between the implant and surrounding bone tissue [C.Y. Yang et al., 1995, 1997]. Therefore, decreasing the impurity phases and ACP is important for the long-term mechanical and biological stability of plasma-sprayed HACs. The ACP is a thermodynamically meta-stable component and impurity calcium phosphate phases are undesirable for the HACs, studies pointed out that performing appropriate thermal treatments, such as air or vacuum heat treatments, spark plasma sintering (SPS) technique, and hydrothermal treatments, etc., are available methods to significantly promote HA crystallization and to improve the mechanical properties and biological responsibility of HACs [Ji & Marquis, 1993; Wang et al., 1995; Lee et al., 2005; Yu et al., 2003; C.W. Yang & Lui, 2008].

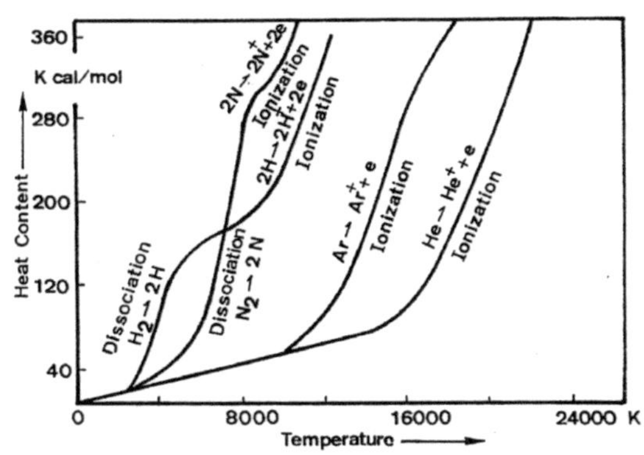

Fig. 2. The variation of heat content and temperature during ionization and dissociation stages of various plasma gases.

Although a thick, tailored HA coating can easily be applied by the plasma spraying process, a limitation of the HACs for applications is its low cohesion and adhesive bonding strength. To solve these problems, it has been generally recognized that performing post-heat treatments is an effective way to improve the bonding strength of HACs. Pure HACs is brittle, thus, some bioinert ceramics or metals, such as dicalcium silicate (β-Ca_2SiO_4) aluminum (Al_2O_3), partially stabilized zirconia (PSZ), titania (TiO_2), titanium and its alloys, have been chosen as the reinforcing additives to fabricate HA/ceramics pre-composite powders [Choi et al., 1998; Zheng et al., 2000; Chou & Chang, 2002a; Y. Yang & Ong, 2003; Sato et al., 2008; Cannillo et al., 2008]. The plasma-sprayed composite coatings made from HA and these reinforcing additives can help to alleviate the brittleness of pure HA and to improve the mechanical properties of HACs, as well as the reinforcements. Since a continuous ACP layer is resulted from rapid solidification of crystalline HA droplets, it is thought of acting a high solubility region and a low energy fracture path. This situation will result in weakening the mechanical integrity of the HA coating/substrate interface and further decreasing the adhesive bonding strength. Therefore, another way to improve the interfacial strength is the application of a stable bioinert intermediate strengthening layer, or so-call as bond coat, at the coating/substrate interface to enhance the adhesion of HACs to metallic substrates. The bond coat can reduce the thermal gradient at the interface to decrease significant thermal decomposition of HA. The bond coat can help to prevent the release of metal ions to the surrounding tissue. It can also provide better mechanical interlocking and even establish a chemical bonding between bond coat and HACs. Many attempts have been made to apply the above-mentioned bioinert ceramics or metals as bond coat materials to improve the performance of plasma-sprayed HACs [Lamy et al., 1996; Chang et al., 1997; Kurzweg et al., 1998; Fu et al., 2001; Lu et al., 2004].

Plasma-sprayed HACs with a better bonding strength can be achieved by adding reinforced additives to form composite coatings and applying heat treatments to acquire a higher HA crystallinity level and fewer coating defects as a result of HA crystallization. With different materials preparation, manufacturing and characterization, or different *in vitro* and *in vivo* examination methods, however, it is difficult to systematically evaluate the relationship between the HA crystallization, interfacial chemical reactions, biological responses and mechanical properties of the coatings. This chapter represents the crystallization effect on influencing the bonding strength of the plasma-sprayed HACs through performing post-vacuum heating and hydrothermal treatments. The benefit of low-temperature hydrothermal crystallization on plasma-sprayed HACs is clarified through the evaluation of crystallization mechanism by the Arrhenius kinetics. Through adding a ceramic and a metallic bond coat, the effects of mechanical interlocking and interfacial chemical reactions at interface on improving the bonding strength of HA/bond coat will be discussed. Since variables associated with implants preparation result in a certain extent of fluctuation for material properties, a strong reliability of implants is required when they are extrapolated to clinical applications. To determine the failure probability and reliability, the failure surface morphologies of HACs and the Weibull model of survival analysis [Weibull, 1951] were used to assess the effects on bonding strength data fluctuation pertaining to microstructural feature and the reliability of HACs.

2. Processing

Medical grade high purity HA (Sulzer Metco XPT-D-701) powder with particle size ranging from 15 to 40 μm were used in the coating process. Commercial yttria-stabilized zirconia (ZrO_2, YSZ, Amdry 142F), pure α-titanium (CP-Ti, Amdry 9182) powders were selected as bond coat materials, and Ti-6Al-4V alloys (ASTM F136 ELI) were selected as substrates. Prior to spraying, substrates were grit-blasted with SiC grit to roughen the surface. The average surface roughness (Ra) of grit-blasted substrate was controlled at about 3.9±0.3 μm. The powders were carried by high purity Ar gas to the plasma torch following the spraying parameters as listed in Table 1. The coating thickness of YSZ and CP-Ti bond coats was controlled at about 30 μm. Total coating thickness of 120±10 μm was prepared for HA coatings with and without intermediate layers. Table 2 lists the surface roughness of YSZ, CP-Ti bond coats and HA top coat of composite coatings.

Post-heat treatments were performed in a vacuum heating chamber (Vacuum industries, System VII) with 1.33×10^{-3} Pa at 600°C (named as V-HACs) with a heating rate of 10°C/min, held for 3h and then furnace cooling. The hydrothermal treatment was carried out in a hermetical autoclave (Parr 4621, Pressure Vessel) at 150°C for 6h (named as HT-HACs). The heating temperature was maintained throughout the experiments using a heater attached to the autoclave and the temperature was precisely controlled by a Parr 4842, PID controller with ± 1°C. The autoclave contained 100 ml deionized water, which was used as the source of steam atmosphere during the hydrothermal treatment, and the saturated steam pressure at 150°C was 0.48 MPa. The specimens were isolated without the immersion in the water. Phase compositions of the YSZ, CP-Ti bond coats and plasma-sprayed HACs were identified by X-ray diffraction (Rigaku D/MAX III. V), using CuKα, operated at 30 kV, 20 mA. A commonly used index of crystallinity (IOC) was adopted for the purpose of further quantitatively evaluating the crystallization state of the vacuum and hydrothermally-treated HACs. The IOC data is a ratio of three strongest HA diffraction peaks ((211), (112), (300), JCPDS 9-432) integral intensity of the HACs (*Ic*) and the as-received HA powder (HAP, *Ip*) according to the relationship IOC = $(Ic/Ip) \times 100\%$. This method supposes that the IOC of as-received HAP is 100% and the calculated IOC value of the as-sprayed HACs is about 20%. To realize the variation of TCP, TP and CaO impurity phases after applying heat treatments, the internal standard method was used to quantitatively determine these phase content within the heat-treated HACs. The integral intensity of known weight percent pure Si powder added in the specimens was taken as internal standard. The calibration curves for impurity phase content have been established by Wang et al. [Wang et al., 1995]. The main peak integral intensity ratio between TCP, TP and CaO phases from various XRD patterns of V-HACs and HT-HACs were compared to the calibration curves and the concentrations (in wt. %) in these specimens were calculated.

Spraying parameters	YSZ bond coat	CP-Ti bond coat	HA top coat
Coating thickness (μm)	30	30	120/90 †
Primary gas (*l*/min)	Ar (41)	Ar (46)	Ar (41)
Secondary gas (*l*/min)	H_2 (10)	H_2 (6)	H_2 (8)
Power (kW)	42.5	38.4	40.2
Powder carrier gas (*l*/min)	Ar (3)	Ar (3)	Ar (3)
Powder feed rate (g/min)	20	20	20
Surface speed (cm/min)	8000	8000	8000

† 90 μm was prepared for the HA top coat of HA/YSZ, HA/CP-Ti composite coatings, and 120 μm was prepared for plasma-sprayed HA coatings without bond coats.

Table 1. Plasma spraying parameters employed for preparing HA composite coatings.

Grit-blasted Ti-6Al-4V	YSZ bond coat	CP-Ti bond coat	Plasma-sprayed HACs	HAC/YSZ coating	HAC/CP-Ti coating
3.9 ± 0.3	6.4 ± 0.4	6.4 ± 0.2	7.6 ± 0.7	8.7 ± 1.1	8.9 ± 0.7

Note: values are given as mean ± S.D., each value was the average of ten tests (*n*=10).

Table 2. Surface roughness (Ra, μm) of the substrate and various plasma-sprayed coatings.

3. Microstructural evolution and biological responses of heat-treated HA coatings

Figure 3(a) shows the phase composition of as-sprayed CP-Ti bond coat. In addition to the diffraction peaks of titanium (α-Ti), the oxidation product of α-Ti within the coating is $TiO_{1.04}$, which represents two different crystal structures: the major oxide is cubic $TiO_{1.04}$ (JCPDS 43-1296) and the main peak another oxide of hexagonal $TiO_{1.04}$ is observed at 2θ = 36° (JCPDS 43-1295). The as-sprayed YSZ bond coat remains a cubic crystal structure (JCPDS 27-0997) as shown in Fig. 3(b). Figure 3(c) shows the phase composition of as-sprayed HACs. A fairly high content (about 49.3 wt. %) of ACP and impurity calcium phosphate phases, including α-$Ca_3(PO_4)_2$ (α-TCP), β-$Ca_3(PO_4)_2$ (β-TCP), $Ca_4P_2O_9$ (TP) and CaO, are identified in the as-sprayed HACs besides the desired HA phase.

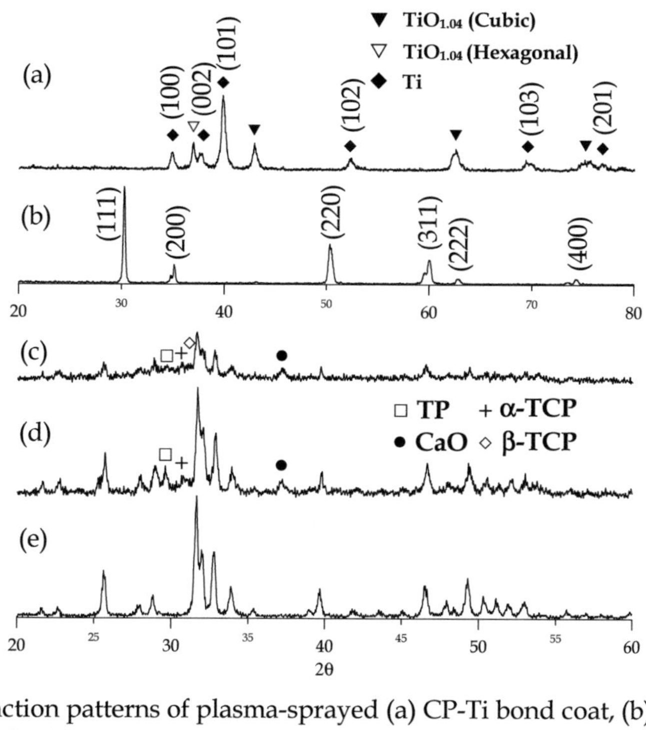

Fig. 3. X-ray diffraction patterns of plasma-sprayed (a) CP-Ti bond coat, (b) YSZ bond coat, (c) as-sprayed HACs, (d) V-HACs and (e) HT-HACs coatings.

Figure 3(d) displays the X-ray diffraction patterns of post-vacuum heat treated coatings (V-HACs). The TCP, TP and CaO impurity phases still remained within coatings after performing 600°C vacuum heat treatment, at which the total impurity phase content within V-HACs is about 20.3 wt. %. The quantitative result of the IOC for V-HACs is about 70%. According to the phase diagram of CaO-P_2O_5 system, since there is a lack of ambient partial water vapor pressure during vacuum heating, TCP and TP phases are stable phases and they cannot be eliminated without the replenishment of hydroxyl groups (OH^-). The CaO remained within the V-HACs because it cannot easily be converted into HA if the ambient heating atmosphere without abundant H_2O molecules [Weng et al., 1996; Cao et al., 1996]. It is worth noting that these TCP, TP and CaO impurity phases are significantly eliminated after hydrothermal treatment as shown in Fig. 3(e). The impurity phase content within HT-HACs is about 12.1 wt. %. The sharpening of three strongest HA main peaks and the flattening of the diffraction background (2θ at about 28° to 34°) mean that the plasma-sprayed HACs further crystallized and the content of ACP significantly decreased by the 150°C hydrothermal treatment in an ambient saturated steam pressure system. The IOC of HT-HACs is about 66%, which is close to the high-temperature vacuum heat treatment. Since the hydroxyl groups promote the reconstitution of ACP into crystalline hydroxyapatite [Tong et al., 1997], therefore, the saturated steam pressure atmosphere of autoclaving hydrothermal treatment can effectively improve HA crystallization and effectively eliminate the ACP and impurity phases of plasma-sprayed HACs with the replenishment of hydroxyl groups.

Figure 4(a) shows the typical surface morphology of the plasma-sprayed HACs, which represents an accumulated molten splats feature with a fair amount of pores and thermal-induced microcracks at the rapid cooling stage. After applying 600°C vacuum het treatment, Fig. 4(b) also shows a surface cracking feature for the V-HACs specimens. Different from the thermal contraction cracking during plasma spraying, however, these cracks are resulted from the significant crystallization-induced contraction effect [C.W. Yang et al., 2006] during high-temperature crystallization of HACs. Figure 4(c) displays the typical coating surface of HT-HACs, which provides evidence in microscopic surface features different from that of as-sprayed HACs and V-HACs. It is worth noting that nano-scale crystalline growth, indicated by the circle in Fig. 4(c), is observed on the surface of the HT-HACs specimen. These particles can be attributed to HA crystallites, which crystallized from the hydroxyl-deficient structure of plasma-sprayed HACs through the replenishment of hydroxyl groups. In addition, since the ACP is more soluble than crystalline HA phase in an aqueous environment, part of the new-growth crystalline HA might have formed through a dissolution-recrystallization process. The nano-scale HA crystalline experiences further grain growth with a larger crystal size in the vicinity of microcracks as indicated by the arrow. The reduction of coating defects for hydrothermally-treated HACs can be recognized as the self-healing effect of the hydrothermal treatment [C.W. Yang & Lui, 2008]. As a result of crystallization during heat treatments, the contraction-induced cracking and the self-healing phenomena will significantly influence the bonding strength and the failure mechanism of HACs.

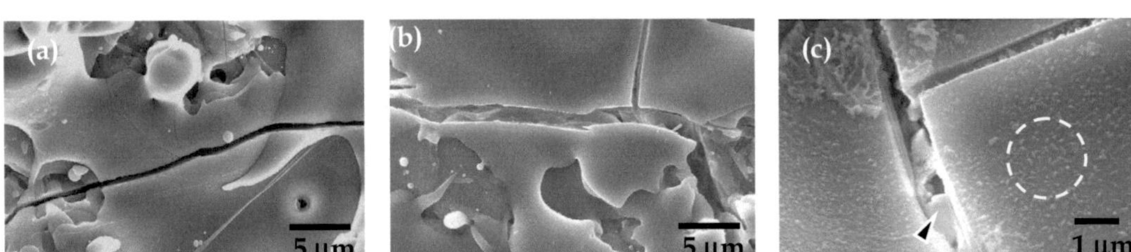

Fig. 4. Surface morphologies of (a) as-sprayed HACs, (b) V-HACs and HT-HACs coatings.

Figure 5 shows the cross-sectional microstructures of the as-sprayed HACs, V-HACs, HT-HACs and various composite coatings. According to the quantitative calculation by an image analyzer, the spraying defects content (in volume %), including pores and thermal-induced microcracks, is about 3.9 % for as-sprayed HACs in the case of Fig. 5(a). Since a significant volume contraction during HA crystallization [C.W. Yang & Lui, 2007], the V-

HACs specimen shows a coating structure with many vertical, apparent contraction-induced cracks as indicated by the arrow in Fig. 5(b). The defects content of V-HACs coating is about 5.3%. As shown in Fig. 5(c), the HT-HACs possess significant fewer microcracks and lower defects content (about 2.6%). It displays a much denser microstructure than the as-sprayed HACs and V-HACs. The self-healing effect of hydrothermal crystallization resulted from the new-growth HA crystallites can be recognized to diminish the coating defects and further increase the densification of plasma-sprayed HACs. Figure 5(d) and 5(e) shows the cross-sectional features of HA/CP-Ti and HA/YSZ composite coatings, respectively. The HA top coat shows a similar microstructure to the as-sprayed HACs. Since the bond coat can reduce the thermal gradient at HA/substrate interface, the HA top coat in Fig. 5(d) and 5(e) shows less thermal-induced microcracks for these composite coatings. The coating thickness of both CP-Ti and YSZ bond coats is about 30 μm, and the Ti-6Al-4V substrate is fully covered by the bond coat. In addition, the CP-Ti and YSZ bond coats provide a rougher surface than the substrate (refer to Table 2) to the HAC top coat. This can help to improve the mechanical interlocking between HA/substrate interface and further increase the adhesive bonding strength of the HA coatings.

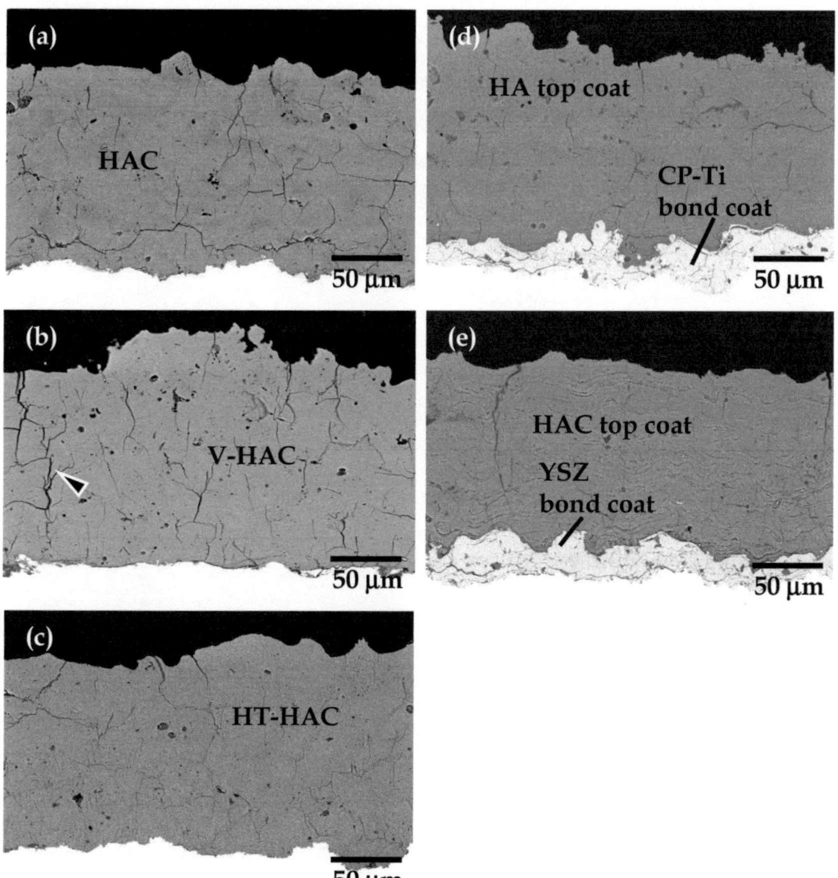

Fig. 5. Cross-sectional images of (a) as-sprayed HACs without bond coat, (b) V-HACs coating, (c) HT-HACs coating, (d) HA/CP-Ti and (e) HA/YSZ composite coatings.

In addition, the biological responses of plasma-sprayed HACs, V-HACs and HT-HACs are quantitatively evaluated *in vivo* using the Chinese coin implant model in the femoral of a goat, and details of the experimental procedure have been well described in the previous reports [C.Y. Yang et al., 2007, 2009]. The osteoconductivity of the implants is evaluated quantitatively in terms of the new bone healing index (NBHI), which defined as the (area of new bone/area of surgical defect region) × 100%. The ability of osseointegration of implants is addressed as apposition index (AI), which defined as the (length of direct bone-implant contact/total length of bone-implant interface) × 100%. This method can help to determine the success or failure of an implant by evaluating the interaction occurring at the bone-biomaterial interface. The mean NBHI and AI data listed in Table 3 indicated that the crystallized HACs with applying heat treatments have a statistically higher extent of new bone healing and apposition index compared to the as-sprayed HACs after 12 weeks of implantation. Figure 6 shows the amount of new bone increased within the surgical defective bone regions of as-sprayed HACs, V-HACs and HT-HACs after 12 weeks of implantation. Since the phase composition and crystallinity of post-heat-treated HACs remain stable at 12 weeks, it provides better treated-HACs-to-bone contact area, which can provide the firm bone/implant fixation compared to as-sprayed HACs. Considering the crystallized coatings of V-HACs and HT-HACs, hydrothermally-crystallized HACs show better *in vivo* biological responses at 12 weeks and show the potentiality to provide biological fixation than the other condition in the present results.

	Plasma-sprayed HACs †	V-HACs †	HT-HACs †
NBHI ‡ (%)	74.7 ± 6.6	77.8 ± 5.1	79.0 ± 7.5
AI ‡ (%)	67.3 ± 7.1	76.8 ± 5.7	78.1 ± 6.4

† Values are given as an average ± standard deviation (SD).
‡ NBHI: new bone healing index. AI: apposition index.

Table 3. NBHI and AI values for plasma-sprayed HACs, V-HACs and HT-HACs after 12 weeks of implantation.

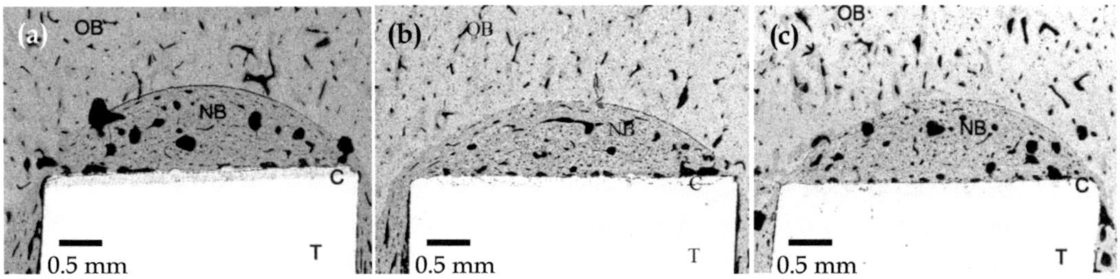

Fig. 6. SEM/BEI of histological section at the surgical defect region for 12 weeks post-implantation: (a) plasma-sprayed HACs, (b) V-HACs and (c) HT-HACs. The new bone is repaired within the surgical defect regions of these coatings. The osseointegration is found at the bone/HA coating interfaces (T: Ti-6Al-4V, C: coating, NB: new bone, OB: old bone).

4. Affected factors on crystallization during heat treatments

The experimental results demonstrate that the autoclaving hydrothermal treatment can actually promote significant crystallization to improve the phase purity, crystallinity, microstructural homogeneity and biological responses of plasma-sprayed HACs. Previous studies have indicated that the kinetics of crystallization and chemical reactions during heat treatments are significantly related to heating temperatures, which is recognized as a main factor for promoting HA crystallization [Chang et al., 1999; Campos et al., 2002; Roeder et al., 2006]. Since the IOC value represents the degree of crystallization for heat-treated HACs, it can be recognized of the conversion ratio from ACP to crystalline HA under different heating conditions. Considering the theory of chemical reaction kinetics and the definition of IOC for crystallized HACs, the HA crystallization process should follow the Arrhenius equation [Chang et al., 1999; Huang et al., 2000; Liu et al., 2001; C.W. Yang & Lui, 2007] as represented in Eq. (1). Based on the reaction kinetics of Arrhenius equation, the rate constant (k) can be thought as the reaction rate, and it represents the crystallization rate during heat treatments. The reaction rate and the activation energy of HA crystallization within vacuum can be quantitatively evaluated by the IOC of each specimens and HA crystallization under the vacuum heating follows the second-order Arrhenius reaction kinetics.

$$r = \frac{dIOC}{dt} = k(1 - IOC)^2 \qquad (1)$$

However, a significant crystallization of hydroxyl-deficient HACs requires at least 600°C [Gross et al., 1998; Feng et al., 2000; Lu et al., 2003], and high heating temperatures tend to undermine the structural integrity and cause phase decomposition of crystalline HA. In addition, the effect of the ambient heating atmosphere is another factor that should be considered to affect the reaction rate for HA crystallization. Referring to the phase diagram of $CaO\text{-}P_2O_5$ at 500 mmHg partial steam pressure (P_{vapor}), HA is a stable phase and water vapor is a significant factor to promote HA crystallization. The low-temperature hydrothermal treatment system with a surrounding saturated steam pressure can help to diminish the ACP, impurity phases within the plasma-sprayed HACs and significantly promote HA crystallization. This is a result of replenishment of missing OH^- groups with surrounding H_2O molecules [Chen et al., 1997; Y. Yang et al., 2003]. Thus, the influence of ambient water vapor during the autoclaving hydrothermal treatment should be considered to evaluate the kinetics of the hydrothermal crystallization at lower heating temperatures.

Since the dehydroxylation is a result of hydroxyl groups (OH^-) broken away from HA crystal structure during plasma spraying process, the ACP with a reduced crystallinity of HA occurred in the coating layers. When the hydrothermal treatment is applied to promote the crystallization of plasma-sprayed HACs, the water is vaporized. The ionized water vapor molecules contain H^+ and OH^- groups within the hermetical autoclave and the content of H^+ and OH^- groups increases with increasing the temperature [Zhang et al., 2001]. The resultant OH^- groups within the water vapor atmosphere are expected to react with ACP and other low-crystalline calcium phosphate components, and convert them into crystalline HA phase through the replenishment of OH^- groups. The previous study has demonstrated that the reaction order under the hydrothermal crystallization not only depends on the effects of heat treatment time and temperatures, but the saturated steam pressure (P_{vapor}) factor is involved at each hydrothermal heating temperature [C.W. Yang &

Lui, 2009]. According to a series of examinations, Eq. (2) concludes the modified form which involves a saturated steam pressure term following the second-order reaction kinetics of Arrhenius equation.

$$r = \frac{dIOC}{dt} = k(1 - IOC)^{3/2} P_{vapor}^{1/2} \qquad (2)$$

Experimental evidence confirmed that the ambient saturated steam pressure plays an important role in lowering heating and reactions temperatures. This results in a significant microstructural self-healing effect through the grain growth of HA nanocrystallite [C.W. Yang & Lui, 2009], as shown in Figs. 4(c), within the hydrothermally-crystallized HACs, which also shows a statistically higher extent of new bone apposition [C.Y. Yang et al., 2007] essential in the initial fixation of implants in clinical applications.

The XPS analysis results clarify the replenishment of OH^- groups and the reduction of the dehydroxylation state of hydroxyl-deficient HACs during the hydrothermal treatment. Figure 7 shows the high resolution XPS O 1*s* spectra of HA coatings with curve-fittings, which resulted from the Gaussian peak-fitting routine. The corresponding O 1*s* band of the as-sprayed HACs presented in Fig. 7(a) consists of three components at about BE = 531.4 eV, BE = 532.4 eV and BE = 533.2 eV, which correspond to the PO (PO_4^{3-}), POH bonds of HA crystal and the surface adsorbed H_2O. The adsorbed H_2O peak represents that the surface of the as-sprayed HACs is easily affected by the surrounding moisture within the air, and H_2O molecules can be physically adsorbed on the HACs.

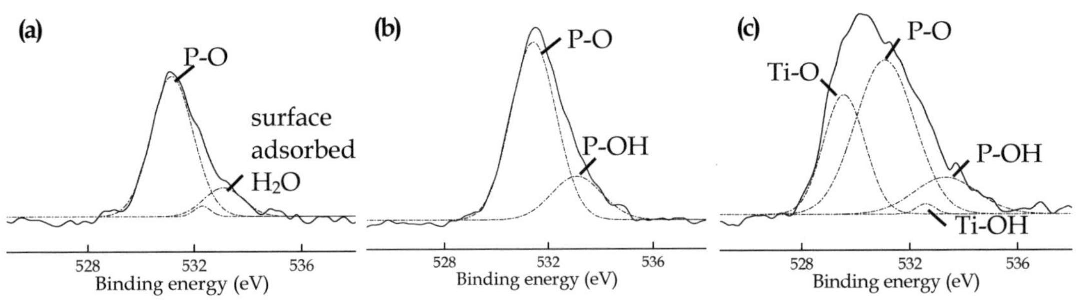

Fig. 7. XPS O 1*s* spectra curve-fitting results of (a) plasma-sprayed HACs, (b) HT-HACs, and (c) the surface close to the HT-HACs/Ti-substrate interface.

In contrast with the as-sprayed HACs, XPS O 1*s* spectra of hydrothermally-treated coatings (HT-HACs) with relatively large integration area of the POH bonding peak but without the adsorbed H_2O peak is shown in Fig. 7(b). The surface residual adsorbed H_2O molecules can be seen to be reduced. The hydroxyl-deficient state of the as-sprayed HACs is significantly improved with the abundant replenished OH^- groups from the hydrothermal treatment, especially under a higher saturated steam pressure atmosphere. In Fig. 7(c), the O 1*s* spectra obtained at the HT-HACs/Ti-substrate interface are fit with four peaks: the above-mentioned PO, POH peaks of HA, the Ti-O peak at 529.6 eV and the Ti-OH peak at 532.5 eV. The Ti-O peak can be attributed to the surface oxide ion of Ti-substrate, and the peak at ΔBE about 3.0 eV from Ti-O peak can be assigned to the chemisorbed OH^- groups of Ti-OH [Healy & Ducheyne, 1992; Takadama et al., 2001]. Since the rapid solidification of molten

HA droplets during plasma spraying induces the formation of ACP within the as-sprayed HACs, XPS analysis results demonstrate that the hydrothermal treatment helps to promote the HA crystallization through the replenishment and the chemisorption of OH^- groups. The presence of Ti-OH bonding can enhance the bioactive properties of the HA coating by promoting the osteointegration process [Massaro et al., 2001].

5. Effect of the strengthening bond coats on the adhesive bonding strength of composite HA coatings

The most commonly used method of determining tensile bonding strength for the plasma-sprayed coatings is the criterion ASTM C633-01. The roughness of substrates is an important factor in achieving high bonding strength of plasma-sprayed HACs because the bonding of the HACs to metallic substrates appears to be mechanical interlocking in nature. There is less degree of chemical bonding in as-sprayed HACs. Ti-6Al-4V cylindrical rods with dimensions of 25.4 mm (ϕ) and 50 mm (l) are used as substrates for the tests. Each test specimen is an assembly composed of a substrate fixture, to which the HACs of 120 ± 10 μm are applied, and a loading fixture. The loading fixtures are also grit-blasted and attached to the surface of the HACs top coat using adhesive glue with an adhesive strength of about 60 MPa. After curing, the assemblies are subjected to tensile tests at a crosshead speed of 1 mm/min until failure. For the statistical significance of the following Weibull analysis, 20 specimens are tested for bonding strength measurements. Fig. 8 shows that the bonding strength of plasma-sprayed HACs is improved (ANOVA statistical analysis, $p < 0.05$) with applying the reinforced bond coats and post-heat treatments. The highest bonding strength about 39.9 ± 2.4 MPa is acquired for the HA/YSZ composite coating, and the HT-HACs shows a fairly high bonding strength of 38.9 ± 1.0 MPa. It can be recognized that the adhesive bonding strength is significant improved with applying the hydrothermal treatment and adding the YSZ bond coat compared to the other conditions.

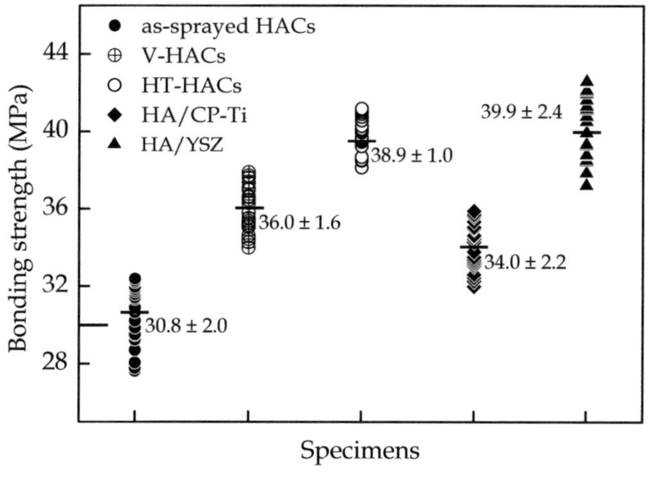

Fig. 8. Results of measuring bonding strength with the average value ± S.D. and the data fluctuation of as-sprayed HACs, V-HACs, HT-HACs, HA/CP-Ti and HA/YSZ composite coatings (S.D. means the standard deviation).

Referring to the cross-sectional features shown in Fig 5(b), the 600°C vacuum heat-treated specimen shows many vertical and apparent large cracks, which are resulted from crystallization-induced contraction of coating by the thermal dilatometry measurements [C.W. Yang& Lui, 2007]. Figure 4(c) represents cracks are obviously healed with the crystalline HA grains within the HT-HACs, which display a dense microstructure as shown in Fig. 5(c). It can be seen that the microstructural homogeneity with a self-healing effect occurred from the hydrothermal crystallization throughout the whole HA coating layers under the abundant saturated steam pressure environment. Although the bonding strength is improved with the crystallization of coating layers during vacuum heat treatments, however, it can be recognized that a detrimental effect of contraction-induced cracks accompanied with HA crystallization may result in the deterioration of bonding strength.

The representative failure surfaces of these coatings are shown in Fig. 9. According to ASTM C633-01, the variation of bonding strength in situ is suggested to be governed by the cohesive strength of coatings and the adhesive strength of a coating to a metal substrate. The affecting factors of the adhesive strength of a coating and substrate interface include the surface roughness of substrate and the residual stress. As for the cohesive strength of coating, the factors include the crystallinity and the densification of a coating, which appearing on the Young's modulus of a coating. A large area fraction of cohesive failure (co) can be commonly observed for high strength coatings [Kweh et al., 2000; C.W. Yang & Lui, 2008]. The cohesive failure is dominated by the microstructural features such as crystallinity, defects and lamellar texture. Compared with the failures of as-sprayed HACs shown in Fig. 9(a), the failure morphologies of HT-HACs (Fig. 9(b)) represent homogeneity and display a larger area fraction of cohesive failure, since strengthened coatings resulted from the self-healing effect of hydrothermal crystallization. In contrast, the decreased area fraction of adhesive failure (ad) indicates that the adhesion of HT-HACs to Ti-6Al-4V substrate is improved. Referring to the evidence from the XPS analysis as shown in Fig. 7(c), the hydrothermal treatment helps to promote the interfacial crystallization through the replenished and the chemisorbed OH^- groups, which results in a significant chemical bonding of HA coating to Ti-substrate interface.

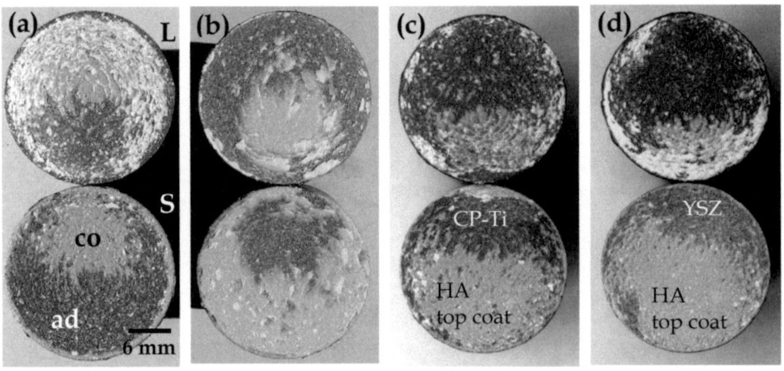

Fig. 9. Failure surfaces of (a) as-sprayed HAC, (b) HT-HACs, (c) HA/CP-Ti and (d) HA/YSZ composite coatings. The bonding strength measured is a manifestation of cohesive (co) and adhesive (ad) strength (L: loading fixture, S: substrate fixture).

Considering the HA/CP-Ti and HA/YSZ composite coatings, the decreased area fraction of adhesive failure (Fig. 9(c) and 9(d)) represents that the adhesion of HA top coat to substrate is also enhanced by adding YSZ and CP-Ti bond coats. The significant improvement of bonding strength for HA/CP-Ti and HA/YSZ composite coatings can be recognized that a higher surface roughness of CP-Ti and YSZ layers than grit-blasted Ti-6Al-4V substrate (Table 2) to provide better interfacial mechanical interlocking. The idea to further increase interlocking is to establish a chemical bonding between HA coating and bond coat. The evident shift of XPS binding energy of Ca 2*p* peak for the HA/YSZ interface compared with the as-sprayed HACs (Fig. 10) indicated that there is a significant interfacial diffusion [Vincent, 2000] for Ca ions at the interface of the HA top coat to the YSZ bond coat. However, there is no interfacial chemical reaction between the HA coating and the CP-Ti bond coat. Therefore, it can be related to the fact that the diffusion of Ca ions from HA matrix into the YSZ bond coat and the formation of chemical bonding of Ca-ZrO_2 [Khor et al., 2000; Chou & Chang, 2002b]. The interfacial chemical bonding can help to improve the bonding of HA/YSZ composite coatings.

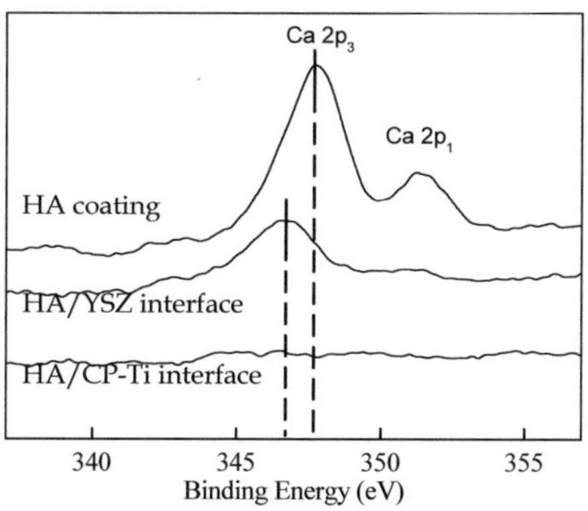

Fig. 10. XPS Ca 2*p* spectra for inter-diffusion analysis resulted from the as-sprayed HACs, the HA/YSZ bond coat interface and the HA/CP-Ti bond coat interface.

6. Evaluation of failure behaviors with statistical techniques

To characterize the strength data fluctuation, reliability, failure probability and failure mechanism of materials, a powerful statistical distribution function, which is called as the Weibull distribution function, was invented by Waloddi Weibull in 1937 and delivered his hallmark American paper in 1951 [Weibull, 1951]. He claimed that this model can be applied to a wide range of problems, and the Weibull models have been used in many different applications for solving a variety of problems from many different disciplines. Equation (3) shows the general form of the Weibull distribution function.

$$F(\sigma_i) = \int_{\sigma=0}^{\sigma=\sigma_i} f(\sigma)d\sigma = 1 - \exp\left[-\left(\frac{\sigma_i - \sigma_o}{\eta}\right)^m\right] \tag{3}$$

In Eq. (3), σ represents the bonding strength, and at least twenty specimens (n=20) are tested for the purpose of statistical significance of the analysis. $F(\sigma_i)$ is the cumulative failure probability corresponding to a bonding strength σ_i (i is the ranking of specimens). The parameter m represents the Weibull modulus, η is the characteristic strength and σ_o is the minimum strength. The failure behavior of materials is determined by m, η and σ_o. The Weibull modulus, which controls the shape of function curves, is a measure of the variability of the data. The characteristic strength η corresponds to the strength at which the cumulative failure is 63.2%. The minimum strength σ_o means that the failure probability of HACs at applied stress below this value is zero. Fitting the bonding strength data into Eq. (3), the failure probability density function $f(\sigma)$ curves of the as-sprayed HACs, HA/CP-Ti and HA/YSZ composite coatings are plotted in Fig. 11(a). The cumulative failure probability $F(\sigma_i)$ is estimated using the Benard's median rank of Eq. (4), which is a very close approximated solution of a statistical function [Faucher & Tyson, 1988]. The reliability function $R(\sigma_i)$ with a relation of $R(\sigma_i) = 1-F(\sigma_i)$ is defined as the survival probability. Fig. 11(b) shows the natural logarithmic (ln) graphs for the cumulative failure probability at each corresponding bonding strength σ_i (i=1-20) of the specimens, it can graphically evaluate the Weibull modulus (m) from the slope of a least-squares fitting method of Eq. (5) at a maximum coefficient of determination (R^2).

$$F(\sigma_i) = \frac{i-0.3}{n+0.4} \tag{4}$$

$$\ln\ln\left(\frac{1}{1-F(\sigma_i)}\right) = m\ln(\sigma_i - \sigma_o) - m\ln\eta \tag{5}$$

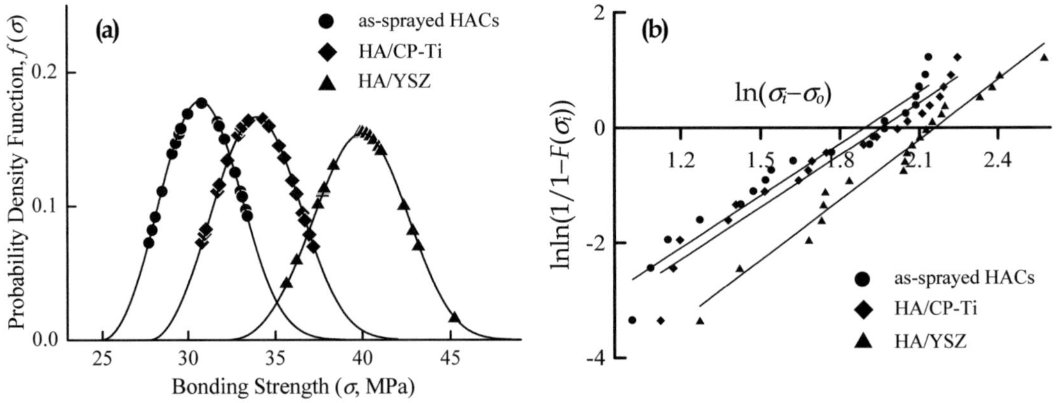

Fig. 11. (a) The failure probability density function $f(\sigma)$ curves, and (b) the Weibull distribution plots of the plasma-sprayed HACs, HA/CP-Ti and HA/YSZ composite coatings.

Since the Weibull distribution function is used to model the reliability and the failure behavior of materials, a failure rate function $\lambda(\sigma_i)$ shown in Eq. (6) at each corresponding bonding strength is defined for evaluating the failure behaviors [Burrow et al., 2004].

$$\lambda(\sigma_i) = \frac{f(\sigma_i)}{R(\sigma_i)} = \frac{m}{\eta^m}(\sigma_i - \sigma_o)^{m-1} \tag{6}$$

The examination of the Weibull modulus listed in Table 4 represents that HACs are reliable materials with a wear-out failure model (m > 1) of increasing failure rate (IFR). Figure 12 shows the failure rate function ($\lambda(\sigma)$) and reliability function ($R(\sigma)$) curves of the as-sprayed HAC, HA/CP-Ti and HA/YSZ composite coatings. These curves start from the minimum strength (σ_o), which implies the failure probability of HACs less than this strength is zero and the reliability of HACs is 1.0. The minimum strength can be recognized as the safe loading level for the plasma-sprayed HACs. Meanwhile, knowledge of the Weibull distribution function can provide further explanation for the strengthening effect of reinforced YSZ and CP-Ti bond coats on the bonding strength of as-sprayed HACs, and it can be used to determine which coating has higher uniformity and reliability. The Weibull modulus is also a measure of the variability of the data, which being larger as the degree of bonding strength fluctuation decreases. It is evident that the failure probability density function (Fig. 11(a)) and the failure rate (Fig. 12(a)) curves shift to the higher bonding strength and produce a concentrated data distribution for the HA/YSZ composite coating. The YSZ bond coat effectively enhances the bonding strength of plasma-sprayed HAC, and helps to acquire more stable HACs with less reliability decrease (Fig. 12(b)) while the loading exceeds the minimum strength.

Samples	Weibull Modulus, m	Minimum strength, σ_o (MPa)	Characteristic strength, η (MPa)
As-sprayed HACs	3.0	24.9	31.8
HA/CP-Ti	3.0	27.7	35.0
HA/YSZ	3.5	32.1	40.9

Table 4. Results of Weibull model analysis for the HACs

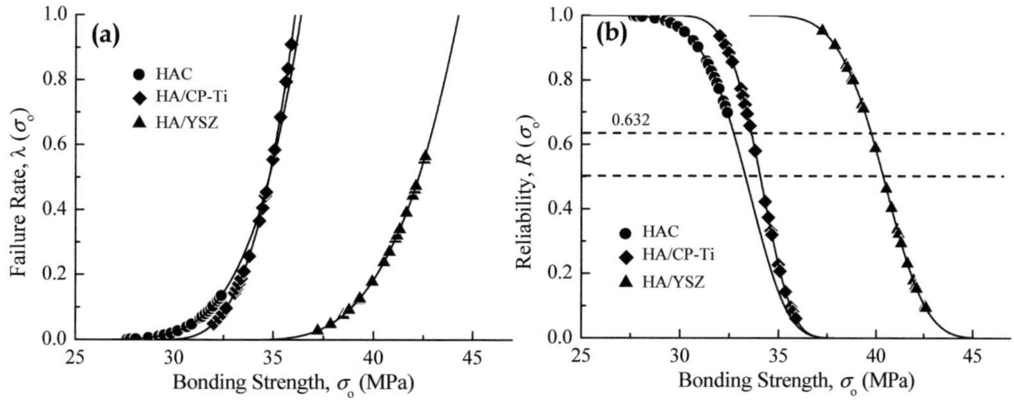

Fig. 12. (a) The failure rate function $\lambda(\sigma)$ curves, and (b) the reliability function $R(\sigma)$ curves.

7. Conclusion

The evolution of microstructural features, biological responses, crystallization kinetics, tensile mechanical properties and failure behaviors of post-spray heat-treated HACs by the vacuum heating and the hydrothermal treatment are evaluated. Applying heat treatments is an effective way to improve the crystallization state, biological responses and adhesive bonding strength of plasma-sprayed HACs. Compared with these two heat treatments, the hydrothermal treatment is more favorable to eliminate the impurity phases and ACP than high temperature heat treatments in vacuum. Hydrothermal crystallization, which proceeded within a saturated steam environment, significantly improves the microstructural homogeneity, coating density and HA/Ti-substrate interfacial reaction of plasma-sprayed HACs through the self-healing effect with the grain growth of the crystalline HA. In addition, hydrothermally-treated HACs display a higher new bone healing and apposition index than the as-sprayed and vacuum heat-treated HACs. The crystallization of plasma-sprayed HACs during heat treatments is second-order Arrhenius reaction kinetics. The effect of ambient heating atmosphere with a saturated steam pressure is an important factor for the hydrothermal treatment to further promote HA crystallization rate at lower heating temperatures. The addition of CP-Ti and YSZ as intermediate strengthening bond coats between HA/Ti-substrate can significantly enhance the adhesive bonding strength of plasma-sprayed HACs. In addition to the higher mechanical interlocking between HA top coat and bond coat interface, a chemical bonding resulted from the interfacial diffusion at the HA/YSZ bond coat interface can further improve the adhesive bonding strength for the HA/YSZ composite coating. Fractures with less area percentage of interfacial failure are indicative of a better adhesion of a coating. According to the results of Weibull model analysis, plasma-sprayed HACs represent a wear-out failure behavior (the Weibull modulus, $m > 1$) with increasing failure rate. The adhesion and survival probability of plasma-sprayed HACs are improved by adding the YSZ bond coat to form a HA/YSZ composite coating.

8. Acknowledgment

This study was financially supported by the National Science and Council of Taiwan (Contract No. NSC 100-2221-E-150-037) for which we are grateful.

9. References

Akahori, T.; Niinomi, M.; Fukui, H. & Suzuki, A. (2004). Fatigue, Fretting Fatigue and Corrosion Characteristics of Biocompatible Beta Type Titanium Alloy Conducted with Various Thermo-mechanical Treatments. *Materials Transactions*, Vol.45, No.5, pp. 1540-1548, ISSN 1345-9678.

Bauer, T.W.; Geesink, R.C.T.; Zimmerman, R. & McMahon, J.T. (1991). Hydroxyapatite-coated Femoral Stems: Histological Analysis of Components Retrieved at Autopsy. *The Journal of Bone and Joint Surgery [Am]*, Vol.73A, pp. 1439-1452, ISSN 1535-1386.

Ben-Nissan, B. & Choi, A.H. (2006). Sol-gel Production of Bioactive Nano-coatings for Medical Applications Part 1: An Introduction. *Nanomedicine*, Vol.1, pp. 311-319, ISSN 1549-9634.

Bourdin, E.; Fauchais, P. & Boulos, M. (1983). Transient Heat Conduction under Plasma Conditions. *International Journal of Heat and Mass Transfer*, Vol.26, pp. 567-582, ISSN 0017-9310.

Bucholz, R.W.; Carlton, A. & Holmes, R.E. (1989). Interporous Hydroxyapatite as a Bone Graft Substitute in Tibial Plateau Fractures. *Clinical Orthopaedics and Related Research*, Vol.240, pp. 53-62, ISSN 0009-921X.

Burrow, M.F.; Thomas, D.; Swain, M.V. & Tyas, M.J. (2004). Analysis of Tensile Strengths Using Weibull Statistics. *Biomaterials*, Vol.25, pp. 5031-5035, ISSN 0142-9612.

Callaghan, J.J. (1993) Current Concepts Review: The Clinical Results and Basic Science of Total Hip Arthroplasty with Porous-coated Prostheses. *The Journal of Bone and Joint Surgery [Am]*, Vol.75A, pp. 299-310, ISSN 1535-1386.

Campos, A.L.; Silva, N.T.; Melo, F.C.L.; Oliverira, M.A.S. & Thim, G.P. (2002). Crystallization Kinetics of Orthorhombic Mullite from Diphasic. *Journal of Non-Crystalline Soilds*, Vol.304, pp. 19-24, ISSN 0022-3093.

Cannillo, V., Lusvarghi, L, & Sola, A. (2008). Production and Characterization of Plasma-sprayed TiO_2-Hydroxyapatite Functionally Graded Coatings. *Journal of the European Ceramic Society*, Vol.28, pp. 2161-2169, ISSN 0955-2219.

Cao, Y.; Weng, J.; Chen, J.; Feng, J.; Yang, Z. & Zhang, X. (1996). Water Vapor-treated Hydroxyapatite Coatings after Plasma Spraying and Their Characteristics. *Biomaterials*, Vol.17, pp. 419-424, ISSN 0142-9612.

Chang, E.; Chang, W.J.; Wang, B.C. & Yang, C.Y. (1997). Plasma Spraying of Zirconia-reinforced Hydroxyapatite Composite Coatings on Titanium: Part I. Phase, Microstructure and Bonding Strength. *Journal of Materials Science: Materials in Medicine*, Vol.8, pp. 193-200, ISSN 0957-4530.

Chang, C.; Huang, J.; Xia, J. & Ding, C. (1999). Study on Crystallization Kinetics of Plasma Sprayed Hydroxyapatite Coating. *Ceramics International*, Vol.25, pp. 479-483, ISSN 0272-8842.

Chen, J.; Tong, W.; Cao, Y.; Feng, J. & Zhang, X. (1997). Effect of Atmosphere on Phase Transformation in Plasma-sprayed Hydroxyapatite Coatings during Heat Treatment. *Journal of Biomedical Materials Research*, Vol.34, pp. 15-20, ISSN 0021-9304.

Choi, J.W.; Kong, Y.M. & Kim, H.E. (1998). Reinforcement of Hydroxyapatite Bioceramic by Addition of Ni_3Al and Al_2O_3. *Journal of the American Ceramic Society*, Vol.81, pp. 1743-1748, ISSN 0002-7820.

Chou, B.Y. & Chang, E. (2002a). Phase Transformation during Plasma Spraying of Hydroxyapatite-10-wt%-Zirconia Composite Coating. *Journal of the American Ceramic Society*, Vol.85, pp. 661-669, ISSN 0002-7820.

Chou, B.Y. & Chang, E. (2002b). Plasma-sprayed Zirconia Bond Coat as an Intermediate Layer for ydroxyapatite Coating on Titanium Alloy Substrate. *Journal of Materials Science: Materials in Medicine*, Vol.13, pp. 589-595, ISSN 0957-4530.

Cook, S.D.; Thomas, K.; Kay, J.F. & Jarcho, M. (1988). Hydroxylapatite Coated Titanium for Orthopaedic Implant Applications. *Clinical Orthopaedics and Related Research*, Vol.232, pp. 225-243, ISSN 0009-921X.

Cook, S.D.; Thomas, K.A.; Delton, J.E.; Volkman, T.K.; Whitecloud, T.S. & Key, J.F. (1992). Hydroxyapatite Coating of Porous Implants Improves Bone Ingrowth and Interface Attachment Strength. *Journal of Biomedical Materials Research*, Vol.26, pp. 989-101, ISSN 0021-9304.

de Groot, K.; Klein, C.P.A.T.; Wolke, J.G.C. & de Bliek-Hogervost, J.M.A. (1990). Chemistry of Calcium Phosphate Bioceramics, In: *CRC Handbook of Bioactive Ceramics. Vol. II*, T. Yamamuro, L.L. Hench & J. Wilson, (Eds.), 3-16. CRC Press Inc., ISBN 978-084-9332-42-5, Boca Raton: Taylor & Francis.

Ducheyne, P.; Radin, S. & King, L. (1993). The Effect of Calcium Phosphate Ceramic Composition and Structure on *In Vitro* Behavior: I. Dissolution. *Journal of Biomedical Materials Research*, Vol.27, pp. 25-34, ISSN 0021-9304.

Engh, C.A.; Bobyn, J.D. & Glassman, A.H. (1987). Porous-coated Hip Replacement. *The Journal of Bone and Joint Surgery [Br]*, Vol.69B, pp. 45-55, ISSN 0301-620X.

Faucher, B. & Tyson, W.R. (1988). On the Determination of Weibull Parameters. *Journal of Materials Science Letters*, Vol.7, pp. 1199-1203, ISSN 0261-8028.

Fauchias, P.; Coudert, J.F.; Vardelle, M.; Vardelle, A. & Denoirjean, A. (1992). Diagnostics of Thermal Spray Plasma Jets. *Journal of Thermal Spray Technology*, Vol.1, pp. 117-128, ISSN 1059-9630.

Feng, C.F.; Khor, K.A.; Liu, E.J. & Cheang, P. (2000). Phase Transformations in Plasma Sprayed Hydroxyapatite Coatings. *Scripta Materialia*, Vol.42, pp. 103-109, ISSN 1359-6462.

Fu, L.; Khor, K.A. & Lim, J.P. (2001). The Evaluation of Powder Processing on Microstructure and Mechanical Properties of Hydroxyapatite (HA) /Yttria Stabilized Zirconia (YSZ) Composite Coatings. *Surface and Coatings Technology*, Vol.140, pp. 263-268, ISSN 0257-8972.

Geesink, R.G.T.; de Groot, K. & Klein, C.P.A.T. (1987). Chemical Implant Fixation Using Hydroxyapatite Coatings. *Clinical Orthopaedics and Related Research*, Vol.225, pp. 147-170, ISSN 0009-921X.

Geesink, R.G.T.; de Groot, K. & Klein, C.P.A.T. (1988). Bonding of Bone to Apatite-coated Implants. *The Journal of Bone and Joint Surgery [Br]*, Vol.70B, pp. 17-22, ISSN 0301-620X.

Gross, K.A. & Berndt, C.C. (1998). Thermal Processing of Hydroxyapatite for Coating Production. *Journal of Biomedical Materials Research*, Vol.39, pp. 580-587, ISSN 0021-9304.

Gross, K.A.; Gross, V. & Berndt, C.C. (1998). Thermal Analysis of Amorphous Phases in Hydroxyapatite Coatings. *Journal of the American Ceramic Society*, Vol.81, pp. 106-112, ISSN 0002-7820.

Healy, K.E. & Ducheyne, P. (1992). The Mechanisms of Passive Dissolution of Titanium in a Model Physiological Environment. *Journal of Biomedical Materials Research*, Vol.26, pp. 319-338, ISSN 0021-9304.

Hench, L.L. (1991). Bioceramics: From Concept to Clinic. *Journal of the American Ceramic Society*, Vol.74, No.7, pp. 1487-1510, ISSN 0002-7820.

Holmes, R.E.; Bucholz, R.W. & Mooney, V. (1986). Porous Hydroxyapatite as a Bone-graft Substitute in Metaphyseal Defects. *The Journal of Bone and Joint Surgery [Am]*, Vol.68A, pp. 904-911, ISSN 1535-1386.

Huang, L.Y.; Xu, K.W. & Lu, J. (2000). A Study of the Process and Kinetics of Electrochemical Deposition and the Hydrothermal Synthesis of Hydroxyapatite Coatings. *Journal of Materials Science: Materials in Medicine*, Vol.11, pp. 667-673, ISSN 0957-4530.

Ingham, H.S. & Shepard, A.P. (1965). *Plasma Flame Process*, METCO Inc., p. 11, Westburg, Long Island, New York, USA.

Jansen, J.A.; van de Waerden, J.P.C.M.; Wolke, J.G.C. & de Groot, K. (1991). Histologic Evaluation of the Osseous Adaptation to Titanium and Hydroxy-apatite-coated Titanium Implants. *Journal of Biomedical Materials Research*, Vol.25, pp. 973-989, ISSN 0021-9304.

Jarcho, M. (1981). Calcium Phosphate Ceramics as Hard Tissue Prosthetics. *Clinical Orthopaedics and Related Research*, Vol.157, pp. 259-278, ISSN 0009-921X.

Ji H. & Marquis, P.M. (1993). Effect of Heat Treatment on the Microstructure of Plasma-sprayed Hydroxyapatite Coating. *Biomaterials*, Vol.14, pp. 64-68, ISSN 0142-9612.

Khor, K.A.; Fu, L.; Lim, J.P. & Cheang, P. The Effects of ZrO2 on the Phase Compositions of Plasma Sprayed HA/YSZ Composite Coatings. (2000). *Materials Science and Engineering A*, Vol.276, pp. 160-166, ISSN 0921-5093.

Kokubo, T.; Hayashi, T.; Sakka, S.; Kitsugi, T. & Yamamuro, T. (1987). Bonding between Bioactive Glasses, Glass-ceramics or Ceramics in a Simulated Body Fluid. *Journal of the Ceramic Society of Japan*, Vol.95, pp. 785-791, ISSN 0372-7718.

Kuroda, D.; Niinomi, M.; Morinaga, M.; Kato, Y. & Yashiro, T. (1998). Design and Mechanical Properties of New Beta Type Titanium Alloys for Implant Materials. *Materials Science and Engineering A*, Vol.243, pp. 244-249, ISSN 0921-5093.

Kurzweg, H.; Heimann, R.B. & Troczynski, T. (1998). Adhesion of Thermally Sprayed Hydroxyapatite Bond-coat Systems Measured by a Novel Peel Test. *Journal of Materials Science: Materials in Medicine*, Vol.9, pp. 9-16, ISSN 0957-4530..

Kweh, S.W.K.; Khor, K.A.; & Cheang, P. (2000). Plasma-sprayed Hydroxyapatite (HA) Coatings with Flame-spheroidized Feedstock: Microstructure and Mechanical Properties. *Biomaterials*, Vol.21, pp. 1223-1234, ISSN 0142-9612.

Lamy, D.; Pierre, A.C. & Heimann, R.B. (1996). Hydroxyapatite Coatings with a Bond Coat of Biomedical Implants by Plasma Projection. *Journal of Materials Research*, Vol.11, pp. 680-686, ISSN 0884-2914.

Lee, Y.P.; Wang, C.K.; Huang, T.H.; Chen, C.C.; Kao, C.T. & Ding, S.J. (2005). *In vitro* Characterization of Post Heat-treated Plasma-sprayed Hydroxyapatite Coatings. *Surface and Coatings Technology*, Vol.197, pp. 367-374, ISSN 0257-8972.

Lintner, F.; Böhm, G.; Huber, M. & Scholz, R. (1994). Histology of Tissue Adjacent to an HAC-coated Femoral Prostheses: A Case Report. *The Journal of Bone and Joint Surgery [Br]*, Vol.76B, pp. 824-830, ISSN 0301-620X.

Liu, C.; Huang, Y.; Shen, W. & Cui, J. (2001). Kinetics of Hydroxyapatite Precipitation at pH 10 to 11. *Biomaterials*, Vol.22, pp. 301-306, ISSN 0142-9612.

Liu, X.; Chu, P.K. & Ding, C. (2007). Formation of Apatite on Hydrogenated Amorphous Silicon (a-Si:H) Film Deposited by Plasma-enhanced Chemical Vapor Deposition. *Materials Chemistry and Physics*, Vol.101, pp. 124-128, ISSN 0254-0584.

Lu, Y.P.; Song, Y.Z.; Zhu, R.F.; Li, M.S. & Lei, T.Q. (2003). Factors Influencing Phase Compositions and Structure of Plasma Sprayed Hydroxyapatite Coatings during Heat Treatment. *Applied Surface Science*, Vol.206, pp. 345-354, ISSN 0169-4332.

Lu, Y.P.; Li, M.S.; Li, S.T.; Wang, Z.G. & Zhu, R.F. (2004). Plasma-sprayed Hydroxyapatite + Titanium Composite Bond Coat for Hydroxyapatite Coating on Titanium Substrate. *Biomaterials*, Vol.25, pp. 4393-4403, ISSN 0142-9612.

Lugscheider, L.; Remer, P. & Nyland, A. (1996). High Velocity Oxy Fuel Spraying: An Alternative to the Established APS-process for Production of Bioactive Coatings, In:

Proceedings of the Tenth International Conference on Surface Modification Technologies, Singapore, T.S. Sudarsan, K.A. Khor, & M. Jeandin, (Eds.), 717-727.

Massaro, C.; Baker, M.A.; Cosentino, F.; Ramires, P.A.; Klose, S. & Milella, E. (2001). Surface and Biological Evaluation of Hydroxyapatite-based Coatings on Titanium Deposited by Different Techniques. *Journal of Biomedical Materials Research*, Vol.58B, pp. 651-657, ISSN 0021-9304.

Munting, E.; Verhelpen, M.; Li, F. & Vincent, A. (1990). Contribution of Hydroxy-apatite Coatings to Implant Fixation, In: *CRC Handbook of Bioactive Ceramics, Vol. II*, T. Yamamuro, L.L. Hench & J. Wilson, (Eds.), 143-148, CRC Press Inc., ISBN 978-084-9332-42-5, Boca Raton: Taylor & Francis.

Niinomi, M.; Kuroda, D.; Fukunaga, K.; Morinaga, M.; Kato, Y.; Yashiro, T. & Suzuki, A. (1999). Corrosion Wear Fracture of New β Type Biomedical Titanium Alloys, *Materials Science and Engineering A*, Vol.263, pp. 193-199, ISSN 0921-5093.

Niinomi, M. (2001). Recent Metallic Materials for Biomedical Applications. *Metallurgical and Materials Transactions A*, Vol.32A, pp. 477-486, ISSN 1073-5623.

Niinomi, M.; Hattori, T.; Morikawa, K.; Kasuga, T.; Suzuki, A.; Fukui, H. & Niwa, S. (2002). Development of Low Rigidity β-type Titanium Alloy for Biomedical Applications. *Materials Transactions*, Vol.43, No.12, pp. 2970-2977, ISSN 1345-9678.

Ozeki, K.; Aoki, H. & Fukui, Y. (2006). Dissolution Behavior and In Vitro Evaluation of Sputtered Hydroxyapatite Films Subject to a Low Temperature Hydrothermal Treatment. *Journal of Biomedical Materials Research*, Vol.76A, pp. 605-613, ISSN 0021-9304.

Peng, P.; Kumar, S.; Voelcker, N.H.; Szili, E.; Smart, R.S.C. & Griesser, H.J. (2006). Thin Calcium Phosphate Coatings on Titanium by Electrochemical Deposition in Modified Simulated Body Fluid. *Journal of Biomedical Materials Research*, Vol.76A, pp. 347-355, ISSN 0021-9304.

Pfender, E. (1994). Plasma Jet behavior and Modeling Associated with the Plasma Spray Process. *Thin Solid Films*, Vol.238, pp. 228-241, ISSN 0040-6090.

Pilliar, P.M.; Cameron, H.U.; Binnington, A.G. & Szivek, J.A. (1979). Bone Ingrowth and Stress Shielding with a Porous Surface Coated Fracture Fixation Plate. *Journal of Biomedical Materials Research*, Vol.13, pp. 799-810, ISSN 0021-9304.

Radin, S.R. & Ducheyne, P. (1993). Effect of Calcium Phosphate Ceramic Composition and Structure on *in vitro* Behavior. II. Precipitation. *Journal of Biomedical Materials Research*, Vol.27, pp. 35-45, ISSN 0021-9304.

Roeder, R.K.; Converse, G.L.; Leng, H. & Yue, W. (2006). Kinetics Effects on Hydroxyapatite Whiskers Synthesized by the Chelate Decomposition Method. *Journal of the American Ceramic Society*, Vol.89, Mo.7, pp. 2096-2104, ISSN 0002-7820.

Sato, M.; Aslani, A.; Sambito, M.A.; Kalkhoran, N.M.; Slamovich, E.B. & Webster, T.J. (2008). Nanocrystalline Hydroxyapatite/Titania Coatings on Titanium Improves Osteoblast Adhesion. *Journal of Biomedical Materials Research*, Vol.84A, pp. 265-272, ISSN 0021-9304.

Schreurs, B.W.; Huiskes, R.; Buma, P. & Slooff, T.J.J.H. (1996). Biomechanical and Histological Evaluation of a Hydroxyapatite-coated Titanium Femoral Stem Fixed with an Intramedullary Morsellized Bone Grafting Technique: An Animal Experiment on Goats. *Biomaterials*, Vol.17, pp. 1177-1186, ISSN 0142-9612.

Standard Test Method for Adhesion and Cohesion Strength of Thermal Spray Coatings. ASTM C633-01, West Conshohocken, PA, USA.

Sturgeon, A.J. & Harvey, M.D.F. (1995). High Velocity Oxyfuel Spraying of Hydroxyapatite. *Proceedings of ITSC'95*, pp. 933-938, Kobe, Japan, May, 1995.

Sunderman, F.W.Jr.; Hopfer, S.M.; Swift, T.; Rezuke, W.N.; Ziebka, L.; Highman, P.; Edwards, B.; Folcik, M. & Gossling, H.R. (1989). Cobalt, Chromium, and Nickel Concentrations in Body Fluids of Patients with Porous-coated Knee or Hip Prostheses. *Journal of Orthopaedic Research*, Vol.7, pp. 307-315, ISSN 1554-527X.

Suryanarayanan, R. (1993). *Plasma Spraying: Theory and Applications*, World Scientific Publishing Co. Pte. Ltd., p. 4, ISBN 978-981-02-1363-3.

Takadama, H.; Kim, H.M.; Kokubo, T. & Nakamura, T. (2001). An X-ray Photoelectron Spectroscopy Study of the Process of Apatite Formation on Bioactive Titanium Metal. *Journal of Biomedical Materials Research*, Vol.55, pp. 185-193, ISSN 0021-9304.

Tong, W.; Chen, J.; Cao, Y.; Lu, L.; Feng, J. & Zhang, X. (1997). Effect of Water Vapor Pressure and Temperature on the Amorphous-to-Crystalline HA Conversion during Heat Treatment of HA Coatings. *Journal of Biomedical Materials Research*, Vo.36, pp. 242-245, ISSN 0021-9304.

van Audekercke, R. & Martens, M. (1984). Mechanical Properties of Cancellous Bone, In: *Natural and Living Biomaterials*, G.W. Hastings & P. Ducheyne, (Eds.), 89-98, CRC Press Inc., ISBN 978-084-9362-64-4, Boca Raton: Taylor & Francis.

Vincent, C.B. (2000). *Handbook of Monochromatic XPS Spectra, The Elements of Native Oxides*, John Wiley and Sons, Ltd., 49-53, ISBN 978-047-1492-65-8, England.

Wang, B.C.; Chang, E.; Yang, C.Y.; & Tu, D. (1993a). A Histomorphometric Study on Osteoconduction and Osseointegration of Titanium Alloy with and without Plasma-sprayed Hydroxyapatite Coating Using Back-scattered Electron Images. *Journal of Materials Science: Materials in Medicine*, Vol.4, pp. 394-403, ISSN 0957-4530.

Wang, B.C.; Chang, E.; Yang, C.Y.; Tu, D. & Tsai, H. (1993b). Characteristics and Osteoconductivity of Three Different Plasma-sprayed Hydroxyapatite Coatings. *Surface and Coatings Technology*, Vol.58, pp. 107-117, ISSN 0257-8972.

Wang, B.C.; Chang, E.; Lee, T.M. & Yang, C.Y. (1995). Changes in Phases and Crystallinity of Plasma-sprayed Hydroxyapatite Coatings under Heat Treatment: A Quantitative Study. *Journal of Biomedical Materials Research*, Vol.29, pp. 1483-1492, ISSN 0021-9304.

Wei, M.; Ruys, A.J.; Milthorpe, B.K. & Sorrell, C.C. (2005). Precipitation of Hydroxyapatite Nano-particles: Effects of Precipitation Method on Electrophoretic Deposition. *Journal of Materials Science: Materials in Medicine*, Vol.16, pp. 319-324, ISSN 0957-4530.

Weibull, W. (1951). A Statistical Distribution Function of Wide Applicability. *Journal of Applied Mechanics*, Vol.18, pp. 293-297, ISSN 0021-8936.

Weng J.; Liu X.; Zhang X. & de Groot, K. (1996). Integrity and Thermal Decomposition of Apatite in Coatings Influenced by Underlying Titanium during Plasma Spraying and Post-heat-treatment. *Journal of Biomedical Materials Research*, Vol.30, pp. 5-11, ISSN 0021-9304.

Wolke, J.G.C.; Klein, C.P.A.T. & de Groot, K. (1992). Bioceramics for Maxillofacial Applications, In *Bioceramics and the Human Body*, A. Ravaglioli & A. Krajewski, (Eds.), 166-180. Elsevier Science Publishers Ltd., ISBN 978-185-1667-48-2, England.

Yang, C.W.; Lee, T.M.; Lui, T.S. & Chang, E. (2006). Effect of Post Vacuum Heating on the Microstructural Feature and Bonding Strength of Plasma-sprayed Hydroxyapatite Coatings. *Materials Science and Engineering C*, Vol.26, pp. 1395-1400, ISSN 0928-4931.

Yang, C.W. & Lui, T.S. (2007). Effect of Crystallization on the Bonding Strength and Failures of Plasma-sprayed Hydroxyapatite. *Materials Transactions*, Vol.48, No.2, pp. 211-218, ISSN 1345-9678.

Yang, C.W. & Lui, T.S. (2008). The Self-healing Effect of Hydrothermal Crystallization on the Mechanical and Failure Properties of Hydroxyapatite Coatings. *Journal of the European Ceramic Society*, Vol.28, pp. 2151-2159, ISSN 0955-2219.

Yang, C.W. & Lui, T.S. (2009). Kinetics of Hydrothermal Crystallization under Saturated Steam Pressure and the Self-healing Effect by Nanocrystallite for Hydroxyapatite Coatings. *Acta Biomaterialia*, Vol.5, pp. 2728-2737, ISSN 1742-7061.

Yang, C.Y.; Wang, B.C.; Chang, E. & Wu, B.C. (1995). Bond Degradation at the Plasma-sprayed HA Coating/Ti-6Al-4V Alloy Interface: An *in vitro* Study. *Journal of Materials Science: Materials in Medicine*, Vol.6, pp. 258-265, ISSN 0957-4530.

Yang, C.Y.; Lin, R.M.; Wang, B.C.; Lee, T.M.; Chang, E.; Hang, Y.S. & Chen, P.Q. (1997). *In Vitro* and *in Vivo* Mechanical Evaluations of Plasma-sprayed Hydroxyapatite Coatings on Titanium Implants: The Effect of Coating Characteristics. *Journal of Biomedical Materials Research*, Vol.37, pp. 335-345, ISSN 0021-9304.

Yang, C.Y.; Lee, T.M.; Yang, C.W.; Chen, L.R.; Wu, M.C. & Lui, T.S. (2007). The *in vitro* and *in vivo* Biological Responses of Plasma-sprayed Hydroxyapatite Coatings with Post-hydrothermal Treatment. *Journal of Biomedical Materials Research*, Vol.83A, pp. 263-271, ISSN 0021-9304.

Yang, C.Y.; Yang, C.W.; Chen, L.R.; Wu, M.C.; Lui, T.S; Kuo, A. & Lee, T.M. (2009). Effect of Vacuum Post-heat Treatment of Plasma-sprayed Hydroxyapatite Coatings on Their *in vitro* and *in vivo* Biological Responses. *Journal of Medical and Biological Engineering*, Vol.29, No.6, pp. 296-302, ISSN 1609-0985.

Yang, Y.; Kim, K.H.; Agrawal, C.M. & Ong, J.L. (2003). Influence of Post-deposition Heating Time and the Presence of Water Vapor on Sputter-coated Calcium Phosphate Crystallinity. *Journal of Dental Research*, Vol.82, pp. 833-837, ISSN 1544-0591.

Yang, Y. & Ong, J.L. (2003). Bond Strength, Compositional, and Structural Properties of Hydroxyapatite Coating on Ti, ZrO_2-coated Ti, and TPS-coated Ti Substrate. *Journal of Biomedical Materials Research*, Vol.64A, pp. 509-516, ISSN 0021-9304.

Yu, L.G.; Khor, K.A.; Li, H. & Cheang, P. (2003). Effect of Spark Plasma Sintering on the Microstructure and *in vitro* Behavior of Plasma Sprayed HA Coatings. *Biomaterials*, Vol.24, pp. 2695-2705, ISSN 0142-9612.

Yuan, H.; Yang, Z.; de Bruijn, J.D.; de Groot, K. & Zhang, X. (2001). Material-dependent Bone Induction by Calcium Phosphate Ceramics: A 2.5-year Study in Dog. *Biomaterials*, Vol.22, pp. 2617-2623, ISSN 0142-9612.

Zhang, H.; Li, S.; & Yan, Y. (2001). Dissolution Behavior of Hydroxyapatite Powder in Hydrothermal Solution. *Ceramics International*, Vol.27, pp. 451-454, ISSN 0272-8842.

Zheng, X.; Huang, M. & Ding, C. (2000). Bond Strength of Plasma-sprayed Hydroxyapatite/Ti Composite Coatings. *Biomaterials*, Vol.21, pp. 841-849, ISSN 0142-9612.

Part 3

Plasma Spray in Nanotechnology Applications

7

Solution and Suspension Plasma Spraying of Nanostructure Coatings

P. Fauchais and A. Vardelle
SPCTS, UMR 6638, University of Limoges, European Center of Ceramics, Limoges, France

1. Introduction

The main motivation for coating industrial parts with a different material lies on the following needs: (1) to improve functional performance, (2) to improve the component life by reducing wear due to abrasion, erosion and/or corrosion, (3) to extend the component life by rebuilding the worn part to its original dimensions, and (4) to improve the functionality of a low-cost material by coating it with a high performance but more expensive coating. Coating technologies can be roughly divided into thin- and thick- film technologies. Thin films, with thickness of less than 20 µm can be produced by dry coating processes like Chemical Vapor Deposition (CVD) or Physical Vapor Deposition (PVD); they offer excellent enhancement of surface properties and are for example used in optical and electronic device and cutting tools, *Davis J.R. (2004)*. However, most of these thin-film technologies require a reduced pressure environment and, therefore, are more expensive with a limit on the size and shape of the substrate.

Thick films have a thickness over 20 µm and can be several millimeters thick. They are required when the functional performance depends on the layer thickness, e.g. in thermal barrier coatings, when high erosion and corrosion conditions result in wear and the component life depends on the layer thickness, or when the original dimensions of worn parts have to be restored. Thick film deposition methods include chemical/electro-chemical plating, brazing, weld overlays, and thermal spray. Thermal spray processes, *Davis J.R. (2004)*, are well-established surface treatments aiming at forming a coating by stacking of lamellae resulting from impact, flattening and solidification of impinging molten particles. "Thermal spraying comprises a group of coating processes in which finely divided metallic or non-metallic materials are deposited in a molten or semi-molten condition to form a coating. The coating material may be in the form of powder, ceramic rod, wire or molten materials, *Hermanek, F.J. (2001)*."

A thermal spray system consists of five subsystems:

1. The high energy, high velocity jet generation that includes the torch, power supply, gas supply, and the associated controls;
2. The coating material preparation, i.e. powder manufacturing that controls particle size distribution and morphology, injection by a carrier gas into the high energy gas jet and transformation into a stream of molten droplets;

3. The surrounding atmosphere, i.e. atmospheric air, controlled atmosphere (including humidity control), low pressure, etc.;
4. The substrate material and surface preparation; and
5. Mechanical equipment for controlling the motion of the torch and the substrate relative to each other.

   The main driving force for R&D on the manufacture of thick coatings by thermal spraying is their high deposition rate (a few kilograms per hour of feedstock can be processed with torches with a few tens kW power level) at a relatively low and limited operating cost. Thermal spray processes include flame spraying, plasma spraying, wire arc spraying, and plasma transferred arc (PTA) deposition. Plasma spraying is probably the most versatile of all thermal spray processes because there are few limitations on materials that can be sprayed, and few limitations on the material, size and shape of the substrate, *Fauchais P. (2004)*. Coatings are used in numerous industrial fields, as aeronautical and land-based turbine industries (e.g., thermal barrier coatings, abradable seals, etc.), biomedical industry (e.g., hydroxyapatite bio-integrable coatings onto orthoprotheses) and paper industry (e.g., abrasion wear resistant and corrosion resistant coatings) *Davis J.R. (2004)*. Coatings are formed by stacking of lamellae resulting from impact, flattening and solidification of impinging molten particles. They are characterized by a highly anisotropic lamellar structure. Moreover, stacking defects generate specific interlamellar features within the structure, mainly voids, which can be, or not, connected to the upper surface of the deposit (i.e., connected and open voids, respectively).

Conventional thermal spray processes use powders with particle size ranging from 10 to 100μm. They result in coatings that mainly present micrometer-sized features, as the lamellae formed by the impact of the particles onto the substrate are a few μm thick with diameter from a few tens to a few hundreds of micrometers. The interest for developing and studying thermal-sprayed coatings that exhibit nanometer-size features and not micrometer-sized features has been important over the past 30 years and very important the last ten years. This interest comes from the enhanced properties of nanometer-sized coatings as compared to micrometer-sized ones. Reducing the structure scale down to nanometer allows, *Gell (1995)*, increasing strength, improving toughness and coefficient of thermal expansion while reducing apparent density, elastic modulus, and apparent thermal conductivity, among other improvements. One of the major drawbacks in processing nanometer-sized particles by thermal spraying is the difficulty in injecting them in the core of the high enthalpy flow, since the particle injection force has to be of the same order as that imparted by the gas flow: $S.\rho_g.v_g^2$, where S is the cross section of the injected particle, ρ_g the specific mass of the plasma and v_g the plasma velocity. Of course the last two terms vary along the particle trajectory in the plasma jet. The force of the injected particle is proportional to its mass (depending on the cube of its diameter) and the velocity of the carrier gas flow, which must be substantially increased as soon as the particle size decreases. However, the carrier gas flow rate disrupts the plasma jet as soon as its mass flow rate is over 1/5 of that of the plasma-forming gas. Thus, it is not practically possible to inject particles with sizes below 5-10 μm.

Four possibilities exist to circumvent this drawback (see the reviews of *Fauchais et al (2008) and (2011), Pawlowski (2008) and Viswanathan et al (2006))*. They consist of spraying:

1. Micrometer-sized agglomerates made of nanometer-sized particles that are injected using conventional injection route based on carrier gas. The operating spray parameters must, then, be adjusted in such a way that the molten front within such agglomerated particles progresses slower compared to the one in fully dense particles of same diameters. Upon solidification, the molten fraction of particles (for example the smallest particles or the outer shell of larger particles made of a single element) generates micrometer-sized zones in the coating ensuring its cohesion while the unmolten fraction of particles (inner core) keep their nanometer-sized structure. Such a coating architecture is usually called as "bimodal".
2. Complex alloys (5 to 10 components) that have low critical cooling rates for metallic glass formation result in the formation of amorphous coatings when thermally deposited. The amorphous coatings, when heated after spraying, to above their crystallization temperature, are devitrified. Since the diffusion rate in the solid state is very low at the transformation temperature (typically 0.4–0.7 T_m for iron alloys, T_m being the melting temperature), nanometer-scale microstructures are formed.
3. A suspension of nanometer-sized particles, with the carrier gas replaced by a liquid ("Suspension Thermal Spraying": STP). The nanometer-sized particles are dispersed into the liquid phase by means of dispersants: the suspension is then injected either as a liquid stream or as drops after nebulization. Depending upon the injection and plasma jet conditions, the liquid stream or droplets are fragmented due to flow shear forces and vaporized.
4. Precursors in solution that will form in-flight nanometer-sized particles ("Solution Precursor Thermal Spraying": SPTS). This significantly limits the safety issues associated with handling of nanometer-sized particles (e.g. see the very complete review by Singh et al. (2009)) and avoids most of the drawbacks associated with suspension stabilization, in particular when dissimilar materials (e.g., metallic alloys and oxides) are mixed together.

The following sections discuss successively: plasma torches and plasma jets used for suspension or solution spraying, interaction of the plasma jet with liquids, the preparation of suspensions and solutions, the nanometer structured coatings obtained and, their characterization and finally potential applications.

2. Plasma torches used for suspension or solution spraying

2.1 Conventional direct current plasma torches

Most spraying processes are carried out in air at atmospheric pressure, except for radio frequency r.f spraying. A conventional d.c. plasma torch (more than 90% of industrial torches are operated at power levels below 50 kW) with a stick type cathode is shown schematically in Figure 1a *Fauchais (2004)*. The cathode is made of thoriated (2 wt %) tungsten and the anode-nozzle of high purity oxygen-free copper encasing a cylindrical insert of sintered tungsten with internal diameter (i.d.) between 5 and 8-mm. The arc column (3 in Figure 1a) develops between the conical cathode tip (Θ in Figure 1a) pumping part of the plasma-forming gas (1 in Figure 1a), the other part flowing along the anode wall (cold boundary layer 2 in Figure 1a). The most commonly used plasma gases are: Ar, Ar-He, Ar-H_2, N_2, N_2-H_2, but more complex mixtures such as Ar-He-H_2 are also used. Ar and N_2 are mostly used for their mass, while the secondary gases (He and H_2) are used for their thermal properties: For example, 25 vol. % of H_2 in Ar increases the mean thermal conductivity of Ar by a factor 8 at 4000 K, *Boulos et al (1994)*. In comparison to H_2, He increases the thermal conductivity of the plasma and, its high viscosity up to about 14000 K delays the mixing of the surrounding air with the plasma jet.

The arc attachment to the anode wall, through the connecting column (4 in Figure 1a), continuously fluctuates in length and position. This is due to the movements induced by the drag force of the gas flowing in the cold boundary layer (2 in Figure 1a), the arc chamber pressure fluctuations, the Helmholtz oscillations in the space upstream of the arc, *Coudert et al (2007)*, and the magneto-hydrodynamic forces, all of which resulting in upstream and downstream short circuits. The corresponding transient voltage, which depends on the cold boundary layer thickness in the arc attachment area, exhibits a restrike (saw tooth shape), take-over (regular periodic variation) or mixed mode, and its value can reach ± 75 % of the time-averaged voltage. The restrike mode is the most probable with plasma forming gases that contain diatomic gases, while the take-over mode mainly occurs with monoatomic gases, these phenomena being drastically enhanced by the Helmoltz oscillations.

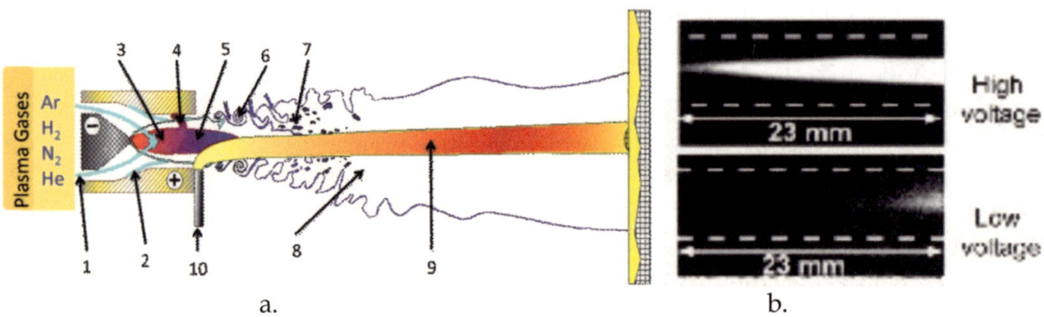

Fig. 1. (a.) schematic of a conventional dc arc spray torch with: Θ—stick type thoriated tungsten cathode, ⊕—anode; nozzle 1—the plasma forming gas injection, 2—the cold boundary layer at the anode wall, 3—the arc column, 4—the connecting arc column, 5—the plasma jet exiting the nozzle, 6—the large scale eddies, 7—the surrounding atmosphere bubbles entrained by the engulfment process, 8—the plasma plume, 9 - particles jet, 10- injector. (b.) plasma jet pictures taken with a shutter time of 10^{-4} s at maximum and minimum voltages: Ar-H_2 (45-15 slm) plasma produced with a PTF4 torch, anode-nozzle internal diameter d = 6 mm, arc current I = 600 A.

The arc root fluctuations help to keep the anode integrity because the heat fluxes at the arc root can be as high as 10^9-10^{10} W.m^{-2} thus limiting the residence time of the arc root to about 150 µs. Voltage fluctuations are characterized by the ratio $\Delta V/V_m$ (ΔV being the fluctuation amplitude and V_m the average voltage) that can vary from 0.25 (in the best conditions of the takeover mode) to 1.5 (in the worst conditions of the restrike mode). Correspondingly, the power dissipated in the arc, and thus the enthalpy, fluctuates with arc voltage (the torch being supplied by a constant current source) resulting in plasma jets continuously fluctuating in length and position (Figure 1a) at frequencies ranging between 2000 and 8000 Hz, depending on the cold boundary layer thickness *Fauchais (2004)*. An example is given in Fig. 1b representing pictures of an Ar-H_2 (25 vol. %) d.c. plasma jet taken at the highest (80

V) and lowest (40 V) voltages, respectively. As the plasma enthalpy fluctuates, the momentum density of the plasma jet ($\rho_g \times v_g^2$) at a given location in the jet also varies substantially with time. For $\Delta V/V_m$=1, the plasma flow average specific mass, ρ_g (depending on T), varies by less than 30%, while v can vary by a factor of up to 2 or 3. The injection force imparted by the carrier gas to a conventional particle (i.e. a few tens of micrometers in diameter), cannot follow the arc root fluctuations that are in the few kilohertz range. Therefore, the average particle trajectory fluctuates accordingly and, the temperature and velocity of the particles fluctuate at the same frequency as that of the arc root *Bisson et al (2003)*. The effect on the micrometer-sized coating morphologies can be rather important: i.e, less dense and more porous coatings are obtained with highly fluctuating plasma jets than those obtained with more stable plasma jets with the same enthalpy.

The effects of these fluctuations are even more drastic when the feedstock material injected in the plasma jet is in liquid form (suspension or solution) (see section 3). Therefore, Ar–He mixtures, which bring about lower fluctuations are often used to spray liquid feedstock, in spite of the fact that the power level is lower than that reached with diatomic gases *Fauchais et al (2011)*. Most of the temperature and velocity measurements of plasma jets have been performed in time ranges that integrate fluctuations. Under these conditions, typical plasma jet temperatures at the torch nozzle exit are between 14000 and 8000 K, irrespective of the plasma-forming gases and anode-nozzle internal diameter, while jet velocities vary with these two parameters between 800 and 2200 m s^{-1}, i.e. sub-sonic velocities at these temperatures and atmospheric pressure.

2.2 Other direct current plasma torches

To spray liquid suspensions or solutions, two other types of d.c. torches were developed in the nineties:

- Plasma torches with three cathodes insulated between them and supplied by independent power sources as the Triplex® system from Sulzer-Metco. The electrical energy is distributed through three parallel arcs striking at a single anode preceded by insulating rings (Figure 2). The internal diameter (i.d.) of the nozzle is between 6 and 8 mm. The generation of arcs that are longer than in conventional dc plasma torches makes it possible to reduce significantly (4 to 5 times) the percentage of voltage fluctuations of these torches compared to conventional dc plasma torches. Indeed if the voltage is more important with the long arc the voltage fluctuations at the anode are similar to those obtained with a conventional d.c. plasma torch, or even smaller with the more limited anode attachment. Figure 2(b) shows an Ar–He plasma jet produced by the Triplex II torch and illustrates its stability (compare with Figure 1(b)). Moreover, the threefold symmetry with three feedstock injectors can be aligned towards the warmest or coldest parts of the plasma jet to enable the optimization of the injection of the feedstock material.
- Plasma torches composed of three cathodes and three anodes operated by three power supplies (total power ranging from 50 to 150kW), such as Axial III® from Mettech. The feedstock material is injected axially between the three plasma jets converging within an interchangeable plasma nozzle. Hence, the particle residence time in the hot zones can be drastically improved.

2.3 Radio frequency plasma torches

Radio frequency torches used for spraying have internal diameters of 35 to 60 mm and power levels below 100 kW. The main differences with d.c. torches are in the torch i.d., resulting in flow velocities below 100 $m.s^{-1}$ and in the axial injection of particles. As can be seen in Figure 3, the injector is positioned almost at the middle of the coil. As the coupling between the coil and the plasma occurs in a ring close to the wall, the gas close to the torch axis is heated only by convection-conduction, and the water-cooled injector can be positioned axially with no coupling to the coil. In the spray RF plasma torches supplied by TECKNA (only industrial supplier of RF plasma spray torches), a ceramic tube with a higher thermal conductivity replaces the quartz tube generally used in RF plasma torches.

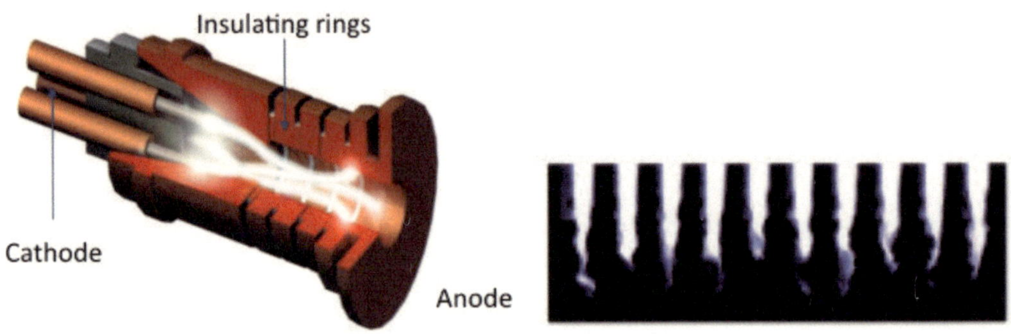

a. b.

Fig. 2. (a) Schematic view of Triplex® Plasma torch, (b) Pictures of the plasma jet: Aperture time: 1.5 μs, time between images: 130 μs.

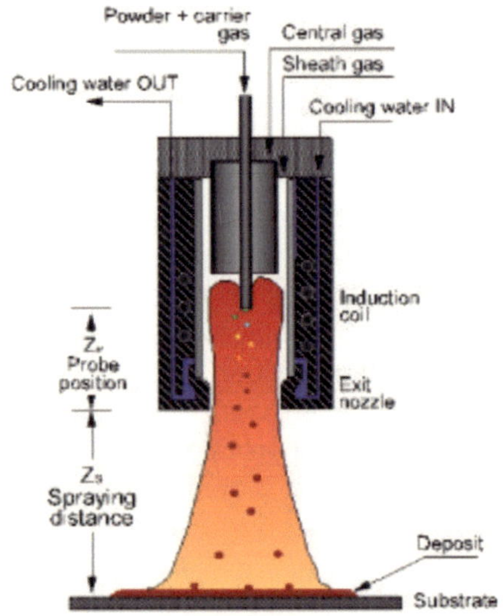

Fig. 3. Principle of radio frequency plasma spraying *Fauchais (2004).*

The coil is inserted in the torch body and it allows a perfect alignment and a smaller separation between the coil and the discharge and, thus, a better coupling.

The combination of these elements with a careful aerodynamic design of gas injectors and laminated high-velocity water-cooling allows reliable functioning at high power density. Spray torches with power levels up to 100 kW generally operate at 3.6 MHz. As the gas velocity is nearly inversely proportional to the square of the torch i.d, that is the plasma gas velocity is below 100 $m.s^{-1}$, corresponding to particle velocities below 60 $m\ s^{-1}$ and correspondingly high residence times (in the tens of milliseconds range). These conditions allow for the melting of metallic particles up to 200 μm with argon despite its low thermal conductivity. Using argon as the plasma forming gas allows for easy coupling at reasonable power levels. However, the sheath gas can also be pure oxygen allowing, for example, to spray materials that are very sensitive to oxygen losses such as perovskites.

3. Plasma–liquid interaction

3.1 Measurements and modeling

The visualization of the interaction between the liquid feedstock and plasma jet is necessary for better understanding the involved phenomena. *Etchart-Salas et al (2007)* were the first to propose a solution by using the fast-shutter camera of a Spray Watch detector from Oseir® coupled with a laser (808 nm wavelength) sheet flash. The image recording (Figure 4a) was triggered when the transient voltage reached a certain threshold. According to the image size and the number of pixels (600x600), one pixel represented an area of about 30 μm^2. It is thus impossible with such a device to visualize droplets of sizes below 5-6μm in diameter. To obtain more information on the plasma jet-liquid stream interactions, 10 images taken under the same conditions (in about 1 s) were superimposed after having eliminated the luminosity of the plasma jet. The resulting final image allowed determining two characteristic angles of the liquid stream penetration in the plasma jet: the dispersion angle (θ) and the deviation angle (α), as illustrated in Figure 4b.

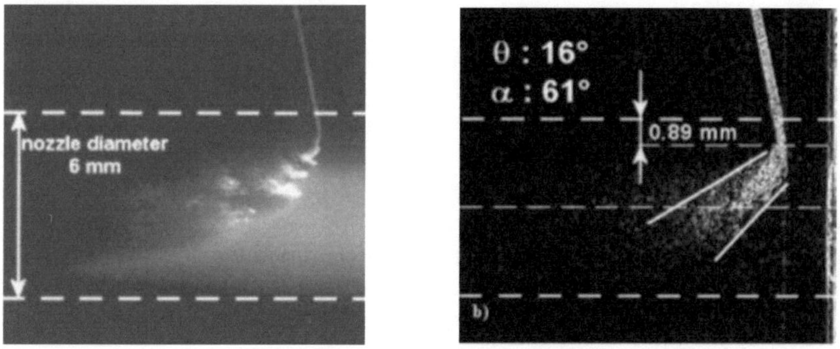

a. b.

Fig. 4. (a) Picture taken with a laser flash of an ethanol suspension mechanically injected at a velocity of 33.5 $m.s^{-1}$ in an Ar-He (30-30 slm) plasma jet (700 A, 40 V, anode-nozzle i.d. 6mm), (b) Image obtained by superposing ten successive images in the same conditions, *Etchart-Salas (2007)*.

More sophisticated devices, e.g., using a shadowgraph technique coupled with particle image velocimetry (PIV), also allowed observing the drops within the plasma jet and measuring their number, size and velocity. A double-pulsed Nd:YAG laser (532 nm wavelength with 8 ns pulse duration) backlight illuminated the liquid material. The detection system consisted of two charge-coupled device (CCD) cameras with 1376x1040 pixels and 12 bits resolution. A programmable hardware-timing unit controlled the synchronization of the laser with the cameras *Fauchais and Vardelle (2011)*. With this system, it was possible to observe the behavior of droplets down to about 3-5 μm, and to measure their velocity and diameter distributions in a relatively small volume ($2.5\times2\times1.5$ mm^3). This technique is very useful under plasma spray conditions to validate model predictions and observe the effect of operating parameter change. However results have to be cautiously interpreted and need to be coupled with pyrometer measurements.

The effect of liquid injection onto the plasma jet temperature can be performed by emission spectroscopy, as shown by *Fazilleau et al (2006)*. However the axial symmetry of the plasma jet is destroyed in the injection zone and recovered only when the liquid has been fragmented and vaporized, thus tomography is mandatory, *Landes (2006)*.

Caruyer et al (2010) have developed the numerical simulation of the interaction between a liquid jet and a plasma flow. They described the dispersion of liquid in order to understand the effect of injection conditions on the surface coating quality. They proposed an original model for dealing with three-dimensional interactions between the d.c. plasma flow and a liquid phase with "volume of fluid" (VOF) method. A compressible model, capable of representing incompressible two-phase flows as well as compressible motions, was used. The first comparisons of predictions with experimental shadography data showed fair agreement during the first moments of the injection.

3.2 Liquid injection

Two main techniques are used: atomization or mechanical injection.

3.2.1 Spray atomization

This method has been used for suspensions and solutions, *Rampon et al (2008)*. Very often, co-axial atomization is used. It consists of injecting a low velocity liquid inside a nozzle where it is fragmented by a gas (mostly Ar because of its high mass density) expanding within the body of the nozzle, *Filkova I. and P. Cedik (1984)*. For liquids of viscosity between a few tenths to a few tens of mPa.s, their break-up into drops depends on the Weber number, which expresses the competitive effects of the force exerted by the flow on the liquid and surface tension forces. This means that for a liquid with a given surface tension atomization depends on both gas velocity and specific mass. Atomization also depends, but to lesser extent, on the Ohnesorge number including the effect of liquid viscosity. Weber, We, and Ohnesorge, Z, dimensionless numbers are defined as follows:

$$We = \rho_g . u_r^2 . d_l / \sigma_l \qquad Z = \mu_l / \sqrt{\rho_l . d_l . \sigma_l}$$

where ρ_g is the gas mass density, μ_l the liquid viscosity, u_r the relative velocity between the gas and the liquid, d_l the diameter of the droplet and σ_l the surface tension of the liquid phase.

However, if the viscosity is too high (> 0.8 mPa.s), difficulties with feeding the liquid may appear. Measurements showed that atomization was affected by the following parameters:

the relative velocity between the liquid and the gas, the ratio of the gas to liquid volume feed rates, called RGS (generally over 100), or the gas-to-liquid mass ratio, called ALR (less than 1), the nozzle design, and the properties of the liquid (density, surface tension, dynamic viscosity). For example, depending on the Ar atomizing flow rate, the mean droplet diameters of alcohol vary between 18 and 110 μm. Also, for the same injection parameters, shifting from ethanol (σ_{eth} = 22.10^{-3} N/m at 293 K) to water (σ_w=72.10^{-3} N/m) modifies the mean diameter from 70 to 200 μm. Increasing the atomizing gas constricts the droplet jet and also perturbs the plasma jet. Similar results have been obtained when considering the influence of RGS, the droplet size is diminished with the increase of RGS. Quadrupling the RGS leads to a decrease in the droplet size by a factor of ten and allows obtaining a narrower Gaussian curve. It is also interesting to note that the weight percentage of solid in the suspension broadens the particle size distribution. For more details the reader is referred to the book of *Lefebvre (1989)*. For example, *Jordan et al (2008)*, have used three different types of atomizers to spray solutions: (1) a narrow angle hydraulic atomizing fan nozzle, (2) an air cap transverse air blast atomizing nozzle with a relatively large spray angle, and (3) a home made capillary atomizer with a liquid exiting one capillary and an air atomizing jet at 90 degrees to the liquid discharge used as a transverse jet atomizer. Of course the drop size distribution is the broadest with the air cap atomizer and the narrowest for the capillary atomizer with which the best coatings are achieved.

3.2.2 Mechanical injection

Two main techniques are possible: either to have the liquid in a pressurized reservoir from where it is forced through a nozzle of given i.d., or to add to the previous set-up a magnetostrictive rod at the back side of the nozzle which superimposes pressure pulses at variable frequencies (up to a few tens of kHz).

For example the first device developed at the SPCTSLaboratory of the University of Limoges, France, *Etchart-Salas (2004)*, consisted of four tanks in which different suspensions and one solvent were stored. Any reservoir or both of them could be connected to an injector consisting of a stainless steel tube with a laser-machined nozzle with a calibrated injection hole. A hole of diameter d_i produced a liquid jet with a velocity v_l (m/s) linked to the incompressible liquid mass flow rate (kg/s) by the following equations:

$$m_l^o = \rho_l.v_l.S_i \qquad \text{and} \qquad \Delta p = f.\rho_l.v_l^2 / 2$$

where ρ_l is the liquid specific mass (kg.m^{-3}) and S_i the cross sectional area of the nozzle hole (m^2). Assuming that the liquid is non-viscous and ideal, v_l depends on the pressure drop Δp between the tank and surrounding atmosphere through the Bernoulli equation where f is a correction factor (0.6 to 0.9) for friction and viscous dissipation. For example with an injector i.d. of 150 μm, the tank pressure was varied between 0.2 and 0.6 MPa to achieve injection velocities between 22 and 34 m.s^{-1}. To achieve the same injection velocity with a d_i of 50 μm as the velocity obtained at 0.5 MPa with d_i = 150 μm, the pressure should be multiplied by 81 and so reached 40.5 MPa requiring adapted equipment and special precautions!. The liquid exiting the nozzle, observed with a CCD camera, flows as a liquid jet, with a diameter between 1.2 and 1.5.d_i depending on the tank pressure and nozzle shape. After a length of about 100-150 times d_i, Rayleigh-Taylor instabilities lead to fragmentation of the jet into drops with a diameter of about 1.3-1.6 times that of the jet. Thus, depending on the position of the injector exit relative to the plasma jet (radial injection), either a liquid jet or droplets are injected in the plasma jet.

Blazdell and Kuroda (2000) used a continuous ink jet printer, which allowed uniformly spaced droplets to be produced by superimposing a periodic disturbance on a high-velocity ink stream. They used a nozzle 50 µm in i.d. (d_i) and a frequency f of 74 MHz producing 64,000 droplets/s. *Oberste-Berghaus et al. (2005)* have used a similar set-up with a magnetostrictive drive rod (Etrema AU-010, Ames, Iowa) at the backside of the nozzle, working up to 30 kHz. They produced 400-µm drops with 10-µs delay between each and with a velocity of 20 m/s.

3.2.3 Liquid penetration into the plasma flow

In conventional particle spraying where the injection force of the particle is nearly that imparted to it by the plasma jet, the optimum particle trajectory is achieved (see Introduction). For liquid jet or drops injection, when they penetrate within the plasma jet they are progressively fragmented and their volume and apparent surface become smaller. Accordingly, the mass of the drops as well as the force imparted to them by the gas jet is reduced. Thus, drop penetration ceases rapidly, and the condition for good penetration implies that $\rho_l.v_l^2 >> \rho_g v_g{}^2$. For example, considering a suspension of zirconia in ethanol injected into an Ar-He d.c. plasma jet, a better penetration of the liquid jet within the plasma jet core is achieved when its velocity is 33.5 $m.s^{-1}$ instead of 27 $m.s^{-1}$. In the first case, the liquid jet momentum density is 0.96 MPa against 0.02 MPa for the mean value of that of the plasma jet (with a 27 $m.s^{-1}$ injection velocity the liquid jet momentum is still 0.6 MPa).

3.2.4 Drops or jet fragmentation and/or vaporization

Upon penetration in the plasma jet, drops or liquid jet are submitted (i) to a strong shear stress due to the plasma flow, which, under conditions described below, fragment them into smaller droplets, and (ii) to a very high heat flux that vaporizes the liquid. Thus, a very important point is the value of the fragmentation time, t_f, relatively to the vaporization time, t_v, in order to see if the two phenomena can be separated. One approach is that of complex models, such as that of *Caruyer et al (2010)*. They considered (i) 3-D transient or stationary description of the turbulent plasma flow and its mixing with the ambient atmosphere, (ii) a suitable description of the liquid feedstock injection as jet, drops or droplets into the plasma jet and (iii) an accurate description of the possible mechanisms that control the treatment of the liquid material in the plasma flow (i.e. mechanical break-up, thermal break-up, coalescence). The fragmentation undergone by drops or jets has been extensively studied for cold gas impacting orthogonally a liquid jet *Lee and Reitz (2001)*. These results have been considered to be a valid first approximation for the interaction of hot gases or plasmas and a liquid even though thermal effects are not considered. However, if results established for cold gases are acceptable in understanding the implied phenomena, experiments should be implemented with hot gases to validate them. The fragmentation depends upon the dimensionless Weber number, We, (section 3.2.1) and according to its value different regimes can be considered:

- for 12 < We < 100, the fragmentation is named 'bag break-up' : it corresponds to the deformation of the drop as a bag-like structure that is stretched and swept off in the flow direction;

- for 100 < We < 350, the fragmentation is named 'stripping break-up' : thin sheets of liquid are drawn from the periphery of the deforming droplets;
- for We > 350, the fragmentation is named 'catastrophic break-up'and corresponds to a multistage breaking process.

The main question is when fragmentation occurs? Indeed, the We number increases with the relative velocity between the drop and the jet, u_r, and with the drop size. The different limit values of this number can be slightly modified according to the Ohnesorge number.

- For r.f. plasmas where gas velocities are below 100 $m.s^{-1}$, where axially injected drops have sizes over 20-30 μm, fragmentation is highly improbable and the main phenomena are the liquid vaporization followed by the heat treatments of the particles contained in the liquid (suspension) or formed in flight (solutions).
- For d.c. plasma jets with radial injection of the liquid in the plasma jet the story is quite different. We > 12, except if the injected drops are less than a few tens of micrometers, and the fragmentation process starts in the plasma jet fringes before drops have reached the hot plasma core, as illustrated in Figure 5a from *Etchart-Salas (2007).* As explained in section 5, the main problem when fragmentation starts in the d.c. plasma jet fringes is that the fragmented droplets are nevertheless vaporized and their content heated enough to stick onto the coating under formation and create stacking defects. Of course when the atomized liquid exhibit rather broad distributions of drop size and velocity, drop penetration within the plasma jet is poorly controlled, as shown in Figure 5b and creates stacking defects within coating.

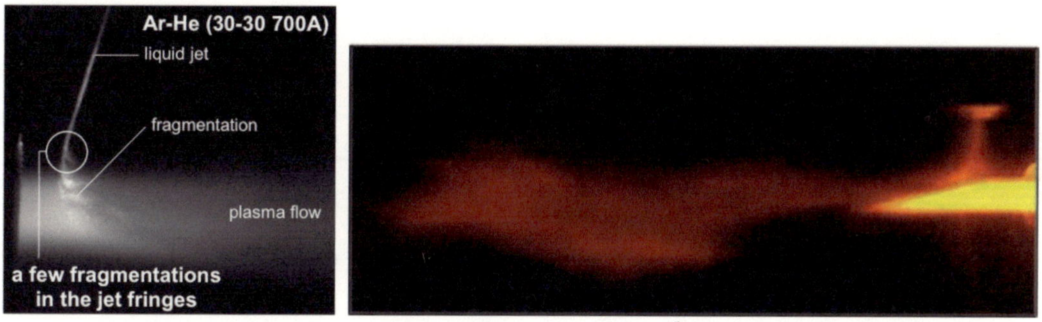

a. b.

Fig. 5. (a) Example of fragmentation of an ethanol jet, about 300 μm in diameter, starting in the fringes of a d.c. plasma jet (Ar-He: 30-30 slm, 700 A, 40 V, Anode-nozzle i.d. 6 mm) *Etchart-Salas et al (2007)*, (b) Interaction of an Ar-H_2 plasma jet with atomized ethanol drops *Fazilleau et al (2006).*

A very important point is the value of the fragmentation time t_f relative to that of vaporization time t_v. The fragmentation time can be roughly calculated by assuming that atomization is completed when the liquid surface tension force is equal to the drag force of the plasma jet resulting in a minimum droplet diameter d_m. *Fazilleau et al (2006)* studied the fragmentation of ethanol drops along their mean trajectory in an Ar–H_2 d.c. plasma jet (operating parameters: Ar-H_2 45-15 slm, 500A, V_m =65 V, anode-nozzle i.d. 6mm). For the calculation of the fragmentation and evaporation times, they considered the plasma temperatures and velocities "seen" by so-called mother drops of various diameters and resulting droplets along their trajectories. They also took into account the buffer effect due to the vaporization of the liquid phase and the Knudsen effect. The results are summarized in Figure 6. It has to be kept in mind that these calculations did not consider the plasma jet cooling resulting from the vaporization of the liquid phase and, thus, the drop diameter decrease along the plasma jet radius was probably overestimated. Nevertheless it can be readily seen that, in the plasma jet fringes, the drop diameter starts to be reduced due to fragmentation and the resulting droplets very rapidly decrease in size because of vaporization. Fragmentation and vaporization times differed by at least two orders of magnitude whatever the considered diameter of the mother drop. It is also important to note that the vaporization time of a drop 300 µm in diameter is about 4 orders of magnitude longer than that of a drop of 3 µm, which is 100 times smaller. The fast (< 1 µs) fragmentation of drops followed by the fast (~1 µs) vaporization of resulting droplets explains the sequence of successive events in an Ar-H_2 plasma jet upon water drop penetration *Fazilleau et al (2006)*. The plasma jet is at first disrupted into two parts distributed on both sides of the plane defined by the torch centerline axis and the injector axis. Then, the axial symmetry is restored 15 mm downstream of the nozzle exit. Once drops are fragmented into much smaller droplets, they are vaporized very fast and the liquid phase vapor is rapidly transformed into plasma. Assuming no evaporation, the predicted residence time of a 2 µm water droplet located on torch centerline at nozzle exit is of the order of ten microseconds for the first 15 mm of trajectory. These characteristic times clearly demonstrate that droplets are fully vaporized in the hot core of the plasma jet as about 1 µs is sufficient to vaporize a droplet 2 µm in diameter, and the water vapor is transformed into plasma.

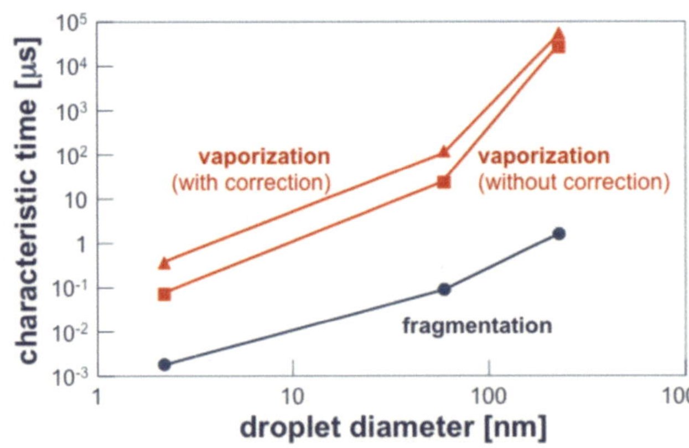

Fig. 6. Variation of fragmentation and vaporization times of ethanol drops with droplet diameter without and with the correction that takes into account the buffer effect of the vapor cloud around droplets for a stationary (V = 65 V, I = 600 A, nozzle i.d. 6 mm) Ar-H_2 plasma jet *Fazilleau et al (2006)*.

3.2.5 Effect of plasma jet fluctuations

Arc root fluctuations have a more important negative effect on liquid injection than they have on powder particle injection. For example with an Ar-H_2 plasma jet (Ar-H_2 45-15 slm, 500 A, V_m =65 V, anode-nozzle i.d. 6mm) with arc operating in the restrike mode, the arc voltage varies between 40 and 80 V, resulting in high variations of the plasma jet length, as illustrated in Fig. 1b. A difference of 40 V in arc voltage corresponds to a variation in the predicted mean velocities of gas of about 800 m.s^{-1}, resulting approximately in a gas momentum density variation ($\rho_g.v_g^2$) of 320%! These variations bring about different penetrations of the suspension stream in the plasma flow, as illustrated in Figure 7. The liquid jet dispersion angle (θ) for an arc voltage of 40 V is about 64° when it is about 33° for an arc voltage of 80 V while the deviation angle (α) is almost constant. Therefore, Ar-He plasma gas that exhibit lower arc voltage fluctuations are often preferred to Ar-H_2 plasma gas, even if the arc power levels are lower, except with the Triplex® torch.

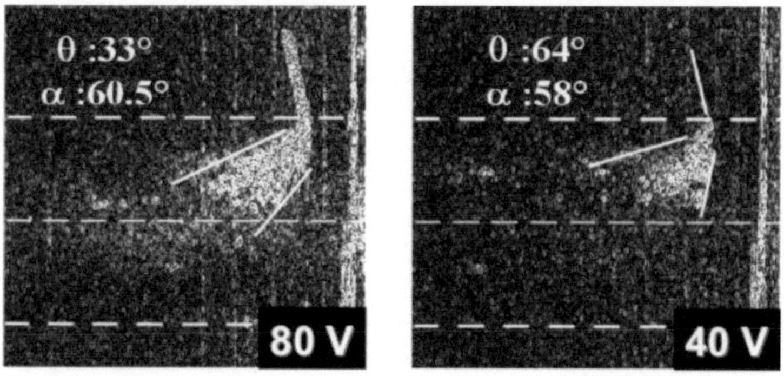

Fig. 7. Plasma jet–suspension stream interactions for arc voltage of (a) 80 V and (b) 40 V (suspension stream injection velocity of 26.6 m.s^{-1}, distance between the injector tip and the torch centerline axis of 20 mm, other working parameters depicted above) *Etchart-Salas et al (2007).*

3.2.6 Effect of liquid injection velocity

To counteract the strong decrease in drop sizes when they penetrate the plasma jet because of break-up mechanisms, the liquid injection velocity must be increased. For example with the Ar-He plasma jet presented in Figure 4b the liquid injected at 33.5 m.s^{-1} travels 0.89 mm in the jet fringes without fragmentation and goes beyond the torch axis while when it is injected at 27 m.s^{-1}, fragmentation starts after 0.58 mm trajectory in the plasma jet fringes and the visible part of the fragmented jet hardly reaches the torch axis.

3.2.7 Effect of liquid vaporization

The liquid vaporization and then the transformation of the vapor into plasma consume energy and cool down the plasma jet. For example when a water jet with a mass flow rate of 3.6×10^{-4} kg/s (typical flow rate for liquid injection) is injected in an Ar-H_2 (25vol.%) plasma jet with a mass flow rate of $1.36\text{x}10^{-4}$ kg/s and a mass enthalpy of 32 MJ.kg^{-1} the latter decreases by almost 9 MJ.kg^{-1}, *Fazilleau et al (2006)*. Correlatively the plasma jet core length where temperature is over 8000 K is reduced by 25%. Spectroscopic measurements showed that about 15 mm downstream of the liquid injection the symmetry of the jet, cut into two parts at the injection location, was restored: a large part of the liquid was evaporated and transformed into plasma, and the remaining droplets were uniformly distributed within the plasma jet core. Of course liquids with lower vaporization energies as ethanol cool down less the plasma. However, generally when the vaporization energy decreases, the surface tension also decreases favoring liquid fragmentation in the plasma jet fringes.

4. Suspensions and solutions preparation

4.1 Suspension preparation

4.1.1 Slurry route

The easiest way to prepare a suspension is to make slurry with particles and liquid phase, particle sizes varying from a few tens of nanometers to a few micrometers. The most used liquid phases are ethanol, water or a mixture of both, *Fauchais (2007)*. After stirring, the suspension stability can be tested by a sedimentation test. Typical values of slurry stability are a few tens of minutes up to a few hours, and the stability increases with the mass load, *Rampon et al (2008)*. However, for loads over 15-20 wt.%, solid particles are less melted when they are sprayed. Slurries with TiO_2, ZrO_2 , Al_2O_3 and ZrO_2-Al_2O_3 have been prepared following this route, *Fauchais et al (2011)*. However, it should be noted that nanometer-sized particles of oxides have the tendency to agglomerate or aggregate, even when stirring the suspension. This drawback can be partially or totally overcome by using a suitable dispersant, which adsorbs on the particle surface and allows an effective dispersion of particles by electrostatic, steric or electro steric repulsions. For example, a phosphate ester that aids particle dispersion by a combination of electrostatic and steric repulsion has been used with zirconia particles, *Fazilleau et al (2006)*. The percentage of dispersant must be adjusted in such a way that it displays the minimum viscosity of the suspension with a shear-thinning behavior, *Fazilleau et al (2006)*. This behavior means that when the shear stress imposed by the plasma flow to the drops of a liquid stream is low, the suspension viscosity is high and it decreases drastically when the shear stress increases as the drop penetrates more deeply within the plasma flow. The pH adjustment is also an important characteristic of the suspension-liquid mixing. The problem becomes complex with WC–Co particles because of the different acid/base properties of both components: *Oberste-Berghaus et al (2005)*. Indeed, WC or, more precisely, WO_3, at its surface is a Lewis acid, while CoO is basic. Thus, a complex equilibrium between the dispersing agent and the suspension pH must be found. For example, the latter must be adjusted to less basic conditions without cobalt dissolution. Similar problems have been observed with Ni particles. When the weight % of powder increases in the suspension, its viscosity increases too.

Different products can be added to the liquid phase to modify its viscosity and/or surface tension, *Rampon et al (2008)*. For example, the addition of viscous ethylene glycol with a boiling point of 200 °C will change the suspension viscosity but at the expense of additional thermal load on the plasma, *Oberste-Berghaus et al (2006)*, while the addition of binders makes it possible to control the suspension viscosity almost independently of the dispersion.

It is also important to (i) adapt the size distribution of particles within the suspension to the potential heat transfer of the hot gases, (ii) limit the width of particle size distribution as in conventional plasma spraying in order to reduce the dispersion of particle trajectories and (iii) avoid powders that have a tendency to agglomerate or aggregate, which is often the case with nanometer sized particles, especially oxides, when prepared by chemical routes.

4.1.2 Chemical route

Another route, called Prosol process, consists in preparing a zirconia sol by neutralization of zirconium oxy-chloride in an aqueous media followed by hydrothermal crystallization, *Wittmann-Ténèze et al (2008)*. When the mixture is heated in a container at a minimum temperature of 170 °C for 20 h at a minimum pressure of 2 MPa, crystalline oxide precipitates as a mixture of monoclinic and tetragonal phases. After washing the excess of ammonium hydroxide, the particles are mixed with water with addition of hydrochloric acid (pH 3) to form a suspension.

4.1.3 Amorphous particles

Chen et al (2009) asserted that the suspension plasma spraying process, using molecularly mixed amorphous powders as feedstock, was an ideal process for the deposition of homogeneously distributed multi-component ceramic coatings.

They produced Al_2O_3 -ZrO_2 molecularly mixed amorphous powders by heat treatment of molecularly mixed chemical solution precursors below their crystallization temperatures. For that, aluminum nitrate and zirconium acetate were dissolved in de-ionized water based on molar volumes to produce a ceramic composition of Al_2O_3–40 wt. % ZrO_2. The resulting solution was heated at 80 °C and stirred continually to get the sol transformed into dried gel. The dried gel powders were heated at 750 °C with a a heating rate of 10 °C.min^{-1}, and then held for 2 h at this temperature. The as-prepared powders were then mixed with ethanol heated to 750 °C with a loading rate of 50 wt. % and then ball-milled by using ZrO_2 balls for 24 h. XRD patterns showed that the powders were amorphous. The Al_2O_3–ZrO_2 particles had a size distribution (d_{10} –d_{90}) of 0.71 μm with an average particle size of 0.5 μm

4.2 Solution preparation

The precursors used in solutions include, *Ravi et al (2006)*, (i) mixture of nitrates in water/ethanol solution; (ii) mixtures of nitrates and metal-organics precursors in isopropanol (hybrid sol); (iii) mixed citrate/nitrate solution (polymeric complex) and (iv) co-precipitation followed by peptization (gel dispersion in water/ethanol).

Compared with other thermal spray techniques, solution plasma spraying using constituent chemicals mixed at the molecular level allows an excellent chemical homogeneity of coatings. It is also worth noting that aqueous solutions permit higher concentrations than organic solutions, are cheaper to produce and are easier and safer to store and handle. For example, water has been used as solvent for solutions of zirconium, yttrium and aluminum salts, *Chen et al (2007)*.

The precursor concentration in solutions can be varied up to the equilibrium saturation. A simple way to determine the equilibrium saturation concentration, is to put the solution in an evaporator at room temperature until precipitation occurs. In their study of 7YSZ solutions, *Chen et al (2008)* have considered two different precursor concentrations: a high molar concentration (2.4 M) and a low molar concentration (0.6 M). When the initial 7YSZ precursor was concentrated four times in water, the solution viscosity increases from 1.4×10^{-3} to 7.0×10^{-3} Pa. s and the surface tension decreases from 5.93×10^{-2} to 4.82×10^{-2} N.m^{-1}. Both precursors pyrolysed below 450 °C and crystallized at about 500 °C, *Chen et al(2007)* showing that the solution precursor concentration had little effect on the precursor pyrolysis and crystallization temperatures. If a variation in precursor concentration has almost no effect on the solution specific mass and surface tension, it however brings about large variations in the solution viscosity. *Chen et al (2010)* also studied the effect of the liquid phase on the treatment of the solution in the plasma jet and got results quite similar to that already got with suspensions, i.e. droplets with high surface tension and high boiling point liquid phase experience incomplete liquid phase evaporation in the plasma jet while droplets with low surface tension and low boiling point liquid phase undergo rapid liquid phase evaporation.

5. Coating formation from suspension and solution droplets

5.1 General remarks about the in-flight treatment of nanometer or sub-micrometer particles

The coating microstructure depends on the interaction between the plasma jet and the original micrometer or sub-micrometer droplets and, then, between the plasma jet and the particles contained in the suspension or formed in the solution. These particles may have sizes in the sub-micrometer or nanometer range and form the coating when they impact on the substrate. The momentum and heat transfers between particles and plasma jet can be estimated from the drag coefficient, C_D that quantifies the resistance of particles in fluid environment and the heat transfer coefficient calculated from the Nusselt number, Nu, that is the ratio of **convective** to **conductive** heat transfer across the particle boundary. Both the drag coefficient and the heat transfer coefficient must be corrected to take into account *Boulos et al (1993)*: (i) the high temperature gradient between the gas and the particle surface. This correction is independent of the particle size and corresponds at the maximum to a 30% decrease in C_D and Nu; (ii) the buffer effect of the vapor issued from droplet and particle. It is generally only considered for the Nusselt number and is also independent of particle size. This correction is particularly important for liquid feedstock that undergoes an intense evaporation in plasma jet; (iii) the rarefaction or Knudsen effect occurring when the ratio of the gas molecules mean free path λ to the particle diameter d_p is smaller than one. This effect is particularly important for nanometer and sub-micrometer sized particles such as particles contained in suspensions or formed in solutions.

5.1.1 Knudsen effect

The non-continuum effect can be important when the mean free path of the plasma molecules, atoms, ions, λ, is of the same order of magnitude or lower than the diameter of the particles, d_p. According to the review of *Boulos et al (1993)*, the Knudsen effect should be taken into account in the Knudsen regime characterized by $0.01 < Kn < 1.0$, where, $Kn = \lambda / d_p$. The Knudsen effect is rather low at room temperature even for particles as small as 40 nm since λ is of the order of 0.4 µm. However, under thermal plasmas conditions, λ is a few micrometers at 10 000 K (in a first approximation $\lambda \sim T/p$) and the Knudsen effect becomes important. For example, for a particle at 1000 K immersed in an Ar-H_2 plasma at 10 000 K, the correction factor for the drag coefficient is divided by a factor close to 3 when the particle diameter decreases from 1 µm to 0.1 µm and the Nusselt number by a factor 10 *Delbos et al (2006)*. The heat and momentum transfer to small particles are, thus, drastically reduced.

Moreover, these small particles that have a very low inertia, decelerate very fast and may follow the gas flow deflected by the substrate. For example in an Ar-H_2 (45–15 slm) plasma produced by a plasma torch operating at 600 A with an internal diameter of the nozzle of 6 mm, the plasma velocity decreases from 2200 m.s^{-1} at the nozzle exit to 1500 m.s^{-1} 15 mm downstream of the torch nozzle exit. At the same distance, the velocity of a 100-nm zirconia particle, injected close to the nozzle exit is at the most 500 m s^{-1}, and it decreases to 350 m.s^{-1} 35 mm downstream of the nozzle exit because of the very low inertia of particle. The particles hit the substrate when certain conditions are fulfilled (see section 5.1.4).

5.1.2 Vaporization effect

The vapor surrounding the droplet or particle has a strong effect on the heat flux brought by the plasma jet as it uses part of it to heat from the particle surface temperature to the gas temperature. This effect is near independent of particle size.

5.1.3 Thermophoresis effect

The drag force is generally by far the most important force acting on particles immersed in a fluid. However, for particles below 0.1 µm, accelerated in a plasma jet, the thermophoresis force can become important *Boulos et al (1993)* in areas where steep temperature gradients exist, such at the limit between the plasma core and its plume. The plasma core corresponds roughly to the jet zone where temperatures are over 8000 K corresponding, for most plasma gases except pure Helium, to an electrical conductivity over 1 kA.V^{-1}.m^{-1}. The thermophoresis force tends to eject small particles into the plasma plume where gas temperatures and velocities are lower, as shown in Figure 8.

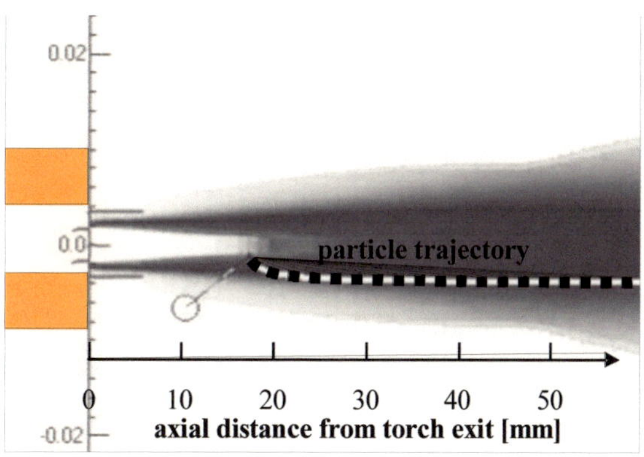

Fig. 8. Schematic of the Thermophoresis effect

5.1.4 Stokes effect

Because of their low inertia, small particles can follow the gas flow that travels parallel to the substrate surface without impacting it. To hit the substrate, particles must cross the boundary layer that develops at the substrate surface. The Stokes number characterizes the behavior of particles **suspended** in a fluid flow. It is defined as follows: $St = \frac{\rho_p . d^2{}_p . v_p}{\mu_g . l_{BL}}$

where indexes p and g are related to particles and gas, respectively, ρ is the specific mass (kg.m^{-3}), d the diameter (m), v the velocity (m s^{-1}), μ the molecular viscosity (Pa.s) and l_{BL} the thickness of the flow boundary layer in front of the substrate (m). When St is higher than 1, that is when particle velocity is high enough, particles can detach from the flow and hit the substrate.

For the Ar–H_2 plasma already considered the Stokes number of particles 0.3 µm in diameter is equal to 1 when their velocity reaches 300 m.s^{-1} assuming a flow boundary layer at the substrate surface in the order of 0.1 mm, *Delbos et al, (2006)*. It is thus of primary importance that the particles have high velocities just prior to impact, especially when their size decreases to sub-micrometer or nanometer values. As the inertia of small particle is very low and the boundary layer thickness in front of the substrate decreases, in a first approximation, as the inverse of the square root of the gas velocity, the substrate must be located at short distances of the torch nozzle exit. For example, with stick-type cathode plasma torch, *Fauchais et al (2008)*, spray distances between 30 and 50 mm are commonly used. Accordingly, heat fluxes imposed by the plasma flow on the forming coating and on the substrate are very high, as illustrated in Figure 9. Compared with micrometer-sized conventional coatings sprayed on substrates located at 100–120 mm and subjected to heat flux below 2 MW.m^{-2}, the very high heat fluxes (up to 40 MW.m^{-2}) imposed on suspension plasma-sprayed coatings contribute to their morphology modification.

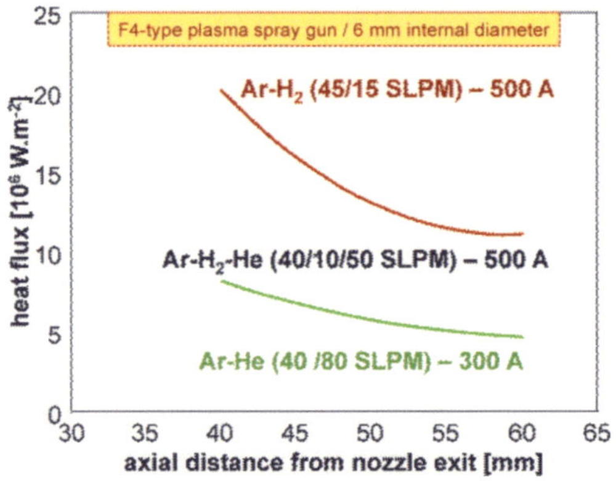

Fig. 9. Variation with the spray distance of the heat flux imparted to a surface by a stick-type cathode plasma torch (6 mm i.d. anode–nozzle) working with Ar–H_2, Ar–He and Ar–He–H_2 plasma-forming gas mixtures *Fauchais et al (2008)*.

5.2 Radio frequency plasma spraying

5.2.1 Suspensions

Bouyer et al (1995 and 1997) were the first to propose the production of hydroxyapatite (HA: $Ca_{10}(PO_4)6(OH))$ coatings using the technique called suspension plasma spraying (SPS). The process involved the injection of the HA material in the form of an atomized colloidal water suspension into the center of an inductively coupled r.f. plasma discharge. This process is a powderless (colloidal suspension) plasma-spraying technique. The HA suspension was brought into the plasma discharge core via a gas-atomizing probe fed with a peristaltic pump. As mentioned previously the relative gas-drops velocity was too small (<50 m.s^{-1}) to fragment the drops (< 100 µm). The latter were successively flash-dried, melted, deposited on the substrate to be coated or collected in-flight as spherical particles. The process took full advantage of the inherent features of the induction plasma, which allows sufficient time for the droplet drying and melting steps. The first parts of the drop trajectories took place in a pure Ar plasma with a rather low heat transfer and it was only downwards that heat transfer increased thanks to the diffusion of the diatomic sheath gas in the torch central part. Figure 10a shows the three main routes for HA deposit preparation, while Figure 10b presents the suspension with needles of HA. The SPS route is, by far, the simplest and least costly, and it also eliminates many potentially contaminating steps in coating preparation. Authors showed the feasibility of suspension plasma spraying for HA coatings with high deposition rates (> 150 µm.min-1.) and also for spheroidized powder production. The use of a r.f. plasma torch made it possible to use oxygen as sheath gas. The decomposition of HA during plasma treatment could be either avoided or at least minimized by using appropriate plasma gases, i.e. a plasma-sheath gas mixture with moderate enthalpy and high oxidizing potential. For example hydrogen acts as harmful gas for HA stabilization, while oxygen is beneficial. In addition, the presence of water in the suspension contributed to the resistance of the apatite structure to decomposition during the high-temperature treatment by maintaining a high partial pressure of water vapor in the deposition reactor.

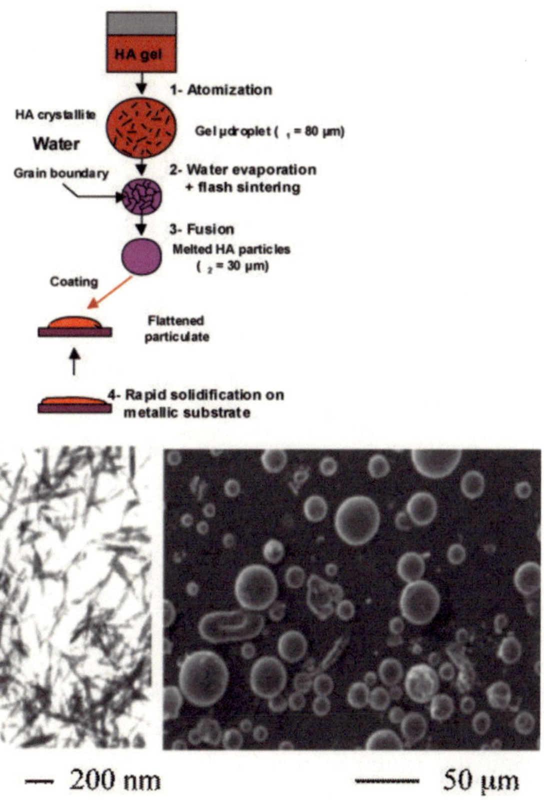

Fig. 10. (a) Schematic routes for HA coating deposition by suspension plasma spraying with r.f. plasma torch, (b) Suspension of HA needle shaped, (c) Spheroidized particles collected in the reactor, *Boyer et al (1995, 1997).*

Schiller et al (1999), using a reactive precursor suspension of MnO_2 powder in an ethanol solution of $LaCl_3$ prepared perovskite powders and coatings. The powder completely melted in the plasma and the $LaMnO_3$ perovskite phase was formed as primary phase. A certain amount of additional phases (La_2O_3, …) were also present in the coating. The purity of the perovskite coatings could be further enhanced through a post-treatment using plasma with a high content of oxygen (80%). However further works are still necessary to completely prevent the formation of La_2O_3, control the coating porosity and dope $LaMnO_3$ with strontium.

This spray technique has also been used by *Bouchard et al (2006)* to produce perovskite cathode materials ($La_{0.8}Sr_{0.2}MnO_{3-\delta}$, $La_{0.8}Sr_{0.2}FeO_{3-\delta}$, and $La_{0.8}Sr_{0.2}CoO_{3-\delta}$), with accurate control of deposit stoichiometry. Most particles of the plasma-synthesized powders were about 63 nm in size, with an average grain size of 20 nm. The plasma-synthesized powders were almost globular in shape, and their BET specific surface areas were ~26 $m^2.g^{-1}$, i.e., about twice that of powders prepared by other routes. *Jia and Gitzhofer (2010)* elaborated coatings by r.f. plasma spraying of suspensions made of gadolinia-doped ceria (GDC) particles with a mean size of 0.6 µm dispersed in distilled water. The resulting coating (Figure. 11a) consisted of layered splats with diameters ranging from 0.2 to 2 µm corresponding to a flattening ratio below 2.7.

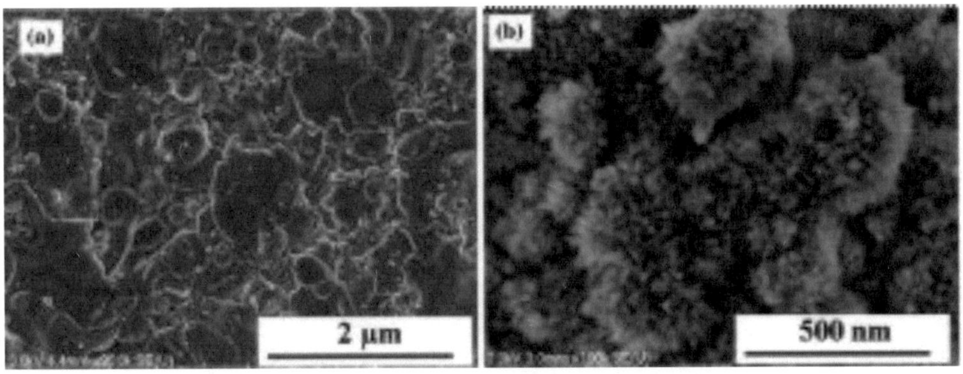

Fig. 11. Field emission scanning electron microscope (FESEM) micrograph of r.f. **suspension** plasma-sprayed gadolinia-doped ceria coating top surface, (b) FESEM micrographs of r.f. **solution** plasma sprayed GDC coating top surface *Jia and Gitzhofer (2010).*

5.2.2 Solutions

In Solution r.f. Plasma Spraying, called SolPS process by *Jia and Gitzhofer (2010)*, GDC solutions of nitrate hexahydrate were completely vaporized in the plasma. Nano-structured GDC particles were synthesized in flight by homogeneous nucleation from the supersaturated precursor vapor and build up a coating with high porosity (Figure 11b) when impacting on a substrate. The formation of globular GDC coatings was observed over the entire experimental range of this study. Similarly *Shen et al (2011)* have r.f. plasma sprayed solutions of Lanthanum strontium cobalt iron oxide (LSCF: $La_xSr_{1-x}Co_yFe_{1-y}O_{3-\delta}$) and GDC. A homogeneously mixed nanometer sized composite GDC/LSCF powder was obtained without a prolonged period of mechanical mixing. The nanometer-sized powders exhibited a perovskite structure, a fluorite structure and separated GDC and LSCF phases.

When using a low solution-feeding rate, nanometer-sized powders with different compositions were easily synthesized by only adjusting the metal nitrate concentrations in the precursor solution. All synthesized powders had spherical particles with diameter between 10 and 60 nm regardless of their composition. The deposited coating had a homogeneous nano-cauliflower structure, similar to that presented in Figure 11b, with an average porosity of 51%.

5.3 Direct current (dc) plasma spraying

In dc plasma spraying, as explained in section 3.2.4, the fragmentation of the liquid feedstock injected in the transverse plasma jet is a few orders of magnitude faster than vaporization and stops, depending on plasma velocities when particles reach diameters ten micrometers or so, corresponding to We < 12.

5.3.1 Suspensions

a. Solid particles in-flight

Once the liquid feedstock in the form of drops or liquid jet penetrates the hot gas flow, if We > 12, it is fragmented into droplets with a size that decreases as they go deeper in the plasma jet. The liquid injection location, either close to the anode-nozzle exit or farther downstream, plays an important role in the fragmentation process as the later depends on the momentum density ($\rho_g.v_g^2$) of the plasma jet at the liquid location and the velocity of the gas, v_g, decreases rather fast with the distance from anode-nozzle exit, because of the jet expansion. The droplets issued from liquid fragmentation have trajectories that are widely dispersed within the plasma jet. Once the liquid phase is completely vaporized, the solid particles contained in each droplet follow different trajectories in the plasma jet fringes or hot core of the jet and can be ejected from the hot core because they travel in area of high temperature gradients as shown in Figure 8. Along these trajectories solid particles are heated and accelerated by the hot gas flow; their initial velocity is that of their 'mother' droplet prior to its total vaporization. When agglomerates are formed, they tend to explode upon complete vaporization of the liquid phase in particular if the nanometer-sized particles of the suspension have been prepared by soft-chemical routes *Delbos et al (2006)* and this phenomenon increases the dispersion of the solid particles constituting the agglomerates. When the sprayed material is collected during a short period (e.g. a few tenths of a second) on a fixed substrate, only splats are observed in the central part of the collecting area while around it only spherical and un-melted particles are observed *Fauchais et al (2008)*. The spherical particles correspond to particles that melt in the plasma jet but are re-solidified before their impact on the substrate while the un-melted particles mainly result from drops fragmented in the jet fringes where the solid particles contained in the drops are insufficiently treated but, often heated enough to stick to the substrate or the coating under formation. Obviously, some additional studies are necessary to further document the organization of the particles upon complete liquid phase vaporization.

When the suspension is injected axially in the plasma jet as with the Mettech plasma torch (see section 2.2) the droplets are also highly dispersed in the whole plasma jet including in the cold boundary layer of the flow because of the flow turbulences, and the thermophoresis effect also takes place. A numerical study of Xiong and Lin (2009) suggested that the optimal diameter of Al_2O_3 particles for axial injection-suspension plasma spraying is about 1.5 μm for 'optimum' coating characteristics. Under these conditions, at the appropriate spray distance, particles can impact at high velocity in a fully molten condition. The small particles (< 500 nm) do not flatten so effectively, because they attain lower momentum and because they may re-solidify before impact, on account of their low inertia. Large particles and agglomerates (>2.5 μm), by contrast, remain partly or entirely un-melted.

b. Bead formation

A line-scan-spray experiment makes it possible to evaluate the degree of melting of particles; it can consider either a simple bead resulting from one pass of the torch in front of the substrate or overlapped beads resulting from successive passes of the torch. The bead thickness depends on the torch operating conditions and on the relative torch to substrate velocity, the number of passes, the suspension flow rate and the injection parameters, the mass loading of powder particles in suspension, the size distribution of particles. The bead profile generally fits rather well with a Gaussian profile.

To study the spray bead manufacturing mechanisms, *Tingaud et al (2008)* made a suspension of alumina of angular single mono-crystalline α-Al_2O_3 (P152 SB, Alcan, Saint-Jean de Maurienne, France) of d_{50} = 0.5 μm dispersed in pure ethanol with an electro-steric dispersant. The powder mass fraction in suspension was 10 wt%. The suspension was sprayed on a 304L stainless steel substrate preheated with the plasma jet at 300 °C before spraying, to eliminate adsorbates and condensates *Chandra and Fauchais (2009)*. As expected, spray parameters and substrate parameters conditioned the spray bead morphology.

In the spray beads shown in Figure 12, two regions could be identified, that corresponded to adherent deposit and powdery deposit. Powdery deposits located at the edges of the spray bead corresponded to precursors traveling in the low-temperature regions of the plasma flow. The central part of the bead was relatively dense and made of splats with a few spherical particles re-solidified in flight and angular un-melted particles that have travelled in colder zones. The splats corresponded to particles adequately (i.e. well-melted) treated in the plasma core. At mid-height of the spray bead, the coating was less dense with many un-melted particles and at its edges it was fully powdery.

To get coatings with good cohesion, the central part of spray beads must contain as much as possible fully or near molten particles at impact.

In the example shown in Figure 12, the densest coating was obtained at a spray distance of 30 mm where the plasma heat flux to substrate reaches 30 $MW.m^{-2}$. The coating density seems to slightly decreases when the mass load of suspension increased from 5 wt. % to 10 wt. %, very likely because of a loading effect. The spray beads on substrate located at 40 mm are thicker but less dense due to the incorporation of more untreated particles.

The particle melting was also deeply affected by the characteristics of the powder used in the suspension, particularly its primary particle size distribution and its agglomeration behavior that can be characterized by the size and strength of agglomerate. With the short spray distances used in suspension plasma spraying, the coating surface temperature reaches 700–800 °C, as measured with infrared pyrometer during two successive passes of the plasma torch on the substrate. It is worth noting that droplets reaching the substrate without being completely vaporized before impact will be vaporized on the surface of the hot coating under formation. Thus, if the solid particles contained in these droplets do not rebound on the surface, they will be incorporated into the coating and create defects. Such temperatures, as shown in section 5.3.1.2 also affect the splat formation and thus coating formation.

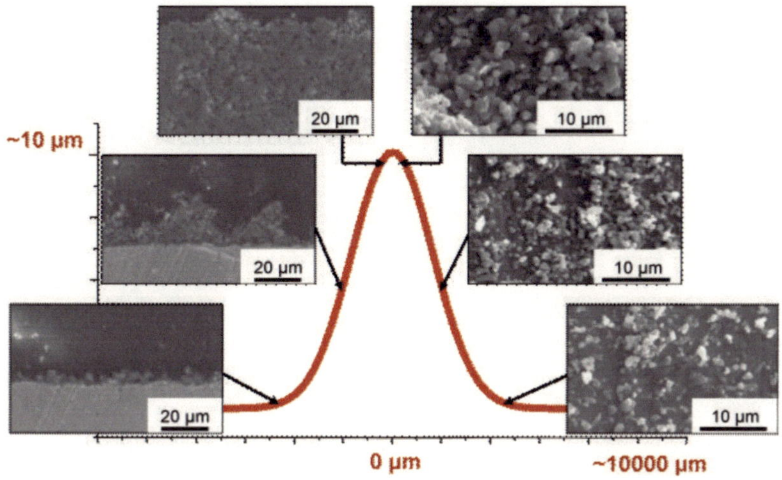

Fig. 12. Al_2O_3 suspension plasma-sprayed coating microstructures for two particle mass loads (5 and 10 wt%) and three spray distances (30, 40, 50 mm). Spray conditions are depicted in Figure 32 caption *Tingaud et al (2008).*

It was also observed that the use of water instead of ethanol as liquid phase in suspension plasma spraying brings about more porous coatings than those achieved with ethanol-based suspensions if the electric power level input to the plasma gas was not increased *Tingaud et al (2008).*

c. Coatings formation

When spray beads overlap during spraying, the poorly treated particles i.e. mainly un-melted, partially molten and re-solidified particles, generally located at the bead edges, are embedded in the coating during deposition. The different passes may thus be separated by porous and poorly cohesive layers that may induce voids and delamination in the upper layer as shown in Figure 13(a). The probable explanation is depicted in Figure 14a, showing how, according to the spray pattern, the poorly treated particles travelling in the jet fringes are deposited at the surface of the hot previous pass and may stick on when its temperature is higher than 800–900 °C. The following well-molten particles, which have travelled in the warm zone of the plasma jet, form the next pass on the powdery layer deposited by the preceding particles travelling in the jet fringes. It is worth noting that the first pass is deposited on the substrate temperature that is generally still below 300 °C and almost no sticking of poorly treated particles occurs. So, the interface between the substrate and the first pass is relatively clean compared with that between next passes. When the spray pattern is adapted, in particular to reduce the mean surface temperature of the coating under construction, it is nevertheless possible to get rid of most of the powdery layer between passes and obtain rather thick and dense coatings, as illustrated in Figure 13b.

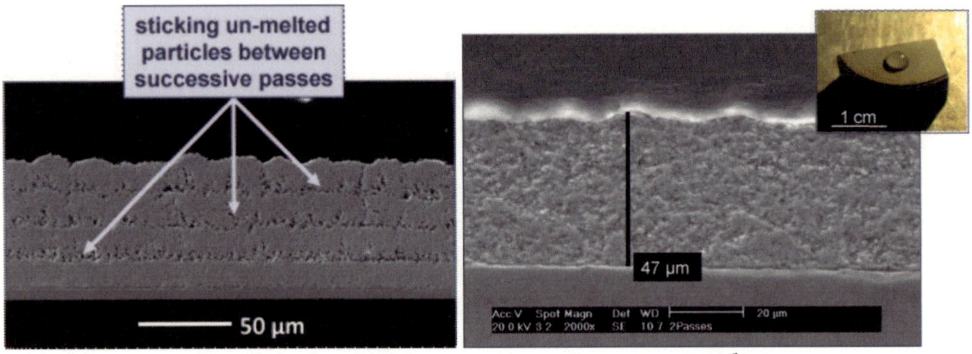

a. b.

Fig. 13. Yttria Partially Stabilized Zirconia (YPSZ) coatings (4–5 passes) obtained with the same Et-OH suspension Unitec 0.02 (Unitec Ceram, Stafford, UK, fused and crushed, d_{50} = 0.39 µm) sprayed under different operating conditions with Plasma-Technik F4-type torch (nozzle i.d. 6 mm). (a) Ar–He (30–30 slm, 700A, v_{inj} = 33.5 $m.s^{-1}$), (b) Dense and thick Y-PSZ (8 wt%) coating deposited under the spray conditions depicted in Figure 13a, the spray pattern being adapted to avoid powdery layers deposited between successive passes *Fauchais et al (2008).*

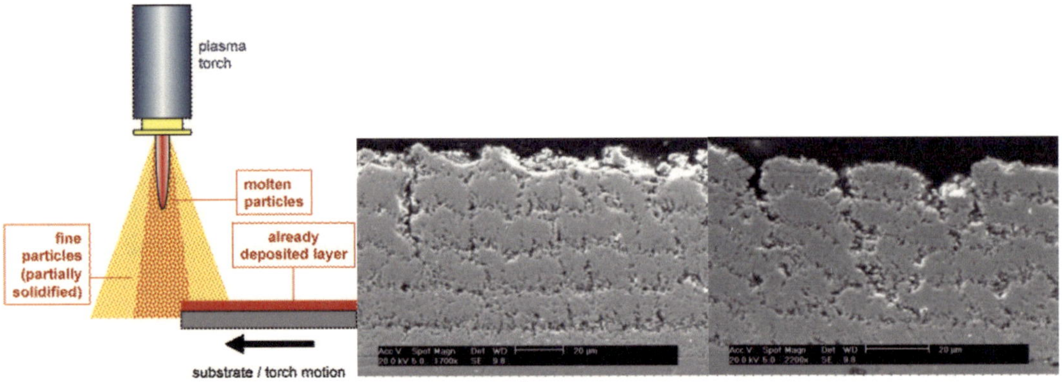

a. b. c.

Fig. 14. (a) Schematic explanation of the deposition on each pass of a powdery layer due to the sticking of poorly treated particles in the jet fringes preceding the deposition of the well-molten particles, (b) YSZ coatings (4-5 passes) obtained with the same suspension as that of Figure 13 Ar–H_2 (45–15 slm, 500A, v_{inj} = 33.5 $m.s^{-1}$), (c) Ar–H_2 (45–15 slm, 500A, v_{inj}= 26.6 $m.s^{-1}$) *Fauchais et al (2008).*

When the suspension is sprayed with highly fluctuating plasma jet as Ar-H_2 plasma ($\Delta V/V_m$ ~ 1 with an Ar-H_2 plasma against 0.25 with an Ar-He plasma), more powdery material is deposited between successive passes and the pore level is higher as shown in Figure 14(b). The situation becomes worse in terms of poorly-treated particle deposition when the suspension is injected in Ar-H_2 plasma with a lower injection velocity that limits the suspension penetration in the plasma jet, as shown in Figure 14c. The situation is even worse when using a suspension containing particles that agglomerate easily with a rather broad distribution of agglomerate sizes between 0.01 and 5 µm. The structure of each pass consists, then, of some sort of columnar structures or ridges, with columns 10–20 μm in diameter made of layered splats and particles *Delbos et al (2006)*.

The high transient heat flux (see Figure 9) imposed by the plasma jet to the substrate and coating in formation affects the coating construction. Indeed, such heat flux leads to coating surface transient temperatures over 1500 °C *Etchart-Salas (2007)*. When Y-PSZ splats are collected on a substrate attached at the extremity of a pendulum crossing the jet at 1 $m.s^{-1}$, the flattening degree is below 2 and is mostly less than 1.5 while it is generally about 4-5 for ceramic particles with size ranging between 10 to 50 μm. This low value is explained by the low inertia and the high surface tension of very small particles. Another important feature to be underlined is that splats with diameter smaller than 2 μm no longer exhibit cracks that would normally be present due to splat quenching stress. According to splat formation mechanisms, it could be expected that suspension plasma-sprayed coatings present a lamellar structure similar to that of conventional coatings made with micrometer-sized particles. In fact the structure of coatings is different, for example, in suspension plasma spraying of Y-PSZ particles, the first deposited layers that have a thickness of about 400 nm, exhibit a columnar structure, as expected from splats layering, while the following layers exhibit a granular structure (Figure 15). This may be explained by a recoil phenomenon experienced by the impinging particle, after flattening and prior to solidification, due to a surface tension effect, which is emphasized for sub-micrometer-sized particles, while, due to the important transient heat flux, the flattening particle cooling is delayed. Similar results were observed with alumina suspensions, where granular particles and splats were observed in coating structures *Darut et al (2009)*. The coating was a mixture of α and γ phases (about 50% of each), in spite of the fact that most particles at impact formed splats. In conventional coatings, mostly made of γ-alumina phase, the transformation of this phase into α occurs when the coating is reheated over 1000 °C (however, the transformation takes a few seconds).

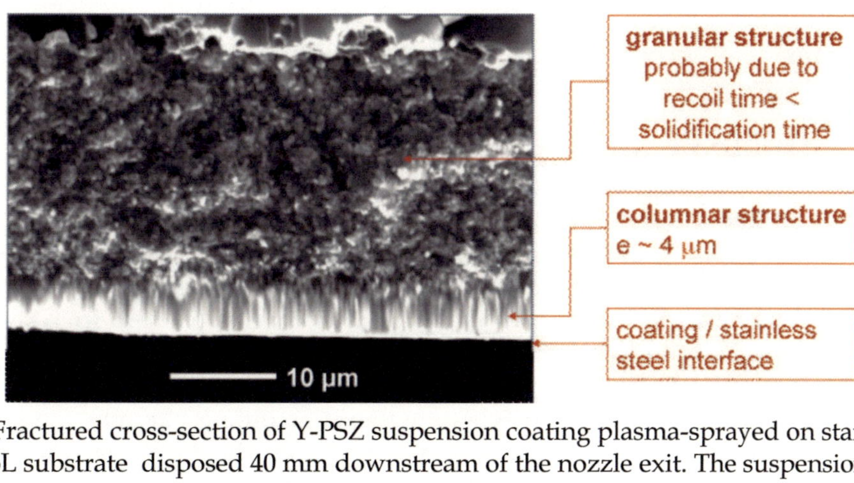

Fig. 15. Fractured cross-section of Y-PSZ suspension coating plasma-sprayed on stainless steel 316L substrate disposed 40 mm downstream of the nozzle exit. The suspension consisted of sub-micrometer-sized attrition milled particles (0.2–3 μm) in ethanol and was sprayed with an Ar–H_2–He plasma *Delbos et al (2006)*.

As in conventional thermal spraying the substrate roughness must be adapted to the sizes of the solid particles of the suspension. Works developed at SPCTS Laboratory of University of

Limoges. *Brousse et al (2008)* showed, that coating architecture was very sensitive to substrate roughness. Large columnar stacking defects developed from the valleys of the surface due to a 'shadow' effect propagating through the coating when the substrate surface roughness was higher than the average diameter of the feedstock particle, as depicted in Figure 16a. Reducing the stacking defects to enhance gas tightness (hermeticity) required spraying onto smooth polished substrates. For example, when decreasing the ratio of the roughness average Ra of the substrate to the feedstock particle average diameter d_{50} from 75 to 2 for Y-PSZ suspension coating, the leakage rate decreased from 0.5 to 0.02 $Mpa.L.s^{-1}.m^{-1}$. However, if no stacking defects can be observed on the SEM picture of Y-PSZ suspension coating corresponding to $Ra/d_{50} = 2$ (Figure 16c), it does not mean that they do not exist, as confirmed by a non-negligible leakage rate. Thus a compromise has to be found between d_{50} and Ra to get rid of these stacking defects.

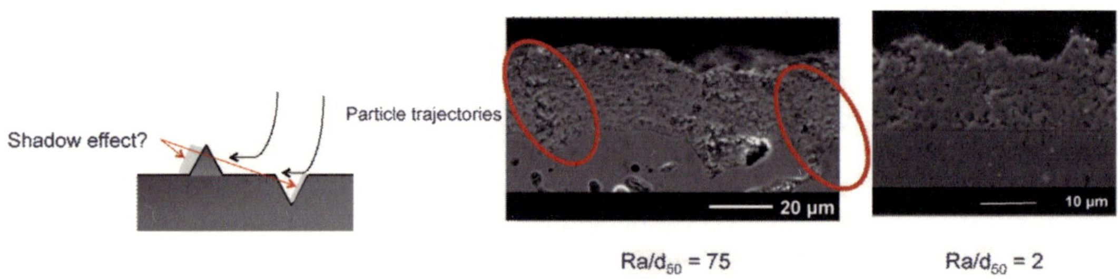

a. b. c.

Fig. 16. (a) Principle of shadow effect, (b) Y-PSZ coating sprayed onto a substrate exhibiting an average substrate roughness 40 times higher than the feedstock particle average diameter (Ra/d_{50}): columnar stacking defects develop through the coating thickness, (c) No shadow effect when $Ra/d_{50} = 2$, by *Brousse et al (2008)*.

5.3.2 Solutions

a. Drops and droplets in flight

Ozturk and Cetegen (2004, 2006) have modeled the behavior of single precursor droplets injected into high-temperature gas jets. The processes undergone by the drops axially or radially injected into the gas flow can be divided into three distinct phases: (i) The first phase is the aerodynamic break-up, which, as for suspension droplets, depends on the Weber and Ohnesorge numbers (see section 3.2.4). (ii) The second phase is droplet heating and surface evaporation in the hot gas jet stream. Loss of liquid phase from the surface concentrates the salt solutes, progressively leading to precipitation of the solute as it reaches a super-saturation level. Precipitation may occur in all regions of the drops or droplets that exceed the equilibrium saturation concentration, according to the homogeneous precipitation assumption, *Saha et al (2009)*. Depending upon the droplet size and the mass transport characteristics within the droplet, different precipitate morphologies can be obtained: shell type morphologies or spherical precipitates. The thickness of the shell can be predicted from the homogeneous nucleation hypothesis unlike the void content. (iii) The third phase is the further heating of the precipitates in the hot gases before their impact on the substrate surface.

Particle morphologies depend on the different processes undergone by the droplets in the hot gas that are linked to their trajectories. According to *Saha et al (2009)*, they include solid particles, hollow shells and fragmented shells as shown in Figures 17(a) to (c). Small droplets with high solute diffusivity exhibit a propensity to precipitate volumetrically and form solid particles as shown in Figure 17(a). Rapid vaporization combined with low solute diffusivity and large droplet sizes can lead to a significant increase in solute concentration near the droplet surface resulting in surface precipitation and formation of a crust around the liquid core of the droplet. The crust/shell may have varied levels of porosity. Shells having low porosity usually rupture due to internal pressurization to form shell fragments (path I in Figure 17b). Shells that are completely impervious rupture and secondary atomization of the trapped liquid core may be observed (path III in Figure 17b). For shells with a high level of porosity, internal pressure rise is counterbalanced by the vapor venting through the pores and resulting in hollow shells (path II in Figure 17b). For particular precursors, elastic inflation and subsequent collapse and rupture of the shell can also be observed (Figure 17c). The particle morphology resulting from droplet processing is hence sensitive to the solute chemistry, mass diffusivity, solute solubility, droplet size, thermal history, injection type and velocity *Saha et al (2009)*. In summary, according to *Saha et al (2009)* the final coating microstructure depends on the size of the droplets, not on whether they follow a trajectory centered on the torch centerline axis or a deviated one after primary precipitation. Globally, droplets in the size range 5 or 10 µm get completely pyrolysed before reaching the substrate, while the 20 µm and larger droplets remain partially pyrolysed. However the initial solution concentration plays a key role in droplet pyrolysis. A high or close to the equilibrium saturation concentration tends to produce volume precipitation *Chen et al (2008)*. Microstructures of the collected solution plasma-sprayed coatings of YSZ on substrates at room temperature from low and high concentration solutions, respectively, are presented in Figures 4 and 5 of the paper of *Chen et al (2008)*. No splats are observed out of the central zone of the deposited bead with the low concentration solution precursor, mainly composed of ruptured bubbles and a small volume fraction of solid spheres (< 0.5 µm). The deposited bead central zone made from the high-concentration solution is mainly composed of overlapped splats, with an average diameter ranging from 0.5 to 2 µm, and a small amount of un-melted solid spheres (< 0.5 µm). The deposited bead edges from both dilute and concentrated solutions are made of un-pyrolysed precursors containing significant amounts of water. The mud-like cracks presented at the edges are the result of shrinkage due to liquid phase evaporation on the substrate.

b. Coating formation

Xie et al (2004) studied a single deposited bead from YSZ solutions resulting from one pass of the torch in front of a polished substrate and overlapped beads resulting from successive passes. The resulting spray beads can be divided into adherent deposits (bead central part) and powdery deposits (bead edges) that correspond to the hot and cold regions of the plasma jet, respectively. *Chen et al (2008)* identified four deposition mechanisms: (i) smaller droplets that undergo further heating to a fully molten state and crystallize upon impact to form ultra-fine (0.5–2 μm average diameter) splats; (ii) at certain spray distances, droplets undergo re-solidification and crystallization before impact upon the substrate to form fine crystallized spheres; (iii) droplets entrained in the cold regions of the thermal jet where they experience sufficient heating to cause solute evaporation leading to the formation of a gel phase, deposited on the substrate. Some droplets also form a pyrolysed shell containing un-pyrolysed solution that fractures during deposition; (iv) some precursor solution droplets can reach the substrate in liquid form, having undergone none of the aforementioned processes.

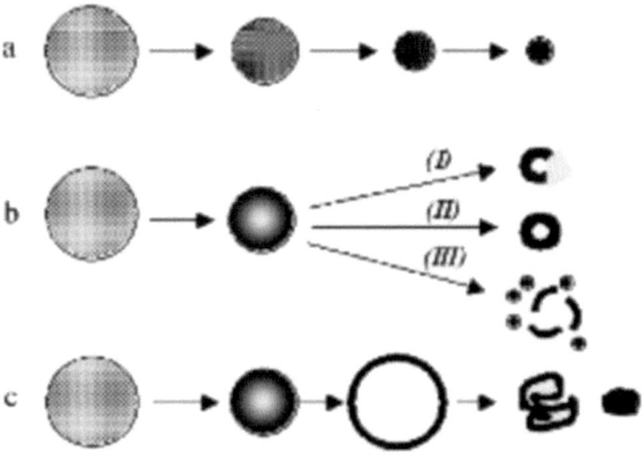

Fig. 17. Different routes for a droplet for vaporization and solid precipitation: (a) uniform concentration leading to solid particles by volume precipitation. (b) Super-saturation near the surface: (I) low permeable shell leading to fragmented shell formation, (II) high permeable shell leading to un-fragmented shell formation, (III) impermeable shell leading to droplet internal heating, pressurization and droplet breakup, secondary atomization. (c) Elastic shell formation causing inflation and deflation by solid consolidation *Saha et al (2009)*.

Most works dealing with solution plasma spraying use atomization to inject solutions into conventional dc plasma jets and, as underlined in section 3.2.1, very different size, velocity and trajectory distributions can be obtained, the best coatings being obtained with narrower droplets, *Jordan et al (2007)*. A high concentration precursor (close to equilibrium saturation) is also important to promote volume precipitation leading to fully melt splat microstructure and high density coating, *Chen et al (2008)*. The type of liquid phase also plays a key role: droplets with a high surface tension and high boiling point liquid phase experience an incomplete liquid phase evaporation process in the thermal jet and forms a mud-like cracked film upon impacting the substrate leading to a porous coating. Droplets created from a low surface tension and low boiling point liquid phase undergo rapid liquid phase evaporation, solute precipitation, pyrolysis, melting process in the plasma jet and form splats upon impact on the substrate, the stacking of which results in a dense coating. At last the substrate and coating temperature during spraying is also very important. Non-pyrolysed materials are pyrolysed at the substrate surface and form aggregates when the substrate temperature is above the precursor pyrolysis temperature. Pyrolysis also occurs when reheating the coating by successive torch passes *Chen et al (2010)*. Under the repeated processing by the high-temperature thermal jet, the mud-like film for example formed when spraying droplets with a high surface tension and high boiling point liquid phase, evaporates, pyrolyses and crystallizes on the substrate and brings about porous coatings. To conclude, the amount of non-decomposed precursor can be controlled by the spray parameters, primarily liquid/gas stream injection momentum densities, spray droplet size dispersion and precursor concentration.

6. Characterization of nanostructures

The knowledge of structural characteristics and functional properties of nanometer-sized coatings is fundamental whatever the envisaged applications and manufacturing processes. The thermo-mechanical behavior of coatings and, in the case of thermal barrier coatings, their thermal insulation performance are mostly related to the void architecture or void networks (for example, see *Rice (1996)*).

The main question is the following: are methods used for conventional micro-structured thermal sprayed coatings suitable for nanostructured coatings? For more details about possible answers the reader is referred to the paper of *Fauchais et al (2010)*. The various possibilities are briefly presented below:

- The observation of coating cross-section mostly by using scanning electron microscopy, SEM, coupled with appropriate image treatment and statistical models makes it possible to quantify the void content as a function of the void size and, with some specific models, of their shape and crack density and orientation. However, the image characteristic dimension should be between 10 and 15 times larger than the objects of the voids to be analyzed in order to have a representative elementary volume (REV) of the structure. Actually the resolution is limited to features larger than 0.1μm, average value, which is by far not sufficient to characterize nanometer-sized coatings. Also cutting and polishing of nanostructured coatings are more difficult because their toughness can be 4 to 8 times larger than that of conventional ones.
- Archimedean porosimetry and electrochemical impedance spectroscopy are based on the penetration of a liquid within the coating. However, the simplistic test of de-ionized water droplet percolation (see the top of Figure 13b) through the deposit permits the determination of the smallest open pore diameter into which the water is able to percolate. For example at atmospheric pressure (about 10^5 Pa), pure water percolates into open voids of equivalent diameter equal to or larger than 1.5 μm, which is by far larger than most of the open voids in nanostructured coatings.
- Mercury intrusion porosimetry (MIP) presents the advantage of covering a large void size range, theoretically from a few nanometers to a few tens of micrometers and has been largely used to quantify void network size distribution of nanostructured coatings (see, for example, *Portinha et al (2005)*).
- Gas pycnometry, generally using helium, is based on gas pressure measurement in an unvarying gas volume measured by implementing a cell, which either does or does not contain the sample. This technique allows the quantification of the open void content of coatings and does not present any relevant drawback. It has been successfully implemented on nanometer-sized suspension plasma-sprayed coatings *Bacciochini et al (2010)*.
- Small angle neutron scattering (SANS) is an advanced technique to assess the void characteristics of thermal spray coatings including surface area distribution and orientation distribution. This technique has been successfully implemented in plasma-sprayed micrometer-sized YSZ thermal barrier coatings. Due to its capabilities

Hammouda (2008), this nondestructive technique is able to probe the nano-void size distribution in coatings. To the best of the authors' knowledge, nevertheless, this technique has not been implemented yet.

- Ultra-Small Angle X-ray Scattering (USAXS) is a nondestructive characterization technique recording elastic scattering of X-rays induced by compositional and structural in-homogeneities *Ilavsky J. et al (2009).* USAXS has been successfully implemented in quantifying void size distribution in YSZ (d_{50}: 50 nm) suspension plasma- sprayed coatings (see, for example, *Bacciochini et al (2010)*) and showed that about 80% of voids, in number, exhibited characteristic dimensions smaller than 30nm, the largest voids in the coatings having characteristic dimensions of a few hundred nanometers. The total void content was between 12.9% and 20.6%, with median diameters ranging from 270 to 400 nm and most of the voids (in number) had characteristic dimension smaller than 20 nm, contrary to micrometer-sized plasma sprayed coatings, which do not contain significant contribution of nano-voids smaller than 20 nm. Such results are difficult, if not impossible, to obtain from other characterization techniques, since they do not have the capability of USAXS to address with a very high resolution the whole set of scatter features (voids in this case) regardless of their characteristics (open, connected or closed).

7. Potential applications

As for all new processes, the mean time for application is between 20 and 25 years, thus it is quite understandable that the solution and suspension plasma spraying processes that are still under development have essentially potential applications. They are mainly in the following fields: wear-resistant, thermal barrier, corrosion-resistant, bioactive, photo-catalytic, and electrochemically functional coatings *Gell et al (2008), Vaßen et al (2010), Fauchais et al (2011), Killinger et al (2011).* A few examples are presented here to illustrate these various fields.

7.1 Thermal barrier coatings (TBC)

Yttria-stabilized or partially stabilized zirconia (YSZ or YPSZ) coatings deposited by solution and/or suspension spraying has been intensively studied. *Ben-Ettouil et al (2009)* investigated the thermal shock resistance of Y-PSZ suspension plasma-sprayed coatings (feedstock particles d_{50} = 50nm) and found that their resistance to isothermal and thermal shocks was higher for coatings with lower crack density. The coating thermal diffusivity, measured at atmospheric pressure, varied from 0.015 at 20°C to 0.025 $mm^2.s^{-1}$ at 250 °C. Such values are 10 times lower than the thermal diffusivity of YSZ coatings with a dual architecture with nanometer- and micrometer- sized features manufactured by air plasma spraying of micrometer-sized agglomerates made of nanometer-sized particles *Lima and Marple (2007).*

For suspension plasma-sprayed YSZ TBCs *Vassen et al (2009)* showed that when applying the optimum process parameters for turbine components, homogeneous microstructures with an evenly distributed pore network could be obtained. Moreover, with appropriate spray conditions, coatings exhibited a high segmentation crack density (approximately 11 $cracks.mm^{-1}$) together with medium porosity (23%) (Figure 18a) and presented the best lifetimes in thermal cycling tests.

Gell et al (2008) studied the properties and resistance to thermal shock test of YSZ coatings made by solution precursor plasma spraying. They tested various atomization systems to inject the solution in the plasma jet. Figure 18b displays a YSZ coating manufactured with a conventional fluid atomizer that yielded a rather broad distribution of droplet sizes ($5 < d_p < 120$ µm) with a relatively large number of drops travelling in the plasma jet (Ar–H_2) fringes. The Vickers hardness of the coating was approximately 450 HV_{3N} and the porosity about 17 %. The vertical cracks were formed as a result of pyrolysation under heating of un-pyrolysed material imbedded within coating, thus resulting in shrinkage inducing tensile stresses. In Figure 18b dense regions of ultra-fine splats, small and uniformly dispersed voids and un-melted particles can be observed. During thermal cycling, the spallation life was improved by a factor of 2.5 compared with APS coatings on the same bond coat and substrate and by a factor of 1.5 compared with high-quality Electron Beam Physical Vapor Deposition (EB-PVD) coatings. The apparent thermal conductivity, as measured by the laser flash technique from 100 to 1000 °C, was about 1.0–1.2 $W.m^{-1}.K^{-1}$, a value lower than that of EB-PVD coatings, but higher than that of conventional air plasma-sprayed (APS) coatings.

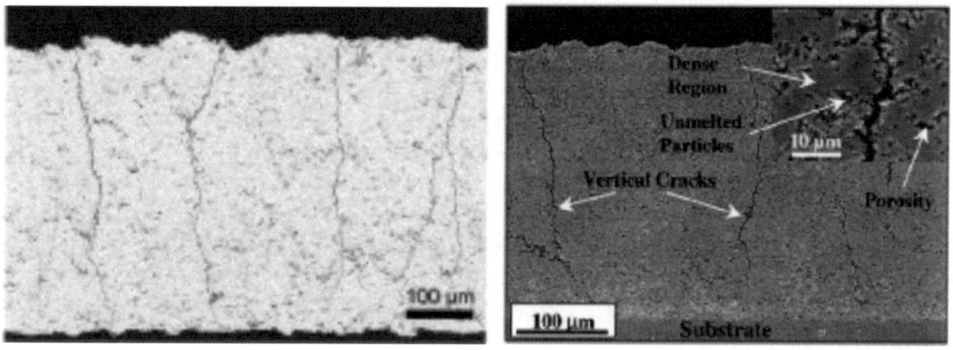

a. b.

Fig. 18. (a) Suspension-plasma-sprayed TBC with medium porosity and high segmentation crack density *Vassen et al (2009)*, (b) Features of solution plasma-sprayed TBCs using an air blast atomizer, details including vertical cracks, dense regions of ultra-fine splats, small and uniformly dispersed porosity, un-melted particles *Gell et al (2008)*.

Another potential application of suspension plasma-sprayed YSZ coating is as a bond coat that provide for the adhesion of a YSZ conventional coating on smooth and thin (1 mm) super-alloy substrates that cannot be prepared by the usual grit blasting method. *Vert et al (2010)*. First results seem promising.

7.2 Solid oxide fuel cells

A large body of studies has been devoted to solution and suspension coatings for solid oxide fuel cells (SOFCs). The objective was to produce the various components of the cell, anode, electrolyte and cathode, by using the same process. The tested coatings involved porous nickel–YSZ cermet coating for the anode, dense and thin (< 15 µm) YSZ coating for the electrolyte and porous perovskite, e.g. lanthanum ferro-cobaltite doped with strontium, (LSCF) or lanthanum manganite doped with strontium (LSM) for the cathode.

For the electrolyte *Brousse et al (2008)* have not succeeded to, but were close to, achieve suspension plasma-sprayed YPSZ coatings with a leakage below the value recommended for an impervious coating. *Jordan et al (2007)* have obtained reasonably dense YSZ coatings by using solution precursor plasma spraying with a specially designed capillary atomizer to atomize and inject the liquid feedstock in the plasma jet. *Marchand et al (2010), Michaux et al (2010), Wang et al (2010)* studied NiO-YSZ coating for the anode deposited by suspension and solution precursor plasma spraying and *Marchand et al (2007)* considered La_2NiO_4 suspension plasma-sprayed coatings for the cathode.

7.3 Wear resistant coatings

The materials tested for wear-resistant coatings are essentially alumina, alumina-titania, and alumina-YSZ. For suspension plasma sprayed alumina coatings *Darut et al (2010)* confirmed that the use of small particles in the suspension (down to d_{50}=0.3 μm) led to a low friction coefficient of about 0.2 against alumina. This coefficient could also be decreased by the addition of SiC in the Al_2O_3 matrix. The additive was not decomposed during suspension plasma spraying, while it is usually partially decomposed when sprayed as micrometer sized particles in conventional air plasma spraying. Al_2O_3-ZrO_2 composite coatings showed higher wear resistance than pure Al_2O_3 coatings, *Tingaud et al (2010)* .

Tarasi et al (2010) investigated the phase formation in Al_2O_3-YSZ sprayed with an Axial III plasma torch. They found that the particle velocity was the key parameter to achieve stable or metastable phases. Stable phases came along with particle low velocity at impact while metastable phases with high particle velocity. *Darut et al (2009)* obtained dense and cohesive coatings of Al_2O_3-TiO_2. *Chen et al (2009)* have first produced an amorphous powder of Al_2O_3-ZrO_2, which was then dispersed in ethanol for suspension spraying. The resulting coating was composed of alpha Al_2O_3 and tetragonal ZrO_2 phases and showed a very homogeneous phase distribution compared to coatings sprayed with crystalline powder by suspension or conventional APS.

7.4 Bio-active coatings

Hydroxyapatite (HA) is a bioactive ceramic material with similar chemical composition and crystal structure to human bone. It is used as a coating material for the integration of prostheses into osseous tissue. *Jaworski et al (2010)* have recently presented the recent developments in suspension plasma sprayed HA coatings with coating thickness between 10 and 50 μm. Coatings phase analysis showed the presence of HA crystals and several phases due to the decomposition of HA during the treatment of HA particles in the plasma jet. These phases and their respective fractions are of great importance since they determine the biological behavior of the coating, such as its dissolution in vivo. *Huang et al (2010)* showed that solution spraying of HA led to lesser degree of decomposition during spraying. This was attributed to the evaporation of the water solvent, reducing the temperature during the process. Moreover coatings exhibited a higher content of OH^- groups, which may lead to a superior structural integrity.

7.5 Photo-catalytic coatings

TiO_2 coatings are used as photo-catalytic surfaces, to degrade organic pollutants but also as electron emitters for light-emitting devices *Jaworski et al (2010)*. The photo-catalytic properties of TiO_2 depend on the phase composition of deposits. Anatase is generally assumed to present a higher photo-catalytic activity than rutile and authors generally consider that 65 vol. % of anastase is necessary to achieve an acceptable photo-catalytic performance. Furthermore, microstructural features such as porosity and specific surface are essential. *Toma et al (2010)* showed that plasma-sprayed suspensions contained up to 67 to 80 vol. % of anatase (most probably un-melted feedstock) for feedstock material containing about 80 vol. % of it.

Vaßen et al (2009) manufactured TiO_2 photovoltaic Graetzel cells by suspension plasma spraying using as feedstock anatase particles with a mean particle size of 60 nm. The coating was deposited on a cold substrate. Highly porous TiO_2 coatings with anatase contents of about 90% and crystallite sizes well below 50 nm were obtained. However, the photovoltaic cell design had still to be optimized to achieve sufficient efficiencies.

8. Conclusions

Plasma spray coatings from liquid feedstock have generated numerous new scientific articles and conference contributions, particularly during the last 6-7 years. All these works are based on results that emerged in the late-1990s making it possible to manufacture nanostructured thick (from about ten to hundred micrometers) coatings exhibiting numerous unique properties, such as good thermal insulation and resistance to thermal shock, excellent wear resistance and improved catalytic behavior. The two processes used are:

i. Suspension thermal spraying with feedstock made of nanometer- or sub-micrometer-sized particles in suspension;
ii. Liquid precursor thermal spraying with feedstock made of a solution.

Compared to conventional plasma spraying, solution and suspension spraying are by far more complex for the following reasons:

- To achieve a good control of drops or liquid jet penetration within the plasma jet they should have velocity and size distributions as narrow as possible, which is difficult to achieve, especially with atomization processes. Moreover liquid properties should be such that fragmentation could not take place in the plasma jet fringes.
- Liquid fragmentation depends strongly upon suspension and solution preparations. For suspensions solvent, dispersant, solid particles mass load, particles size distribution, particle manufacturing route, particles crystallographic state (crystallized or amorphous)… play a key role. For solutions the main parameters are the droplet size, the surface tension and boiling point of the liquid phase, the solute chemistry, its solubility and its mass diffusivity.
- The behavior of small particles generated into the plasma jet depends very much on their sizes. With nanometer-sized particles the Knudsen effect, increases drastically and their inertia is very low. Their initial velocity is thus very important for particles to hit the substrate (according to the Stokes effect) and depends on that of mother droplets. At last, with the thermophoresis effect, small particles are ejected from the hot gases jet as soon as they reach area where temperature gradients are very important.
- The heat transfer to the substrate can be very important due to spray distances that can be as short as 30 mm, resulting in heat fluxes up to 40 $MW.m^{-2}$ and must be controlled (cooling systems and spray pattern).

- The substrate roughness must be adapted to the size of the sprayed particles, which means that the roughness limit is around a ratio Ra/d_{50} below 2 to avoid stacking defects.
- At last it must be underlined that deposition efficiencies are lower (4 to 8 times) than those of conventional coatings.

Two problems are still pending: in-flight measurements and nano-structured coatings characterization.

In spite of them suspensions and solutions plasma sprayed coatings present new microstructures and often improved properties compared to conventional sprayed coatings. Possible applications are mainly in the field of energy conversion systems: TBCs, SOFCs, photo catalytic coatings, bioactive coatings and wear and corrosion resistant coatings. Further research works are needed to achieve a better understanding of phenomena controlling the deposit formation and properties before such coatings appear in commercial applications. This could/should require the development of new, dedicated plasma torches, with higher power to compensate for energy lost in liquid vaporization, and specifically adapted to the processing of liquids (more stable plasma flow, dedicated liquid injection systems and on-line control of liquid injection, safety issues…).

9. Nomenclature

d_l	liquid jet, drop or droplet diameter (m)
d_p	particle diameter (m)
l_{BL}	boundary layer thickness (m)
m_l	liquid mass flow rate ($kg.s^{-1}$)
St	Stokes number ($St = (\rho_p \times d_p^2 \times v_p).(\mu_g \times l_{BL})^{-1}$) (-)
u_r	relative velocity hot gases–liquid ($m.s^{-1}$)
v_g	gas velocity ($m.s^{-1}$)
v_l	liquid velocity ($m.s^{-1}$)
v_p	particle velocity ($m.s^{-1}$)
We	Weber number ($We = (\rho_g \times u_r^2 \times d_l)/\sigma_l$) (-)
Z	Ohnesorge number ($Z = \mu_l.(\rho_l .d_l .\sigma_l)^{-0.5}$) (-)

Greek symbols

μ_l	Liquid viscosity (Pa.s)
ρ_g	Plasma specific mass ($kg.m^{-3}$)
ρ_l	Liquid specific mass ($kg.m^{-3}$)
σ_l	Liquid surface tension ($N.m^{-1}$ or $J.m^{-2}$)

Abbreviations

BL	boundary layer
d.c.	direct current
i.d.	internal diameter (m)
PIV	Particle Image Velocimetry
r.f.	radio frequency
SOFC	Solid Oxide Fuel Cell

SPTS Solution Precursor Thermal Spraying
STS Suspension Thermal Spraying
TBC Thermal Barrier Coating
YPSZ Yttria-Partially Stabilized Zirconia
YSZ Yttria Stabilized Zirconia

10. References

Bacciochini A., Montavon G., Ilavsky J., Denoirjean A. and Fauchais P., (2010), Porous architecture of SPS thick YSZ coatings structured at the nanometer scale (~50 nm), J. Therm. Spray Technol. 19, 198–206

Bacciochini A., Ben-Ettouil F., Brousse E., Ilavsky J., Montavon G., Denoirjean A., Valette S. and Fauchais P. (2010) Quantification of void network architectures of as-sprayed and aged nanostructured yttria-stabilized zirconia (YSZ) deposits, Surface and Coatings Technology 205, 683–9

Basu S., Jordan E.H. and Cetegen B.M., (2008), Fluidmechanics and heat transfer of liquid precursor droplets injected into high-temperature plasmas J. Therm. Spray Technol. 17 60–72

Ben-Ettouil F., Denoirjean A., Grimaud A., Montavon G. and Fauchais P., (2009), Sub-micrometer-sized YSZ thermal barrier coatings manufactured by suspension plasma spraying: process, structure and some functional properties, in Thermal Spray 2009: Proc. Int. Thermal Spray Conf. (Las Vegas, NV, USA) (eds.) B. R. Marple et al (pub.) Materials Park, OH: ASM International) 193–9

Bisson J.-F., Gauthier B. and Moreau C., (2003), Effect of plasma fluctuations on in-flight particle parameters J. Therm. Spray Technol. 12, 38–43

Blazdell P.and Kuroda S., (2000), Plasma Spraying of Submicron Ceramic Suspensions Using a Continuous Ink Jet Printer, Surface and Coatings Technology 123(2-3) 239-246.

Bouchard D., Sun L., Gitzhofer F., and Brisard G.M., (2006), Synthesis and Characterization of $La_{0.8}Sr_{0.2}MO_{3-\delta}$ (M = Mn, Fe, or Co) Cathode Materials by Induction Plasma Technology, Journal of Thermal Spray Technology 15(1) 37-45

Boulos. M.I., Fauchais P., Vardelle A. and Pfender E., (1993), Fundamentals of Plasma Particle Momentum and Heat Transfer, in Plasma Spraying Theory and Applications (ed.) R. Suryanarayanan (pub.) World Scientific Singapore.

Boulos M., Fauchais P. and Pfender E., (1994), Thermal plasmas, Fundamentals and Applications (pub.) Plenum Press NY and London

Bouyer E., Gitzhofer F., and Boulos M.I., (1995), Suspension Plasma Spraying of Hydroxyapatite, Proceedings of the 12 th International Symposium of Plasma Chemistry, (ed.) J.V. Heberlein et al., (Minneapolis, MN: Organizing Committees of the 12th Int. Chem.,) 865-870,

Bouyer E., Gitzhofer F., and Boulos M.I., (1997), The Suspension Plasma Spraying of Bioceramics by Induction Plasma, JOM, February 58-62

Brousse E., Montavon G., Fauchais P., Denoirjean A., Rat V., Coudert J.-F. and Ageorges H., (2008), Thin and dense yttria-partially stabilized zirconia electrolytes for IT-SOFC manufactured by suspension plasma spraying, in Thermal Spray Crossing Borders (ed.) E. Lugscheider (pub.) DVS Düsseldorf, Germany, 547–52

Caruyer C., Vincent S., Meillot E., Caltagirone J.-P., (2010), Modeling the first instant of the interaction between a liquid and a plasma jet with a compressible approach, Surface & Coatings Technology 205, 974–979

Chandra S. and Fauchais P., (2009) Formation of solid splats during thermal spray deposition, J. Therm. Spray Technol. 18, 148–80

Chen D., Jordan E. H. and Gell M., (2010), The solution precursor plasma spray coatings: influence of solvent type, Plasma Chem. Plasma Process. 30, 111–9

Chen D., Jordan E.H., and Gell M., (2009) Suspension Plasma Sprayed Composite Coating Using Amorphous Powder Feedstock, Appl. Surf. Sci. 255(11) 5935-5938

Chen D., E.H. Jordan and M. Gell, (2009) Microstructure of suspension plasma spray and air plasma spray Al_2O_3–ZrO_2 composite coatings, J. Therm. Spray Technol. 18, 421–6

Chen D., Jordan E. H. and Gell M., (2008), Effect of solution concentration on splat formation and coating microstructure using the solution precursor plasma spray process, Surf. Coat. Technol. 202, 2132–8

Chen D., Jordan E. and Gell M., (2007), Thermal and crystallization behavior of zirconia precursor used in the solution precursor plasma spray process, J. Mater. Sci. 42, 5576–80

Coudert J-F, Rat V. and Rigot D., (2007), Influence of Helmholtz oscillations on arc voltage fluctuations in a dc plasma spraying torch, J. Phys. D: Appl. Phys. 40, 7357–66

Darut G., Ben-Ettouil F., Denoirjean A., Montavon G., Ageorges H., and Fauchais P., (2010), Dry Sliding Behavior of Sub- Micrometer-Sized Suspension Plasma Sprayed Ceramic Oxide Coatings, J. Therm. Spray Technol, , 19(1-2) 275-285

Darut G., Valette S., Montavon G., Ageorges H., Denoirjean A., Fauchais P., Klyatskina E., Segova F., (2010) and Salvador M. D., Comparison of Al_2O_3 and Al_2O_3-TiO_2 Coatings Manufactured by Aqueous and Alcoholic Suspension Plasma Spraying, in Thermal Spray: Global Solutions for Future Application, Singapore (2010), (pub.) DVS, Düsseldorf, Germany, 212-217

Darut G., Ageorges H., Denoirjean A., Montavon G. and Fauchais P., (2009) Dry sliding behavior of sub-micrometer-sized suspension plasma sprayed ceramic oxide coatings, J. Therm. Spray Technol. 19, 275–85

Davis J.R. (ed.), (2004), Handbook of thermal spray technology, Pub. ASM International, Materials Park, OH, USA, 338 pages

Delbos C., Fazilleau J., Rat V., Coudert J.-F., Fauchais P.and Pateyron B., (2006) Phenomena involved in suspension plasma spraying: II, Plasma Chem. Plasma Process. 26, 393–414

Etchart-Salas R., (2007), Direct current plasma spraying of suspensions of sub-micrometer sized particles. Analytical and experimental approach on phenomenon controlling coatings reproducibility and quality PhD Thesis University of Limoges (in French)

Etchart-Salas R., Rat V., Coudert J.-F., Fauchais P., Caron N., Wittman K., Alexandre S., (2007), Influence of plasma instabilities in ceramic suspension plasma spraying, Journal of Thermal Spray Technology, 16(5-6) 857-865

Fauchais P., (2004), Understanding plasma spraying: an invited review, Journal of Physics D: Applied Physics, 37, R86-R108

Fauchais P., Etchart-Salas R., Rat V., Coudert J.-F., Caron N., Wittmann-Téneze K., (2008) Parameters controlling liquid plasma spraying: solutions, sols, or suspensions, Journal of Thermal Spray Technology, 17(1) 31-59

Fauchais P., Montavon G., Lima R. and Marple B., (2011), Engineering a new class of thermal spray nano-based microstructures from agglomerated nanostructured particles, suspensions and solutions: an invited review, Journal of Physics D: Applied Physics 44, 093001 (53p)

Fauchais P. and Vardelle A., (2011), Innovative and emerging processes in plasma spraying: from micro- to nano-structured coatings, J. Phys. D: Appl. Phys 44,

Fazilleau J., Delbos C., Rat V., Coudert J. F., Fauchais P., Pateyron B., (2006) Phenomena involved in suspension plasma spraying part 1, Plasma Chemistry Plasma Processes, 26, 371-391

Filkova I. and Cedik P., (1984), Nozzle Atomization in Spray Drying, Advances Drying, A.S. Mujumdar, Ed., Hemisphere Pub. Corp., 3 181-215

Gell M., (1995), Application opportunities for nanostructured materials and coatings, Materials Science Engineering, 204(1) 246-251

Gell M., Jordan E.H., Teicholz M., Cetegen B. M., Padture N., Xie L., Chen D., Ma X. and Roth J., (2008) Thermal barrier coatings made by the solution precursor plasma spray process J. Therm. Spray Technol. 17, 124–35

Hammouda B., (2008) The SANS Toolbox (Gaithesburg, MD: Center for Neutron Research, National Institute of Standards and Technology) 657p

Hermanek, F.J., Thermal Spray Terminology and Company Origins. (2001) ASM International, Materials Park, Ohio.

Huang Y., Song L., Huang T., Liu X., Xiao Y., Wu Y., Wu F., and Gu Z., (2010), Characterization and Formation Mechanism of Nano- Structured Hydroxyapatite Coatings Deposited by the Liquid Precursor Plasma Spraying Process, Biomed. Mater. 5, 054113-120

Ilavsky J., Jemian P. R., Allen A. J., Zhang F., Levine L. E. and Long G. G., (2009) Ultra-small-angle Y-ray scattering at the advanced photon source, J. Appl. Crystallogr. 42, 1–11

Jaworski R., Pawlowski L., Pierlot C., Roudet F., Kozerski S., and Petit F., (2010), Recent Developments in Suspension Plasma Sprayed Titanium Oxide and Hydroxyapatite Coatings, J. Therm. Spray Technol. 19(1-2) 240-247

Jia L. and Gitzhofer F. (2010), Induction Plasma Synthesis of Nano-Structured SOFCs Electrolyte Using Solution and Suspension Plasma Spraying: A Comparative Study, Journal of Thermal Spray Technology 19(3) 566-574

Jordan E.H., Gell M., Bonzani P., Chen D., Basu S., Cetegen B., Wu F., and Ma X. (2007), Making Dense Coatings with the Solution Precursor Plasma Spray Process, Thermal Spray 2007: Global Coating Solution, B.R. Marple, M.M. Hyland, Y.-C. Lau, C.-J. Li, R.S. Lima, and G. Montavon, Eds.), ASM International Materials Park, OH, USA 463-470, e-proceedings

Killinger A., Gadow R., Mauer G., Guignard A., Vaßen R., and Stöver D., (2011) Review of New Developments in Suspension and Solution Precursor Thermal Spray Processes, Journal of Thermal Spray Technology 20(4) 677-695

Landes K., (2006) Diagnostics in Plasma Spraying Techniques, Surf. Coat. Technol., 201, 1948-1954

Lee C.S. and Reitz R.D. (2001), Effect of liquid properties on the break-up mechanism of high-speed liquid drops Atomization Sprays 11, 1–18

Lefebvre A.H., (1989), Atomizations and Sprays, Hemisphere Pub. Corp.

Lima R. S. and Marple B.R. (2007) Thermal spray coatings engineered from nano-structured ceramic agglomerated powders for structural, thermal barrier and biomedical applications: a review, Journal of Thermal Spray Technology 16, 40–63

Marchand O., Bertrand P., Mougin J., Comminges C., Planche M.-P., Bertrand G., (2010) Characterization of suspension plasma-sprayed solid oxide fuel cell electrodes, Surface & Coatings Technology 205 993–998

Marchand C., Vardelle A., Mariaux G., Lefort P., (2008), Modelling of the plasma spray process with liquid feedstock injection, Surface and Coatings Technology, 202, 4458-4464

Mauer G., Guignard A., Vaßen R., Stöver D., (2010), Process diagnostics in suspension plasma spraying, Surface & Coatings Technology 205 961–966

Michaux P., Montavon G., Grimaud A., Denoirjean A., and Fauchais P., (2010) Elaboration of Porous NiO/8YSZ Layers by Several SPS and SPPS Routes, J. Therm. Spray Technol., 19(1-2) 317-327

Oberste-Berghaus J., Legoux J.-G., and Moreau C., (2005) Injection Conditions and In-Flight Particle States in Suspension Plasma Spraying of Aluminia and Zirconia Nano-ceramics, ITSC 2005 (Düsseldorf, Germany), DVS e-proceedings

Oberste Berghaus J., Bouaricha S., Legoux J.-G. and Moreau C., (2005), Suspension plasma spraying of nanoceramics using an axial injection torch, in Thermal Spray Connects: Explore Its Surfacing Potential (eds.) E. Lugscheider and C. C. Berndt (Düsseldorf, Germany: DVS-Verlag) e-proceedings

Oberste-Berghaus J., Marple B.and Moreau C., (2006), Suspension plasma spraying of nanostructured WC-12Co coatings, J. Therm. Spray Technol. 15, 676–81

Ozturk A. and Cetegen B. M., (2005), Modeling of axially and transversely injected precursor droplets into a plasma environment, Int. J. Heat Mass Transfers 48, 4367–83

Ozturk A. and Cetegen B. M., (2006), Modeling of axial injection of ceramic, Mater. Sci. Eng. A 422, 163–75

Pawlowski L., (2008), Finely grained nanometric and submicrometric coatings by thermal spraying: a review, Surface and Coatings Technology, 202(18), 4318-4328

Portinha A., Teixeira V., Carneiro J., Martins J., Costa M. F., Vassen R. and Stöver D., (2005), Characterization of thermal barrier coatings with a gradient in porosity Surf. Coat. Technol. 195, 245–51

Rampon R., Filiatre C. and Bertrand G., (2008), Suspension plasma spraying of YSZ coatings: suspension atomization and injection, J. Therm. Spray Technol. 17, 105–14

Rampon P., Filiatre C., and Bertrand G., (2008), Suspension Plasma Spraying of YPSZ Coatings for SOFC: Suspension Atomization and Injection, J. Therm. Spray Technol., 17(1), 105-114

Ravi B.G., Sampath S., Gambino R., Devi P. S. and. Parise J.B, (2006), Plasma spray synthesis from precursors: progress, issues and considerations, J. Therm. Spray Technol. 15, 701–7

Rice R. W., (1996), Porosity dependence of physical properties of materials: a summary review, Key Eng. Mater. 115, 1–19

Saha A., Seal S., Cetegen B., Jordan E., Ozturk A. and Basu S., (2009), Thermo-physical processes in cerium nitrate precursor droplets injected into high temperature plasma Surf. Coat. Technol. 203 2081–91

Schiller G., Müller M., and Gitzhofer F., (1999), Preparation of Perovskite Powders and Coatings by Radio Frequency Suspension Plasma Spraying, Journal of Thermal Spray Technology 8(3), 389-398

Shan Y., Coyle T.W., and Mostaghimi J., (2010), Modeling the Influence of Injection Modes on the Evolution of Solution Sprays in a Plasma Jet, Journal of Thermal Spray Technology 19(1-2) 248-254

Shan Y., Coyle T. W. and Mostaghimi J., (2007) Numerical simulation of droptlet break-up and collision in solution precursor plasma spraying J. Therm. Spray Technol. 16 698–704

Shen Y., Almeida V. A. B., and Gitzhofer F., (2011), Preparation of Nano-composite GDC/LSCF Cathode Material for IT-SOFC by Induction Plasma Spraying, Journal of Thermal Spray Technology 20(1-2), 145-153

Singh N.,Manshian B,.,Jenkins G.J.S., Griffiths S.M., Williams P. M., Maffeis T. G. G., Wright, C. J., and Doak S. H., (2009) Biomaterials 30, 3891–914

Tarasi F., Medraj M., Dolatabadi A., Oberste-Berghaus J., and Moreau C., (2010), Phase Formation and Transformation in Alumina/YSZ Nanocomposite Coating Deposited by Suspension Plasma Spray Process, J. Therm. Spray Technol. 19(4) 787-795

Tingaud O., Bertrand P., and Bertrand G., (2010), Microstructure and Tribological Behavior of Suspension Plasma Sprayed Al_2O_3 and Al_2O_3-YSZ Composite Coatings, Surf. Coat. Technol. 205(4) 1004-1008

Tingaud O., Grimaud A., Denoirjean A., Montavon G., Rat V., Coudert J.-F., Fauchais P. and Chartier T., (2008), Suspension plasma-sprayed alumina coating structures: operating parameters versus coating architecture J. Therm. Spray Technol. 17, 662–70

Toma F.-L., Berger L.-M., Stahr C.C., Naumann T., and Langner S., (2010) Microstructures and Functional Properties of Suspension-Sprayed Al_2O_3 and TiO_2 Coatings: An Overview, J. Therm. Spray Technol. 19(1-2) 262-274

Vaßen R., Stuke A., and Stöver D., (2009), Recent Developments in the Field of Thermal Barrier Coatings, J. Therm. Spray Technol. 18(2), 181-186

Vaßen R., Kaßner H., Mauer G., and Stöver D., (2010) Suspension Plasma Spraying:Process Characteristics and Applications, Journalof Thermal Spray Technology 19(1-2) 219-225

Vaßen R., Yi Z., Kaßner H., and Stöver D., (2009), Suspension Plasma Spraying of TiO_2 for the Manufacture of Photovoltaic Cells, Surf. Coat. Technol. 203(15) 2146-2149

Viswanathan V., Laha T., Balani K., Agarwal A., Seal S., (2006), Challenges and Advances in Nanocomposite Processing Techniques, Mat. Sc. and Eng., R54, 121–285

Wang Y., Legoux J.-G., Neagu R., Hui R., Maric R., and Marple B. R., (2010), Deposition of NiO/YSZ Composite and YSZ by Suspension Plasma Spray on Porous Metal, in Thermal Spray: Global Solutions for Future Application, 2010 Singapore (pub.) DVS-Berichte, Düsseldorf, Germany, p 446-453

Vert R., Chicot D., Dublanche-Tixier C., Meillot E., Vardelle A., and Mariaux G., (2010), Adhesion of YSZ Suspension Plasma -Sprayed Coating on Smooth and Thin Substrates, Surf. Coat. Technol. 205(4) 999-1003

Wittmann-Ténèze K., Vallé K., Bianchi L., Belleville P. and Caron N., (2008), Nanostructured zirconia coatings processed by PROSOL deposition, Surf. Coat. Technol. 202, 4349–54

Xie L., Ma X., Jordan E. H., Padture N. P., Xiao D. T. and Gell M., (2004), Deposition mechanisms of thermal barrier coatings in the solution precursor plasma spray process, Surf. Coat. Technol. 177–178, 103–7

Xiong H.-B. and. Lin J.-Z, (2009),Nano particles modeling in axially injection suspension plasma spray of zirconia and alumina ceramics, J. Therm. Spray Technol. 18(4), 887–95

8

A Solid State Approach to Synthesis Advanced Nanomaterials for Thermal Spray Applications

Behrooz Movahedi
Department of Nanotechnology Engineering,
Faculty of Advanced Sciences and Technologies, University of Isfahan,
Iran

1. Introduction

Preparation of feedstock powders is the first step for synthesis of thermal spray coatings. A number of techniques that are capable of producing these materials include gas/water atomization, mechanical alloying/milling, thermo-chemical method, spray drying, agglomeration and sintering, plasma fusion, and sol–gel processing techniques (Berndt, 1992). Solid state synthesis is attributed to the chemical reaction and alloying performed at temperature which reactants are solid. As a result, the kinetics of solid state reactions are limited by the rate at which reactant species are able to diffuse across phase boundaries and through intervening product layers. Hence, the conventional solid state technique invariably require the use of high processing temperatures to ensure that diffusion rate is maintained at a high level (Schmalzried, 1995; Stein et al., 1993). More recently, the mechanical milling process has attracted considerable interest, primarily as a result of its potential to generate nanocrystalline and other non-equilibrium structures in large quantities at low temperature. This process is considered as a means to mechanically induced solid state reaction that occur in feedstock powder mixture during collision in the grinding media (Suryanarayana, 2001).

The aim of this chapter is to describe the fundamental, mechanisms and recent developments of solid state approach to synthesis thermal spray advanced materials and related coatings.

2. Solid state synthesis of thermal spray powders

Mechanical milling as a solid state synthesis usually performed using ball milling equipments that generally divided to "low energy" and "high energy" category based on the value of induced the mechanical energy to the powder mixture. The ball milling equipments used for mechanical grinding or mixing are low energy such as Horizontal mill (Tumbler). The speed of the low energy rod or ball mill is quite critical with regards to the efficiency of the process (Fig. 1). It is necessary for the balls (or rods) to drop from the top of the mill onto the feedstock material that is being ground (Fig. 1b). If the mill speed is too fast then the media will not fall at all due to centrifugal forces or it will fall directly onto the media near the bottom of the mill (Fig. 1c). At low speeds the media does not drop at all, whereas at the optimum speed the media continuously "cascades" onto the feedstock material that is being crushed (Brendt, 1992).

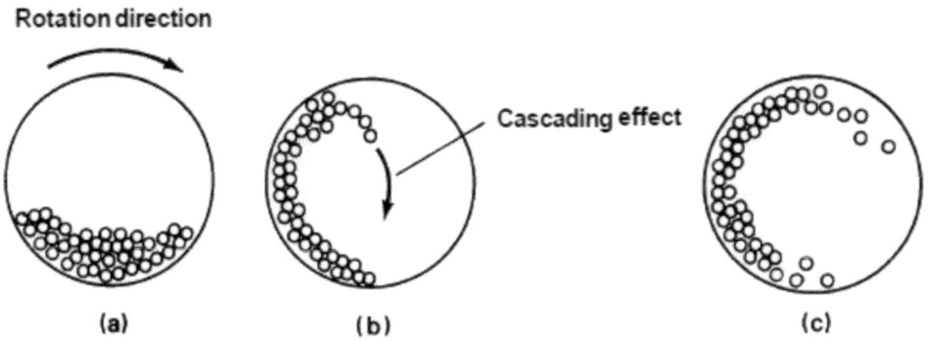

Fig. 1. Low energy mechanical milling variable that control particle grinding and efficiency: (a) low speed, (b) optimum speed, and (c) high speed (Berndt, 1992).

In mechanical milling processes that utilize to change the chemical composition of precursors, the high energy ball milling equipments is generally used. This phenomenon can be performed in various types of high energy ball mills, including attrition, planetary, and vibratory mills that schematically shown in Fig. 2. In an attrition mill, the rotating impeller cause to relative movement between balls and powders. In a planetary ball mill, a rotating disc and vials revolve in opposite direction in order of several hundred rpm. In a vibratory mill that also known as a shaker mill, the vessel is set in 1D or 3D vertical oscillatory motion. Spex 8000 is a commercial type of 3D vibratory mills. (Suryanarayana, 2001, 2004). Of the above types of mills, only the attritor mill has the highest capacity of powder charge. Accordingly, the attritor milling is employed to synthesize thermal spray feedstock powders used for the fabrication of nanostructured coatings.

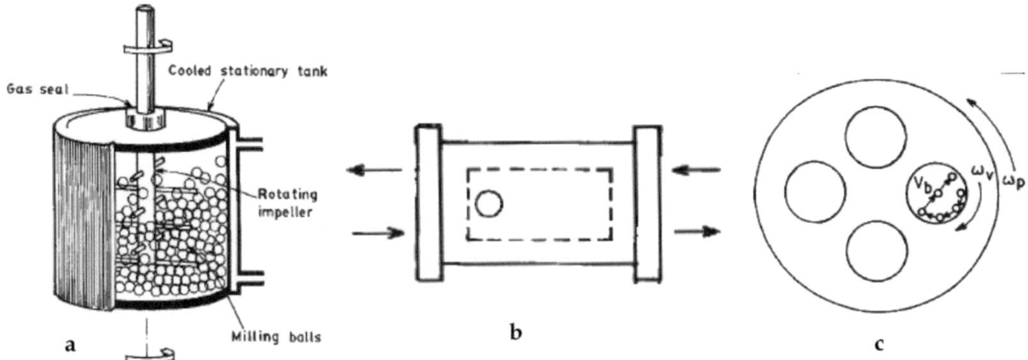

Fig. 2. Various types of high energy ball mills: (a) attrition mill, (b) vibratory mill, and (c) planetary mill (Suryanarayana, 2004).

High-energy mechanical milling is a low-cost process for the production of nanostructured powders and applicable to a variety of advanced materials (Koch, 1997). In such process the elemental blended powders are continuously welded and fractured to achieve alloying at the atomic level (Suryanarayana, 2004). By milling ceramic hard phase and metallic binder, refined composite powders are obtained (He et al; 1998, 2002). By varying the milling conditions, different sizes of the hard phase can be adjusted. Two different terms are most commonly used to describe the processing of powder particles in high-energy ball mills. Mechanical alloying describes the process when mixtures of powders (of different metals or alloys/compounds) are milled together. Thus, if powders of pure metals A and B are milled together to produce a solid solution (either equilibrium or supersaturated), intermetallic, or amorphous phase, the process is referred to as MA. Material transfer is involved in this process to obtain a homogeneous alloy. When powders of uniform (often stoichiometric) composition, such as pure metals, intermetallics, or pre-alloyed powders, are milled in a high-energy ball mill, and material transfer is not required for homogenization, the process has been termed mechanical milling (MM).

Mechanical alloying is a complex process involving optimization of a number of process variables to achieve the desired product phase, microstructure, and/or properties. For a given composition of the powder, some of the important variables that have an important effect on the final constitution of the milled powder are as: type of mill, milling container, milling energy/speed, milling time, size distribution of grinding medium, ball-to-powder weight ratio (BPR), extent of vial filling, milling atmosphere, process control agent (PCA), and temperature of milling. For more information about the milling parameters and the effect of these variables on the final product, refer to the book entitled "*Mechanical Alloying and Milling*" was written by C. Suryanarayana (Suryanarayana, 2004).

Powder particle size and morphology of the feedstock powder influence the melting conditions during spraying and therefore determine coating and microstructure formation. To obtain particles with dissimilar morphologies, powders can be produced by different processing routes including high-energy milling, spray drying and sintering, as depicted by Eigen et al; in Fig. 3. Route A includes milling under agglomerating conditions. If sufficient agglomeration is obtained directly in the milling process, further powder refinement only requires sieving to cut off larger size fractions. In an effort to minimize milling time, milling can therefore, also be stopped at an earlier state followed by a classifying process to yield particle sizes in the sprayable range, thus saving time and energy (route B). In a third route C, high-energy ball milling may be combined with spray drying and sintering, leading to spherical and more open powder morphologies with a finer and more homogeneous phase distribution as compared to conventional feedstock powders (Eigen et al; 2003, 2005).

3. Advanced materials and related thermal spray coatings

In thermal spraying technology, molten or semi-molten powders are deposited onto a substrate to produce a coating. The microstructure and properties of the materials depend on the thermal and momentum characteristics of the impinging particulate (Pawlowski, 2008), which are determined by both the spraying methodology and the type of feedstock materials employed. Various coatings are deposited on the surface of a substrate to either provide or improve the performance of materials in industrial applications. Nanostructured and advanced materials are characterized by a microstructural length scale in the 1–100 nm. More than 50 vol.% of atoms are associated with grain boundaries or interfacial boundaries when the grain is small enough. Thus, a significant amount of interfacial component between neighboring atoms associated with grain boundaries contributes to the physical and mechanical properties of nanostructured materials. Using nanostructured or advanced feedstock powders, thermal spraying has allowed researchers to generate coatings having higher hardness, strength, and corrosion resistance than the conventional counterparts (He et al; 2002).

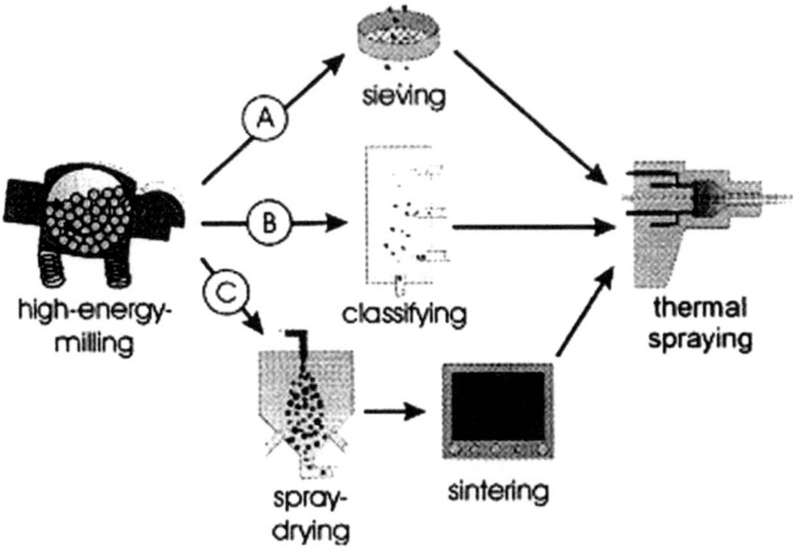

Fig. 3. Schematic of process routes for the production of thermal spray feedstock powders (Eigen et al; 2003).

3.1 Intermetallic compounds

Self-bonding materials are widely used in the thermal spray industry. Because the coating to substrate bond is often the weakest link in thermal spray coating systems, the ability of self-bonding materials to give a tenacious, reliable bond to the substrate greatly enhances the adhesion and therefore the performance of entire coating systems. While a number of these materials exist, the most popular and widely used is the nickel-aluminium powders (Pawlowski, 2008; Steenkiste et al; 2002). The reason for the good adhesion of these powders on non-pretreated surfaces (preheated only at 80–100°C to eliminate condensation), is the nascent heat evolved by the exothermic reaction of Ni and Al which form nickel-aluminide during spraying (Deevi et al; 1997). Different results are available concerning the completeness of the aluminide reaction during the spraying, and the type of aluminide formed. Phase composition in coatings sprayed using Ni–Al powders of different compositions with vacuum plasma spray (VPS) and air plasma spray (APS) methods were studied by Sampath et al. (Sampath et al; 1990). They described that the VPS coatings have a microstructure that results from the reaction between Ni and Al, and APS coatings contain the Ni and Al oxides from air or contain Al_2O_3 together with a solid solution of Al in Ni (α-Ni). It might be concluded that the spraying atmosphere (especially its oxygen content) has a major effect on final microstructure. The chemical reactions resulting directly in the coating phase composition take place during the particle flight in a flame (Chung et al; 2002; Hearley et al; 2000).

Movahedi et al; synthesised a thermal spray Ni–10 wt%Al powder consists of an aggregate containing the two components consisting of an alternative nickel and aluminium layers by low energy ball mills, such as tumbler, whereby the components are milled together for extended periods to form homogeneous powders (Movahedi et al; 2005a, 2009). Such a layered structure can be readily produced by mechanical alloying of elemental powders (Eigen et al; 2003). A rigid bonding between the particles is caused by a cold welding when mechanical energy is applied to different kinds of powders (Chen et al; 1999). Such binderless composite powders, which can be thermally sprayed to form coatings on various substrates, are suitable for using in a thermal spray process (Maric et al; 1996).

Fig. 4. shows XRD patterns of Ni–10 wt%Al powders after different low energy milling times. X-ray diffraction patterns include only elemental Ni and Al peaks without any identification of oxides or intermetallic phases. In contrast, some researchers (Enayati et al; 2004) reported that ball milling of Ni75Al25 powder mixture in a high-energy ball mill (i.e., planetary), led to the formation of a Ni(Al) solid solution that transformed to nickel aluminide on further milling. The extent of plastic deformation of powder particles, the local increase of temperature, and also the increase in the density of lattice defects in low energy ball mills and therefore the mass transport by diffusion are smaller compared to those in high-energy ball mills, making it impossible to obtain nickel aluminide phase (Boldyrev & Tkacova, 2000; El-Eskandarany, 2001).

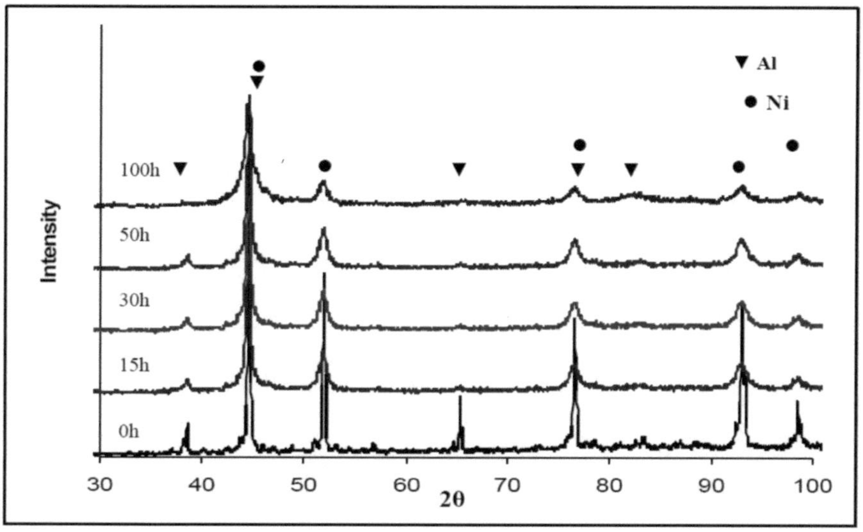

Fig. 4. X-ray diffraction patterns of Ni–10 wt%Al composite powders as received and after different milling times (Movahedi et al; 2009).

It is noted that Ni and Al XRD peaks have a lower intensity and higher width than those of initial powders due to the refinement of the crystallite size as well as an increase in the non-uniform internal strain of Ni and Al crystal lattices. Fig. 5. shows cross-sectional images of powder particles after different milling times. At the early stage of ball milling (10h), the Al particles were flattened and cold welding to the Ni particles. After 20h of low energy mechanical milling, the Ni particles also deformed plastically, and a typical lamellar structure consisting of pure Al and Ni layers with a layer thickness of ~10μm was formed (white areas are Ni and black areas are Al). On continuous ball milling, the layered structure refined so that after 35h of milling time the average layer thickness was ~5μm.

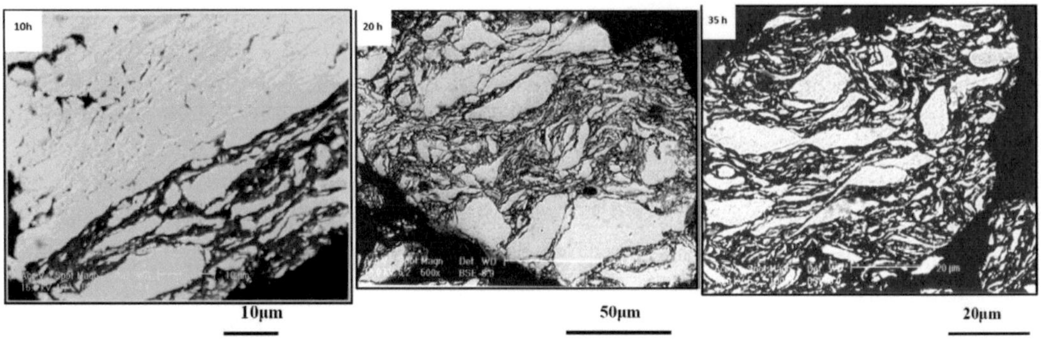

Fig. 5. Cross-sectional SEM images of Ni–10 wt-%Al composite powders after different milling times (Movahedi et al; 2005a, 2009).

Fig. 6. plots the average particle size of the powders on milling time. As milling time increased to 20h, the average Ni–10 wt-%Al particle size increased and reached to the maximum value of 300μm. For milling times longer than 20h, the average particle size decreased and finally approached a constant value of 5μm after 100 h of milling time. In addition, the powder particle size becomes more uniform by increasing the milling time.

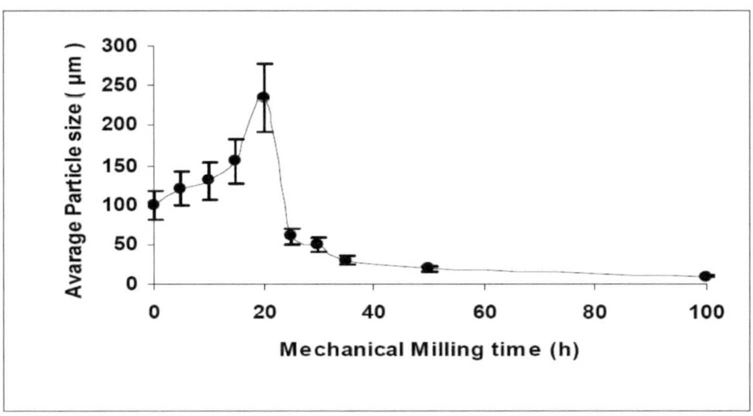

Fig. 6. Average particle size of Ni–10 wt%Al composite powders versus milling time (Movahedi et al; 2009).

The changes of powder particle size during ball milling are caused by fracture and cold welding of Ni and Al powder particles during milling. The smaller particles grow while the larger ones fracture. A rigid bonding between the particles is caused by a welding when mechanical energy is applied to different kinds of powders (Chen et al; 1999). During milling the ingredients of the powder, mixtures are reduced in size and brought into intimate contact by flattening and crushing the particles, welding them together, and repeating the process again and again. The resultant powders essentially consist of a homogeneous and uniform distribution of the initial component within the powder particles (Maric et al; 1996). The particle size distribution of the powders has a major effect on quality and morphology of thermal spray coatings. It is generally accepted that the optimum particle size for thermal spray powders is within the range 35–100μm. These powders are large enough to be easily fed from simple hoppers but small enough for efficient melting to occur. Under a given set of plasma conditions, there is likely to be an optimum size of particle that will be melted in the plasma, transported with sufficient momentum, and subsequently deposited. Smaller particles will lose velocity or may be vaporised with subsequent loss of spraying efficiency, whereas large particles will be incompletely melted and will produce a coating with large pore and low strength. In addition, the particle size significantly affects particle temperature and speed during flight, which subsequently influences the properties of the coating (He & Schoenung, 2002). Movahedi et al; reported that the Ni–10 wt%Al powders made by low energy ball milling have a microstructure consisting of alternative Ni and Al layers and include high defect density such as dislocations and vacancies. Therefore, the exothermic reaction between Ni and Al layers during flight in plasma spray process, readily occurs which subsequently enhances the bonding strength.

Fig. 7. shows the XRD pattern of plasma spray coating. It can be seen that plasma spray coating include NiAl intermetallic phase along with α-Ni solid solution. Significant feature of powders produced by mechanical milling is that a lamellar structure consisting of pure Al and Ni layers forms during milling process. This structure provides a large Ni/Al interfaces for exothermic reaction to occur during flight in plasma spraying. The incomplete Ni and Al reaction is due to the rapid heat losses of powder particles during thermal spraying (Movahedi et al; 2005b, 2009).

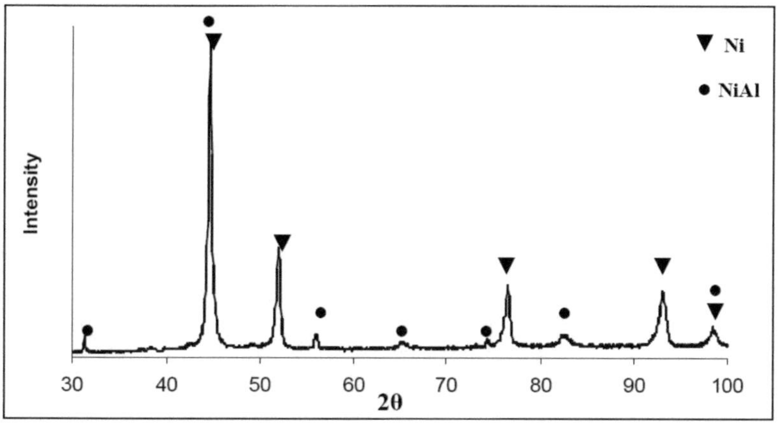

Fig. 7. X-ray diffraction patterns of Ni–10 wt%Al plasma spray coatings (Movahedi et al; 2009).

Fig. 8. shows the cross-sectional microstructure images of the plasma spray coatings. The shape of the pores suggests that they were formed due to the expansion of trapped air when the impacting particles were still molten. Cavities and pores are formed between the individual particles in the sprayed coatings because inter-particle diffusion is limited and particle flow is hindered during the splat cooling process (Movahedi et al; 2009).

Ni-Al intermetallics are being recognized as high temperature structural materials because of their excellent oxidation resistance, high thermal conductivity, low density, and high melting point (Stoloff etal; 2000). Research on nickel aluminides has been expanded during the last 30 years, not only as bulk materials, but also as coatings. Studies have indicated that nickel aluminide alloys have significant potential in wear applications as wear properties of carbon steel parts can be significantly improved by applying nickel aluminide coating (Goldenstein et al; 2004; Houben et al; 1973; Knotek et al; 1973; Rickerby et al; 1991). The possible applications of Ni-Al include; gas turbine engine, rotor blades, and stator vanes, however, low ductility at ambient temperature is the major limitation of this material. A number of attempts have been made to overcome this drawback such as nanocrystallization of structure which may transform nominally brittle compound into the ductile material (Morsi, 2001). One of the synthesis methods of nanocrystalline Ni-Al intermetallic compounds is high energy mechanical alloying. Two different mechanisms, via a rapid explosive reaction or through a gradual diffusion, were found for Ni-Al formation during MA (Atzmon, 1990; Enayati et al; 2008; Mashreghi et al; 2009). The powder prepared by MA can be deposited on surfaces of engineering parts using different thermal spraying techniques including plasma spray and, high velocity oxy fuel (HVOF) process. There are limited investigations in the literature concerning thermally sprayed coatings of nickel aluminides. Hearley et al. used inert gas atomized and reaction sintered Ni-30 wt.%Al powders to prepare NiAl intermetallic coatings by HVOF thermal spraying. They reported that a spherical inert gas atomized powder with narrow particle size range between 15 and 45 μm produced coatings of better quality. Their results also showed that both the fuel and oxygen flow rates influence the coating deposition characteristics and properties (Hearley et al; 1999, 2000).

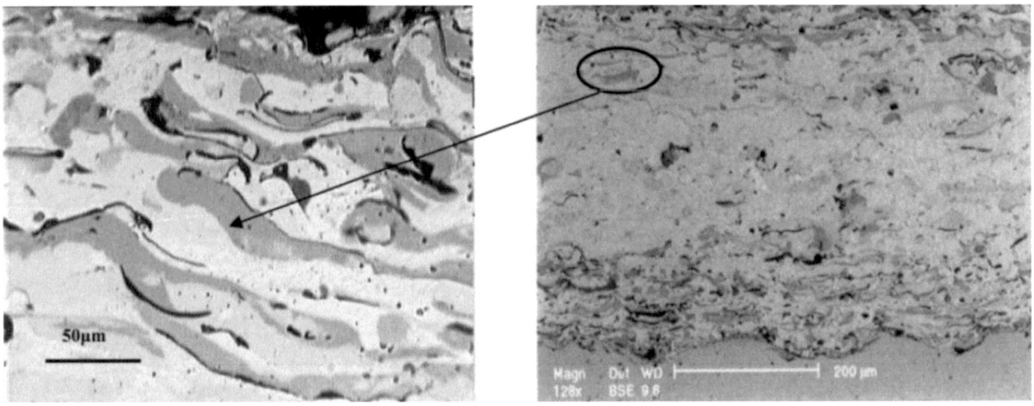

Fig. 8. Cross-sectional SEM images of Ni–10 wt%Al plasma spray coating at different magnifications (Movahedi et al; 2005b, 2009).

Fig. 9. shows the XRD patterns of Ni50Al50 powder mixture as-received and after 60, 90, and 120 min of high energy milling times. The XRD patterns of the as-received Ni50Al50 powder showed diffraction peaks of the crystalline Ni and Al. Increasing milling time to 60 min led to the disappearance of the Ni and Al peaks, while NiAl peaks began to appear. Complete transformation of elemental Ni and Al powder mixture to the NiAl intermetallic phase appeared to occur after 90 min of high energy ball milling. This result shows that the reaction between Al and Ni is promoted by the extensive Ni/Al interface areas as well as the short circuit diffusion paths provided by the large number of defects such as dislocations and grain boundaries introduced during high energy ball milling (Enayati et al; 2011). Hu et al. reported that complete transformation of Ni +Al to NiAl compound during MA occurred after 240h which is much longer than MA time obtained in the work of Enayati, et al. This discrepancy can be due to the different mill machines used.

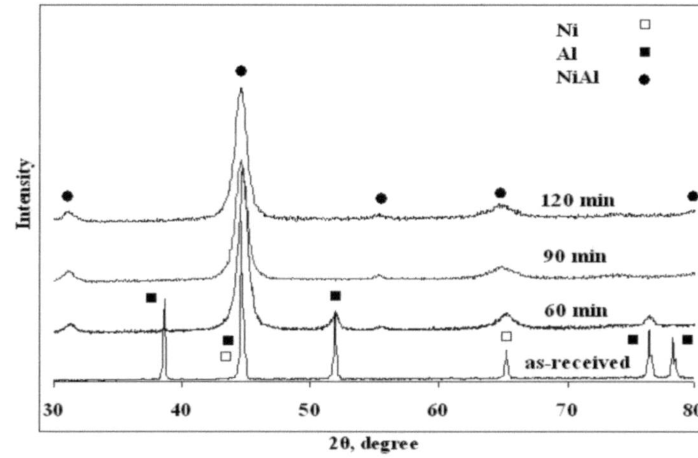

Fig. 9. The development of NiAl intermetallic compound from Ni50Al50 powder mixture during mechanical alloying (Enayati et al; 2011).

The XRD patterns of the as-milled powder for 90 min and as-deposited coatings prepared with two sets of spray parameters (Table 1) are presented in Fig. 10. Besides NiAl main peaks several additional small, broad peaks can be observed on the XRD patterns of the coatings. These broad peaks were identified as Ni and Al_2O_3 phases. Enayati, et al. suggested that oxidizing of Al and subsequent separation of Ni occurred as NiAl particles are subjected to the high temperature (typically 3000 °C) during HVOF spraying. The intensity of Ni and Al_2O_3 peaks are higher in coating II as the fuel/oxygen ratio increases. It means that a higher fuel/oxygen ratio results a higher flame temperature and therefore more oxidation during HVOF processing.

HVOF Parameters	Condition	
	I	II
Oxygen flow rate (l/min)	830	830
Fuel flow rate (ml/min)	210	240
Fuel/oxygen volume ratio	0.025	0.029
Spray distance (mm)	360	360
Powder rate (g/min)	80	80
Number of passes	3	3

Table 1. HVOF Spraying parameters for NiAl coatings (Enayati et al; 2011).

Fig. 11. shows the cross-sectional SEM images of the coatings at several magnifications. The coating exhibits a typical splat-like and layered morphology due to the deposition and re-solidification of molten or semi-molten droplets. The light and dark gray layers are Ni-rich and Al-rich phases, respectively, which are consistent with the work of Movahedi, et al. (Movahedi et al; 2005B, 2009). Enayati, et al. suggested that an improved in uniformity of microstructure was observed for coating II due to the higher fuel flow rate and flame temperature (Enayati et al; 2011).

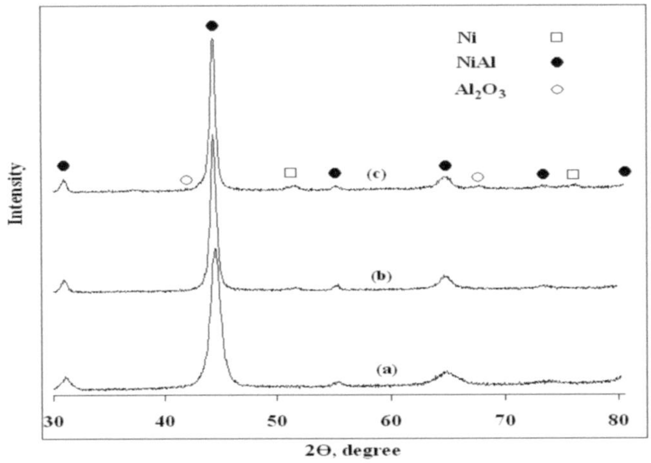

Fig. 10. The XRD patterns of as-milled NiAl powder for 90 min (a) and as-deposited coatings prepared at conditions I (b) and II (c) (Enayati et al; 2011).

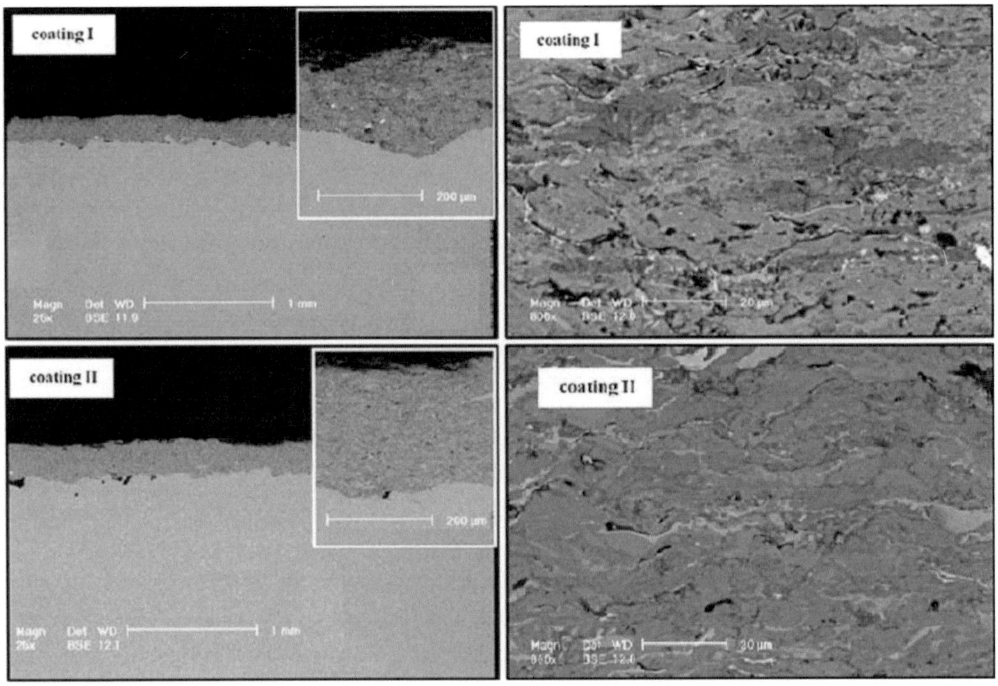

Fig. 11. Cross-sectional microstructure of the HVOF coatings (Enayati et al; 2011).

A dense Ni/Al alloy coating was deposited by cold spraying using a mechanically alloyed powder was reported by Zhang, et al (Zhang et al; 2008). Fig. 12. shows typical SEM images of cross-sectional microstructure of cold-sprayed Ni/Al alloy coating using milled Ni/Al alloy powder. It was observed that the coating exhibited a dense microstructure and some apparent thick layers with a white contrast appeared on the coating microstructure.

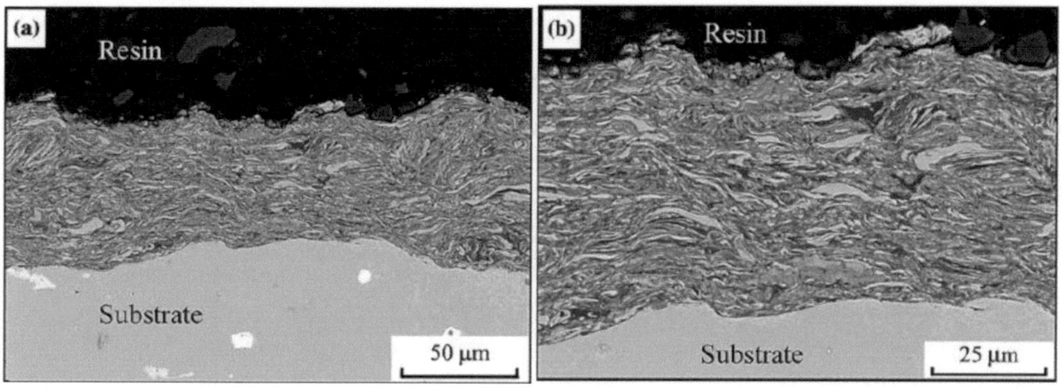

Fig. 12. Cross-sectional microstructure of cold-sprayed Ni/Al alloy coating observed at different magnifications (Zhang et al; 2008).

XRD patterns of the as-sprayed coating and feedstock powder are shown in Fig. 13. Only the peaks of nickel and aluminum were identified in XRD pattern of the ball-milled powder. It is clear that the XRD pattern of the cold spray coating is almost the same as that of the milled powder. This fact indicates that the coating and feedstock exhibited the same phase structure and no oxide was identified in the powder and the coating by XRD. In cold spraying, the particle deposition takes place in a solid state. Consequently, the lamellar structure of the milled powder will be completely retained in the coating, giving a unique effect on the microstructure and properties of the cold-sprayed coating. According to EDS analysis of the coating, was mentioned by Zhang, et al. the thicker layer in a white contrast was a Ni-rich phase and the fine lamella was a Ni-Al solid solution with high-Al content.

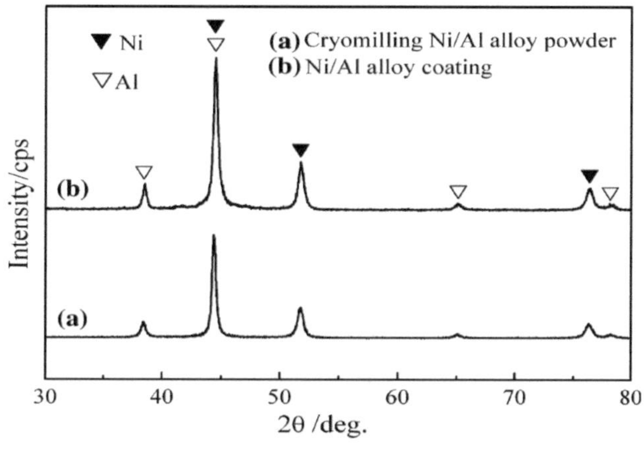

Fig. 13. XRD patterns of the Ni/Al feedstock powder fabricated by ball milling and the as-cold spray coating (Zhang et al; 2008).

The XRD patterns of annealed Ni/Al cold spray coatings are shown in Fig. 14. After annealing treatment at 500°C for 3h, Ni and Al peaks completely disappeared and peaks corresponding to Ni_2Al_3 and NiAl appeared (Fig. 14a). With annealing temperature rising to 600°C, it can be found from Fig. 14(b) that NiAl became the main phase, only minor Ni_2Al_3 exists. As the temperature was raised to 850°C, the diffraction peaks of Ni_2Al_3 disappeared completely, and only diffraction peaks of NiAl phase were present in the XRD pattern, as shown in Fig. 14(c). This fact indicates that the annealing at temperature of higher than 850°C completely converts Ni/Al alloy to NiAl intermetallic compound. As the temperature reached 1050°C, no additional reaction was detected, and the NiAl phase was present in the coating (Fig. 14d) (Zhang et al; 2008).

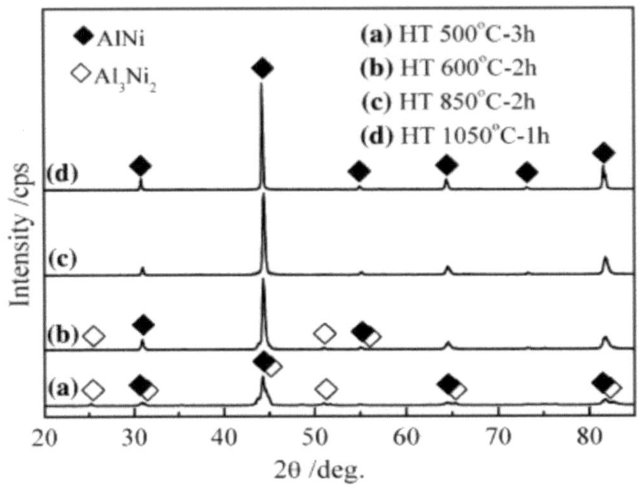

Fig. 14. XRD patterns of the Ni/Al cold spray coatings annealed at different temperatures (Zhang et al; 2008).

3.2 Intermetallic-ceramic and nanocomposites

As mentioned earlier, intermetallic compounds are an important class of materials because of a combination of their high tensile strength, low density, good wear, and creep resistance. These properties have led to the identification of several potential usages including structural applications and protective coatings (Sauthoff, 1995). Two major problems that restrict the application of intermetallic compounds are poor low-temperature ductility and inadequate high temperature creep resistance. These limitations can be overcome by introducing ceramic particles as reinforcements (Morris, 1998). Originally, reinforcement phase can be introduced in the matrix by two routs namely ex-situ addition of reinforcement particles and in-situ formation of reinforcement phase via a displacement reaction which both phase (Intermetallic and ceramic) are formed during ball milling. A rigid bond between the particles is created by cold welding when mechanical energy is applied to powder particles (Chen et al; 1999). Incorporation of hard second phases into an intermetallic composite (IC) matrix is a strategy for effective high-temperature strengthening, creating an intermetallic matrix composite (IMC). Recently, a great deal of work has been done on intermetallic matrix composites (IMCs). Various continuous or discontinuous ceramic reinforcements such as SiC, Al_2O_3, TiB_2, and TiC were explored to obtain increased high temperature strength and better creep resistance, together with adequate ductility and toughness (Inoue et al; 2000). Among these reinforcements, SiC fibers were commercialized for use in IMCs. The SiC reinforcements were added into different nickel aluminide matrices by reaction synthesis, mechanically alloying, and sintering (powder metallurgy) (Lee et al; 2001; Zhang et al; 2004) to improve oxidation and mechanical properties and the workability of the matrices. Hashemi et al. reported the synthesis of nickel aluminide matrix composite coating reinforced by SiC particulates that was fabricated by the plasma spraying of Ni-Al-SiC powder prepared by low energy ball milling. The cross-sectional SEM image of the powder particles after 15h of milling time is presented in Fig. 15. As it can be shown the SiC particles were incorporated into the Ni/Al powder particles.

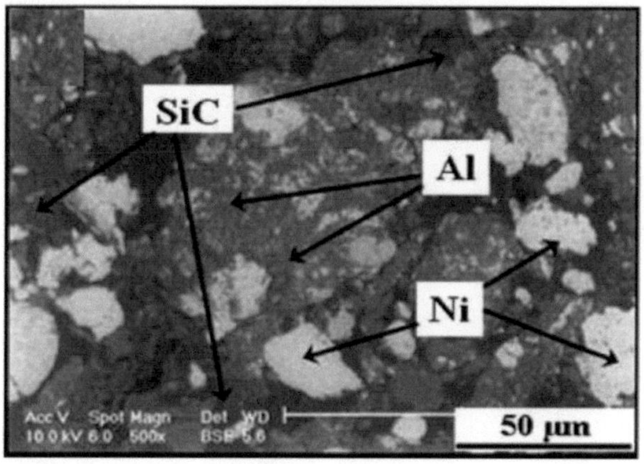

Fig. 15. Cross-sectional SEM images of Ni-Al-SiC powder particles produced by low energy ball milling (Hashemi et al; 2009).

The formation of Ni-Al intermetallic compounds required a long time and a high temperature for the diffusion of Al and Ni. In thermal spray processes, the temperature is high enough for diffusion, but the exposure time of powders to plasma flame is too short. After powder deposition on the substrate, the high-temperature diffusion and therefore the Ni-Al reaction is stopped (Houben et al; 1973; Sampath et al; 2004). Hashemi et al, suggested that by increasing the plasma spray distance the diffusion time increased, while by increasing current density the plasma flame temperature increased. These two parameters determine the content of intermetallic compound in the coatings. They also mentioned that by increasing the current intensity from 600 to 700 A, the relative amount of Ni decreased while that of Ni_2Al_3 phase increased, whereas by increasing the current intensity from 700 to 800 A, a reverse trend was observed. On the other hand, the increased current density will increase the velocity of the powder flow rate, thus declining the dwelling time of the powder particles and further decreasing the amount of Ni_2Al_3. In this approach, powder particles experience lower heat, which eliminates the diffusion of Ni and Al and therefore the development of Ni-Al compounds. The cross-sectional microstructure of as-sprayed coating in optimum spray condition is shown in Fig. 16. (Hashemi et al; 2009).

Horlock, et al. synthesised a reactive powder having a nominal composition of 50wt.%Ni(Cr)-40wt.%Ti-10wt.%C with the planar-type ball milling. They reported that the reaction to produce TiC is initiated within individual powder particles during HVOF spraying, leading to the formation of a coating containing TiC particles within a nanocomposite Ni-rich matrix. The TiC particles were found in the coating (Fig. 17) on the order of 50 to 200 nm in size. The Low-magnification SEM image of a cross section through an HVOF-sprayed coating shows the characteristic layered morphology with little porosity as well as the nanoscale TiC grains (arrowed), which are embedded in a light contrast metallic matrix.

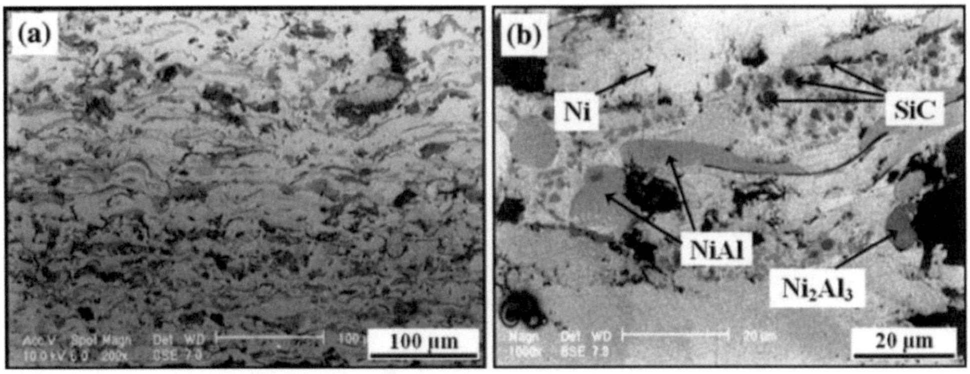

Fig. 16. Cross-sectional SEM images of plasma spray Ni-Al-SiC coating at different magnification (Hashemi, 2009).

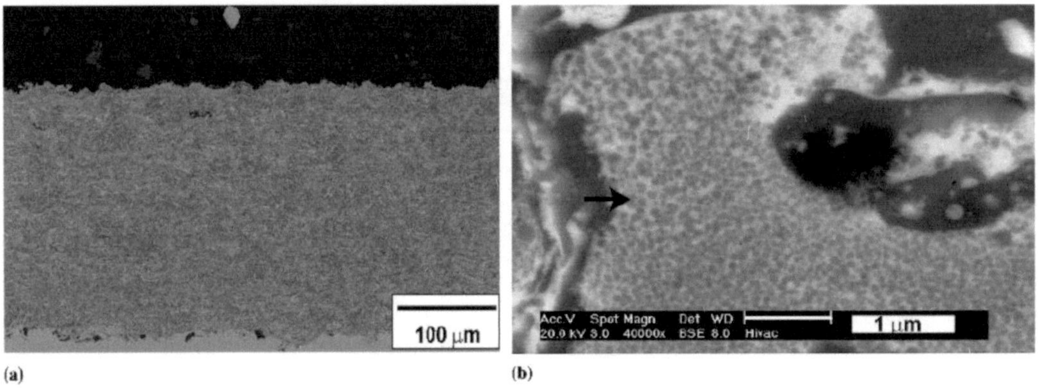

Fig. 17. (a) Low-magnification (b) high-magnification BSE image of a region from (a) of a cross section through an HVOF-sprayed Ni(Cr)-TiC coating (Horlock et al; 2005).

In Horlock's work a pre-alloyed Ni(Cr) powder was used to produce a metallic matrix with the potential to resist corrosion and high-temperature oxidation, while TiC is a ceramic with high hardness and chemical stability. They also purposed that the final powder microstructure contains Ti, Ni(Cr), and C all embedded in the same particle, there is a decreased possibility of fresh Ti surfaces being in contact with free C. Consequently, the possibility of Ti oxidation of TiC is greatly diminished during spraying. The XRD pattern from the as-deposited coating, shown in Fig. 18. has major peaks that can be identified as a Ni-rich solid solution phase and TiC as well as the smaller peaks corresponding to NiTi, TiO_2, and $NiTiO_3$ (Horlock et al; 2004).

TiC-Ni based nanocomposite powders for thermal spraying were produced by high-energy attrition and vibration mills in scale up were studied by Eigen, et al. X-ray diffraction analysis shows (Fig. 19) that the crystallite sizes of both hard phase and binder are refined with increasing milling time reaching a steady state after about 20 h.

Correspondingly, the mixing and refining of the phases can be observed by SEM (Fig. 20). At early stage of milling (Fig. 20a), the ductile binder particles are deformed to plate-like particles while hard particles are initially pressed into the surface of the binder platelets and cover them. In later milling stages, densification takes place, i.e. hard phase particles are fully embedded into the matrix and the matrix is completely cold-welded (Fig. 20b). In the final stage in Fig. 20c, microstructures are characterized by carbide phase dimensions ranging from less than 20 nm (see Fig. 19) to about 500 nm (Eigen et al; 2003).

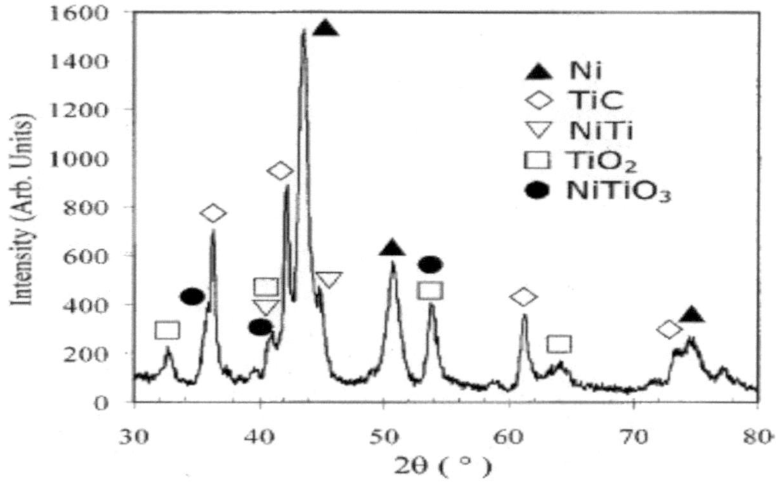

Fig. 18. The XRD pattern obtained from the HVOF as-sprayed Ni(Cr)-TiC coating (Horlock et al; 2004).

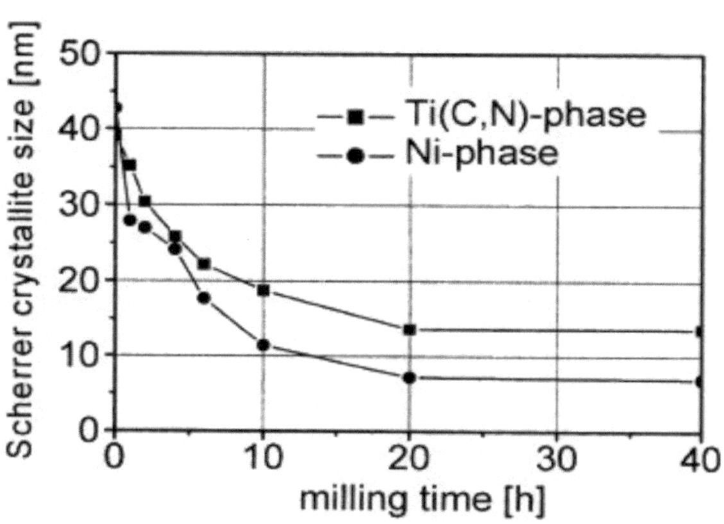

Fig. 19. Refinement of hard phase and binder phase with milling time (vibration mill, BPR 23:1). Crystallite sizes were determined using the Scherrer method (Eigen et al; 2003).

The development of a nanocomposite thermal spray powder is also studied by He, et al. as schematically summarized in Fig. 21. In a nanocomposite powder, such as Cr_3C_2-NiCr and WC-Co, there are hard and brittle carbide particles and a tough metal binder constituent. Hard and brittle carbide particles are fractured into sharp fragments and embedded into the metal binder. The metal binder, with lower hardness, is subjected to enhanced milling from both the balls and hard carbide particles. As milling time increases, carbide fragments are continually embedded into the metal binder. The metal binder and the polycrystalline composite, experience continuous overlapping, cold welding, and fracturing. Finally, a polycrystalline nanocomposite powder system, in which round nanoscale carbide particles are uniformly distributed in a metal binder, is formed. As an example, Fig. 22 shows such a Cr_3C_2-25 (Ni20Cr) polycrystalline nanocomposite powder. Clearly shown that the large proportions of carbides, in the form of round particles, are uniformly distributing themselves in the NiCr solid solution. It is possible to use mechanical milling to synthesize other nanocomposite powder systems with a hard particle and tough binder duplex structure; examples of such systems are WC–NiCr, TiC–NiCr, TiC–Ti, and SiC–Al. The microstructures of conventional and nanostructured Cr_3C_2-25 (Ni20Cr) coatings, examined using SEM, are shown in Fig. 23. A uniform and dense microstructure is observed in the nanostructured coatings, compared to the conventional Cr_3C_2-25 (Ni20Cr) coating that is observed to have an inhomogeneous microstructure (He et al; 2002).

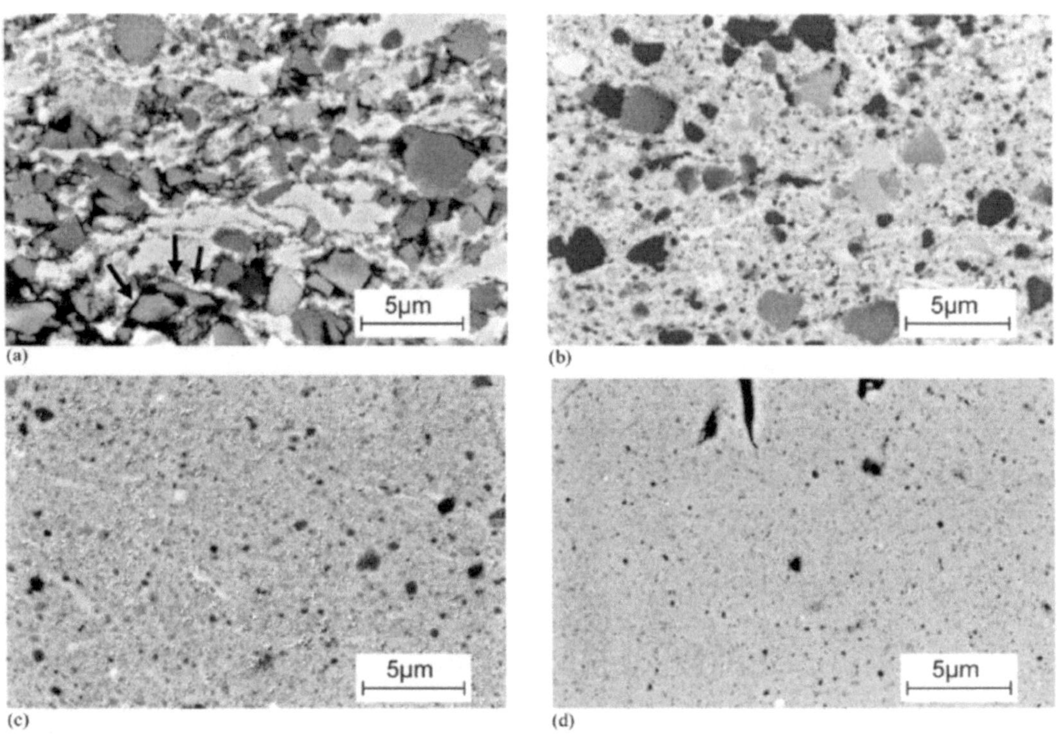

Fig. 20. Refinement of phase distribution after: (a) 2 h; (b) 10 h; and (c) 40 h (BPR 23:1) in the vibration mill; and (d) after 20 h in the attrition mill (BPR 20:1). Black arrows in (a) indicate fragments of one broken hard phase particle (Eigen et al; 2003).

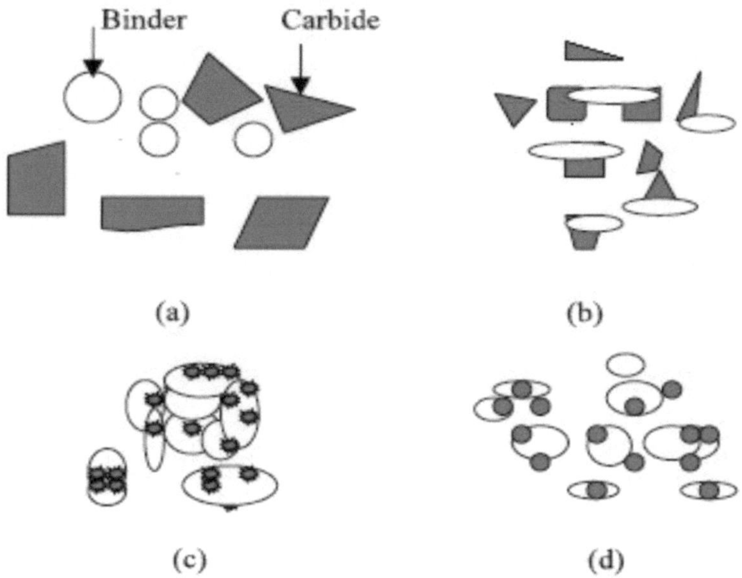

Fig. 21. Schematic diagram of milling mechanism for duplex structure powder: (a) initial stage; (b) NiCr matrix overlaps and deforms, Cr_3C_2 fractures and embed into NiCr; (c) binders deform, fracture, and weld, carbide fracture further; and (d) nanocomposite powder (He et al; 2002).

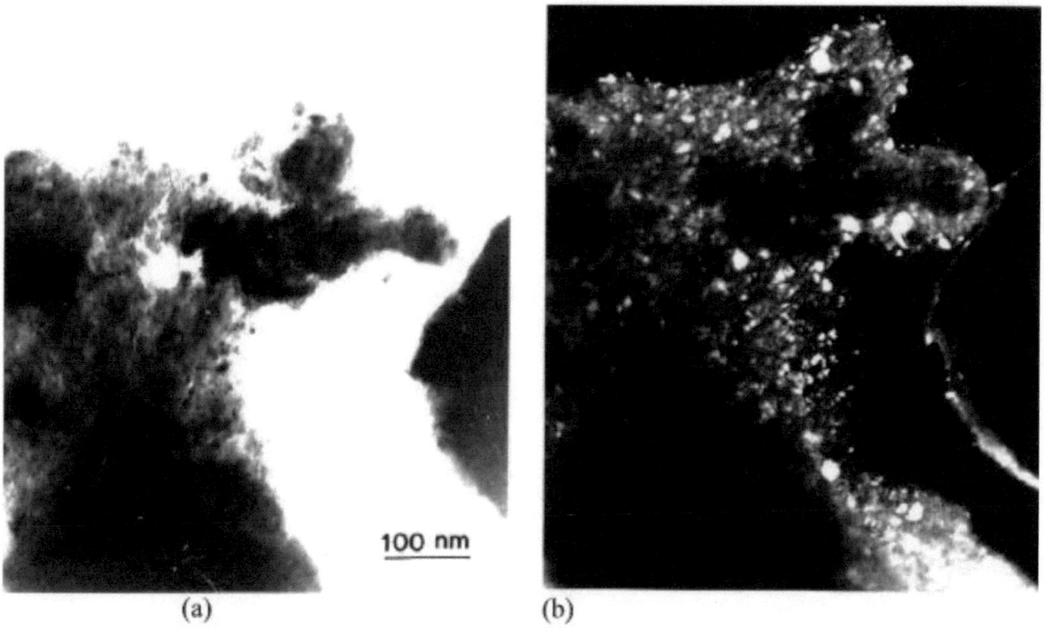

Fig. 22. Cr_3C_2–NiCr powder milled for 20 h: (a) Bright field image; and (b) dark field image. White particles are Cr_3C_2, and dark matrix is NiCr (He et al; 2002).

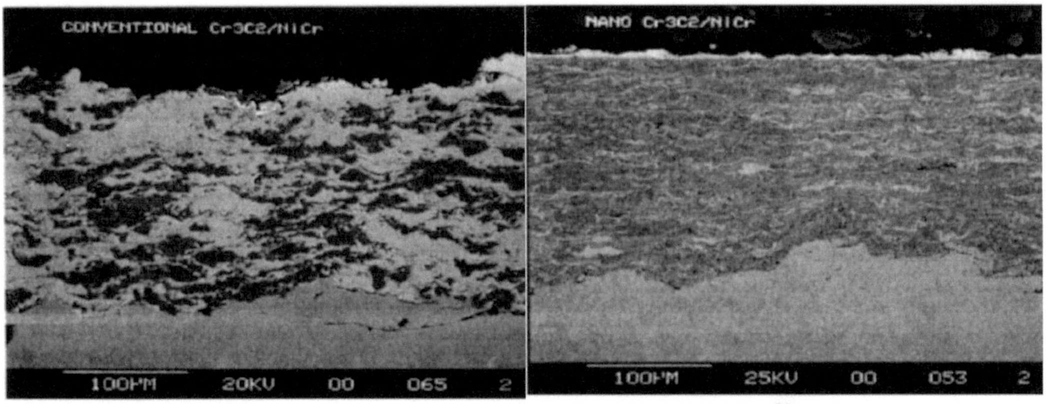

Fig. 23. Microstructure of Cr_3C_2-25 (Ni20Cr): (a) conventional coating; (b) nanostructure coating (He et al; 2002).

3.3 Amorphous-nanocrystalline materials

Amorphous metallic alloys have been of interest not only for fundamental studies, but also for potential applications for over 40 years. Amorphous structures have been made in many alloy systems and show a variety of unique properties compared to their crystalline counterparts. These properties are associated with the amorphous atomic structure and include high yield strength, large elastic limit, high corrosion resistance, good wear resistance, and low elastic modulus (Greer et al; 2002; Schuh et al; 2007). Amorphization by high energy mechanical alloying of elemental powders occurs by an inter-diffusion reaction at relatively low temperature along constituent interfaces. The formation of amorphous phase by MA process depends on the energy provided by the milling machine and thermodynamic properties of the alloy system. The thermodynamic and kinetic principles for amorphization through solid state synthesis or mechanical alloying are discussed by Schwarz and Johnson (Schwarz & Johnson, 1883). They identified two rules for the formation of amorphous alloy by MA in an A–B binary system: (1) A large negative heat of mixing, ΔH_{mix}, between the elemental constituents. (2) A large asymmetry in the diffusion coefficients of the constituents. An amorphous phase is kinetically obtained only if the amorphization reaction is much faster than that for the crystalline phases. It is also believed that during mechanical milling of a homogeneous crystalline alloy, the internal energy of lattice increases due to the introduction of crystal defects. When the free energy of the crystalline structure exceeds the free energy of the amorphous phases, crystalline structure can transform to an amorphous phase (Suryanarayana, 2001, 2004).

Fe-based amorphous alloys and thermal spray coatings are perhaps the most important system for possible applications because of the low cost of iron, and the relatively high strength and hardness of Fe-based amorphous alloys (Chen et al; 2006; Sunol et al; 2001). Extensive research has been carried out on the mechanical alloying of Fe-based amorphous alloys including binary Fe-B, Fe–Cr, and Fe–Zr (Schuh et al; 2007), ternary Fe-Zr-B (Suryanarayana, 2001, 2004) and Fe-Si-B (Chen et al; 2006) and multi-component Fe-Ni-Si(P)-B (Chen et al; 2005), Fe-Al-P-C-B (Minic et al; 2009) and Fe-Nb-Cu-Si-B (Inoue et al; 2000) alloys. Movahedi et al, reported the MA amorphization of 70Fe–15Cr–4Mo–5P–1C–

1Si–4B (wt.%) elemental powder which includes four types of elements: late transition metal (Fe), early transition metals (Cr, Mo), metalloids (B, P, Si), and graphite. The atomic sizes of these elements are in the order of Mo (0.139 nm) > Si (0.132 nm) > P (0.128 nm) > Cr (0.127 nm) > Fe (0.126 nm) > B (0.098 nm) > C (0.091 nm). This composition yields new atomic pairs of Fe–(Cr, Mo), (Cr, Mo)–(B, P, Si) and Fe–(B, P, Si) with various negative heats of mixing. These properties suggest that Fe–Cr–Mo–B–P–C–Si composition has a high glass forming ability (GFA) and appropriate thermal stability (Movahedi et al; 2010a & b).

Fig. 24. shows the XRD patterns of powder mixture as a function of high energy milling (i.e., Retch PM100) times. As-received powder mixture shows sharp crystalline peaks of elemental Fe, Cr, Mo, B, C and Si. Red phosphorus is absent on XRD pattern because of its amorphous nature. As milling progresses, the XRD peaks of the elemental constituents are broadened with a corresponding decrease in their intensities. These effects are caused by a continuous decrease effective crystallite size and an increase of the atomic level strain, as a result of the induced-plastic deformation during MA (Filho et al; 2000; Zhang, 4004). After 15h of milling time, the Cr, Mo, Si, B and C peaks vanished. This may be due to the dissolution of these elements into Fe matrix and/or to their ultra fine crystallite size. There is no significant reaction between the elemental powders at this stage of milling as no new phase was detected. On continued milling a broad peak was developed on the XRD pattern, owing to the formation of an amorphous phase. Meanwhile the crystalline peaks of Fe remain distinct. This structure transforms to an amorphous phase on further milling till 80h of milling time (Movahedi et al; 2010a).

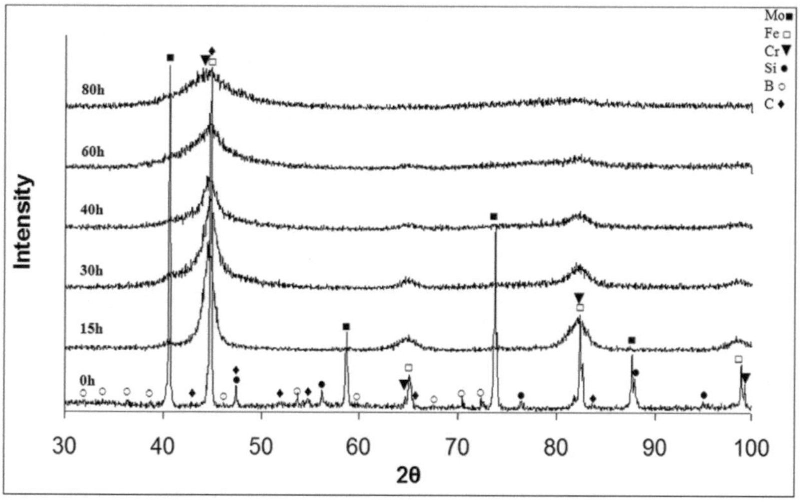

Fig. 24. XRD patterns of 70Fe–15Cr–4Mo–5P–1C–1Si–4B powders as-received and after different milling times (Movahedi et al; 2010a).

Movahedi et al. suggested that the transformation of supersaturated solid solution of Fe to amorphous structure during milling is believed to occur as a result of the internal energy increase of the crystalline structure due to the creation of a high density of lattice defects as well as the dissolution of a great amount of solute atoms with different sizes in Fe lattice. When the free energy of the crystalline solid solution exceeds the free energy of the amorphous state, the crystalline structure thermodynamically becomes unstable and transform to the amorphous structure. This behavior can be accounted for by considering the interaction parameters, proposed by Inoue et al. (Inoue et al; 2000) which describe the difference in bonding energy between the atomic pairs Fe–P–C and Fe–B–C in the ternary Fe–A–C system. The effect of B addition is presumably to generate attractive bonding among the constituent elements. In the Fe–A–C system with a negative interaction parameter, the enthalpy of mixing is also negative. In such a case, the formation of the Fe–A–C solution decreases the free energy of the system by lowering the system mixing enthalpy. These interaction parameters characterize the effect of a third element on amorphization reaction of Fe–C binary alloys processed by mechanical alloying (Inoue et al; 2000; Olofinjana et al; 2007; Suryanarayana, 2001).

Fig. 25. shows cross-sectional SEM images of the powder particles after 2, 4, 10, 15, 30 and 80h of milling. During the early stage of milling, the powder particles are flattened by compressive forces due to the collision of the balls. Thus, micro-forging action deforms the powder particles plastically leading to work hardening and fracturing. The creation of new surfaces enable the particle to weld together and thus to produce a typical lamellar structure consisting of pure elements (Movahedi et al; 2009). On continued ball milling, the layered structure is progressively refined (Fig. 25d–f). At longer milling times the structure became featureless on SEM as a result of development of Fe-base solid solution and subsequent amorphous phase.

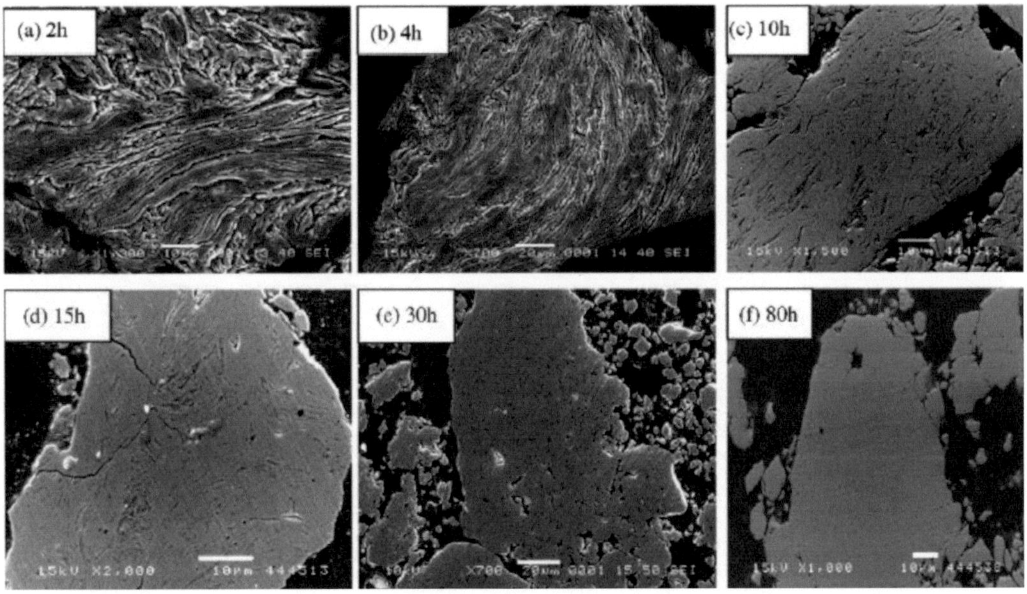

Fig. 25. cross-sectional SEM images of 70Fe–15Cr–4Mo–5P–1C–1Si–4B powders after different milling times (Movahedi et al; 2010a).

High resolution transition electron microscopy (HRTEM) and selected-area diffraction pattern (SADP) of powders milled for 15h confirmed the formation of a nanocrystalline structures (Fig. 26a). After 40 h of milling time amorphous and nanocrystalline phases co-existed in the milled powders. Fig. 26b shows that most amorphous phase is developed at the edge of powder particles indicating that the amorphization reaction starts at edge of particles and progress into the internal regions as MA proceeds. The SADP in Fig. 26b shows some diffraction spots within the amorphous diffuse ring. Furthermore the fast Fourier transform (FFT) images showed a broad diffuse ring at the edge and crystalline diffraction spots at the center of particles (arrows in Fig. 26b) suggesting that the sample contains both amorphous and nanocrystalline phases. Fig. 26c is the HRTEM image and SADP of mechanically alloyed powder after 80h of milling time, showing a fully amorphous microstructure (Movahedi et al; 2010a, 2011).

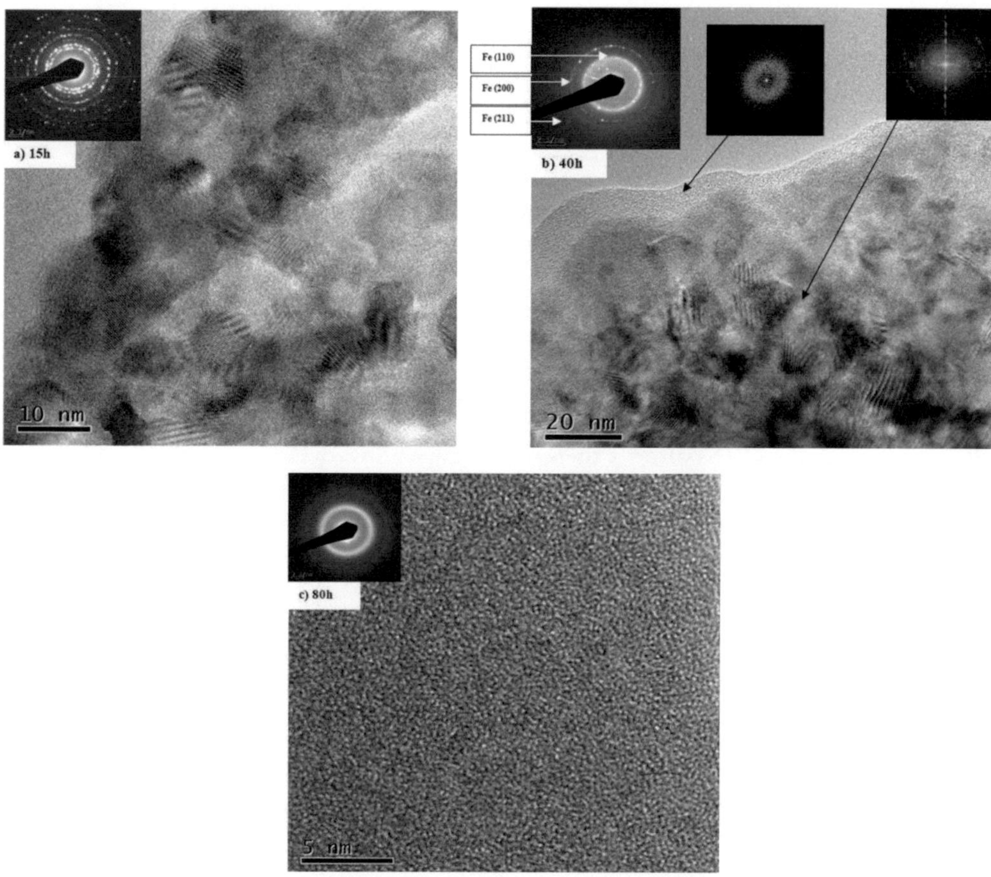

Fig. 26. HRTEM micrographs, SADP and FFT patterns of 70Fe-15Cr-4Mo-5P-1C-1Si-4B amorphous powder after different milling times (Movahedi et al; 2010a, 2011).

Synthesizing amorphous and/or nanocrystalline layers on metal substrates can be utilized to improve surface performance such as wear and corrosion resistance (Kim et al; 2007). Greer et al. reported that amorphous alloys can have very good resistance to sliding and abrasive wear and the coatings can have low friction coefficient (Greer et al; 2002). Thermal spraying process is one of the techniques to deposit amorphous coatings on surfaces, where the amorphous structure is retained due to the sufficiently rapid cooling that inhibits long-range diffusion and crystallization. On impact with the substrate, droplet spreading occurs to give lamellar morphologies with cooling rates of 10^7-10^8 K/s (Wu et al; 2006). A number of researchers have investigated the use of air plasma spraying (APS), low pressure plasma spraying (LPPS) and vacuum plasma spraying (VPS) to deposit alloys, which are capable of solidifying as metallic glasses. Kishitake et al. reported that mixed amorphous and crystalline structures are obtained in APS and LPPS coatings (Kishitake et al; 1996). In recent years, there has been an increasing interest in the use of HVOF for depositing protective coatings. The general methodology involves designing alloys that have low critical cooling rates for glass formation. Alloys with high glass forming ability (GFA) would be favorable for forming fully amorphous phase coating by HVOF process.

The amorphous coatings, while exhibiting interesting properties, can be heated up above their crystallization temperature to initiate devitrification and yield amorphous-nanocrystalline mixture. Since the driving force for the crystallization is extremely high and the diffusion rate in the solid state, at the crystallization temperature, is very low, an extremely high nucleation frequency results. There is limited time for growth before impingement between neighbouring crystallites occurs, resulting in the formation of nanoscale microstructures. Some researchers (Branagan et al; 2001; Kishitake et al; 1996; Otsubo et al; 2000), mentioned that the formation of Fe-base amorphous coatings by LPPS, high-energy plasma spraying (HPS), and HVOF processes with using atomized feedstock powder. In recent years, as mentioned earlier, Movahedi, et al. first synthesised the amorphous mechanical alloying feedstock powder in a new composition 70Fe-15Cr-4Mo-5P-4B-1C-1Si (wt.%) with high GFA and then developed them to amorphous-nanocrystalline coatings with HVOF and plasma spray process (Movahedi, 2010c, 2011).

Fig. 27. illustrates the XRD patterns of mechanically alloyed Fe-Cr-Mo-P-B-C-Si feedstock powder and the as-sprayed HVOF coatings. The presence of halo on XRD pattern confirms that the feedstock MA powder used for HVOF spraying has an amorphous structure. The XRD pattern of HVOF-G1 coating also shows a halo characteristic indicating that this coating has an amorphous structure similar to feedstock MA powder. However, in HVOF-G2 there is an emergent crystalline peak on the top of the amorphous hub suggesting that this coating is a mixture of amorphous and crystalline phases. Structure of HVOF-G3 coating mainly consists of crystalline phases such as α-Fe, $Fe_{23}(C, B)_6$, and Fe_5C_2.

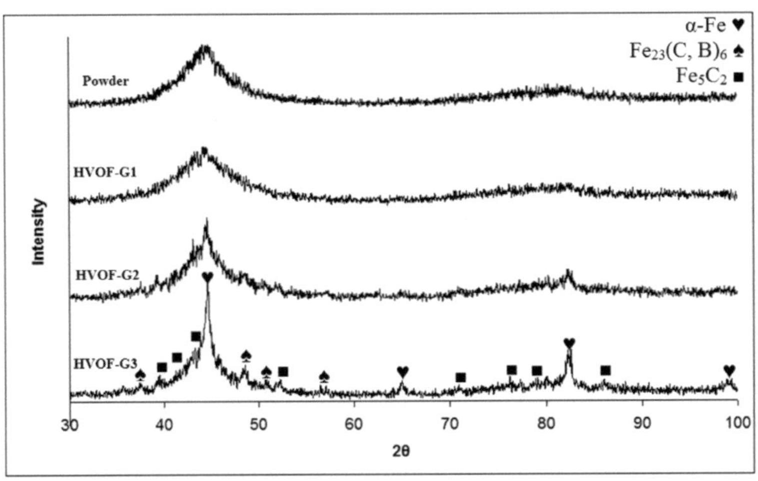

Fig. 27. XRD patterns of mechanically alloyed feedstock powder and Fe-Cr-Mo-P-B-C-Si HVOF coatings (Movahedi et al; 2010c, 2011).

Movahedi, et al. inferred from the diffraction patterns that a full range of amorphous to fully crystalline microstructures can be obtained by adjusting of HVOF parameters especially fuel/oxygen ratio (Table 2). They suggested that the difference in the glassy fraction is related to the mechanism of the amount of cooling rate and remelting of individual particles in HVOF flame at various fuel/oxygen ratios. By increasing fuel/oxygen ratio, the flame temperature becomes too high and the velocity of the powder particles is much higher. Thus, the powder particles were completely remelted in the HVOF flame and then were rapidly solidified and quenched on the cold substrate forming an amorphous structure. By decreasing the fuel/oxygen ratio the particles did not remelt to a significant extent as well as the velocity of them becomes lower, thus the conditions appear to crystallize the amorphous feedstock powder in flame which explains the higher percentage of crystalline phase in the HVOF-G3 (Movahedi, 2010c, 2011; Shin et al; 2007).

Kishitake et al. reported that for Fe-17Cr-38Mo-4C gas atomized powder, amorphous coatings are obtained by the APS while a mixture of amorphous and crystalline phases are formed by HVOF (Kishitake et al; 1996). They suggested that this difference may result from the difference of the cooling rate between the APS and HVOF processes. It is regarded that the composition chosen by Movahedi, et al. satisfies the three empirical rules (Pang et al; 2002) for the stabilization of the super-cooled liquid during HVOF and plasma spraying, leading to highly dense random packed atomic configurations, higher viscosity, and lower atomic diffusivity (Kobayashi et al; 2008) which is primarily attributed to the high GFA.

Parameters	Microstructure		
	Amorphous (HVOF-G1)	**Amorphous-Nanocrystalline (HVOF-G2)**	**Nanocrystalline (HVOF-G3)**
Oxygen gas flow rate (SLPM)	833	682	560
Fuel (Kerosene) flow rate (SLPM)	0.37	0.21	0.14
Fuel/Oxygen (Vol%)	0.044	0.031	0.025
Powder feed rate (g/min)	35	35	35
Spray distance (mm)	300	300	300
Scanning Velocity (mm/s)	50	50	50
Deposit thickness (μm)	300	300	300
Nozzle length (mm)	100	100	100
Compress air cooling	yes	yes	yes

Table 2. HVOF spraying parameters for advanced structures (Movahedi et al; 2010c, 2011).

Typical SEM cross section image of HVOF coatings is shown in Fig. 28. As it can be seen, the microstructure of coatings includes very fine lamella structure which is smooth and dense, adhering well with the substrate with no cracking. Moreover, some pores are rarely observed in this microstructure as indicated by arrows can be seen from the images. The big pores located between flattened droplets are mainly caused by the loose packed layer structure or gas porosity phenomenon, while the small pores within the flattened particles originate from the shrinkage porosity (Totemeier, 2005). Obviously, the porosity of the coatings reduces in the order of HVOF-G3, G2, and G1, and it is believed that increasing the fuel/oxygen ratio, increases both the thermal and kinetic energy of the gas flow, so that the majority of the powder particles are better melted and also accelerated to higher velocities and deformed extensively on impact to form elongated lamella (Ji, 2005). Some unmelted particles are clearly visible in HVOF-G3 (Fig. 28c) coating because of lower flame temperature (minimum fuel/oxygen ratio).

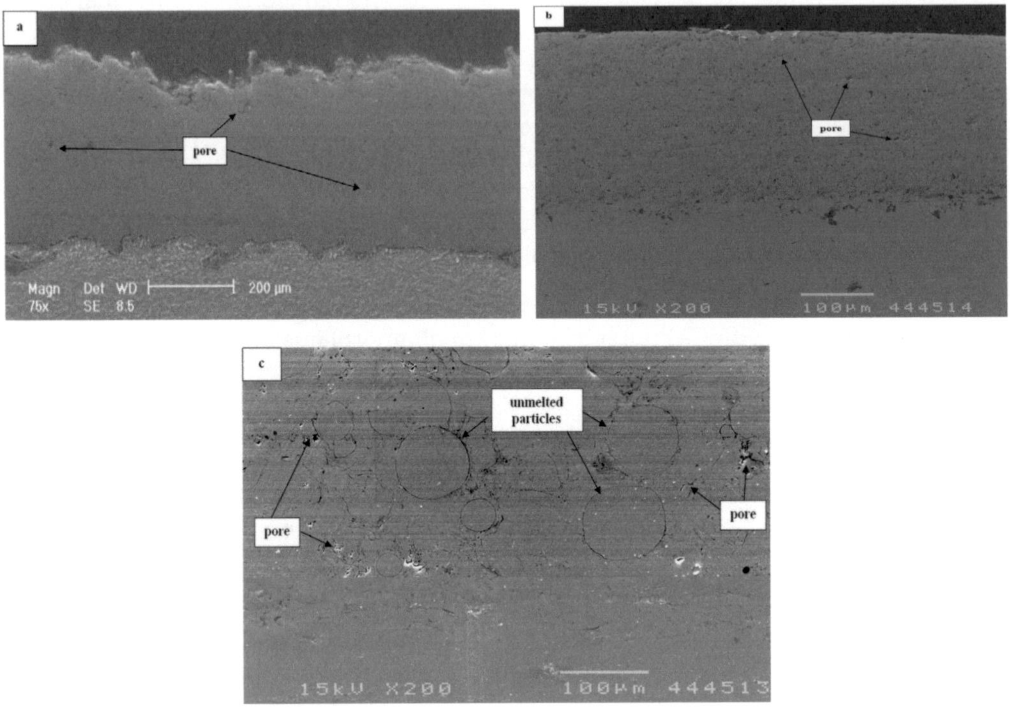

Fig. 28. SEM cross-sectional images of Fe-Cr-Mo-P-B-C-Si coatings (a) HVOF-G1, (b) HVOF-G2, and (c) HVOF-G3 (Movahedi et al; 2010c).

HRTEM image has shown in Fig. 29. confirms that HVOF-G1 coating is completely amorphous. As it can be seen in Table 2, this microstructure appears when the fuel/oxygen ratio has a maximum value. As shown in Fig. 30. the HVOF-G2 coating consists of amorphous phase and nanocrystalline grains. The electron diffraction pattern, in Fig. 30(a) was taken with the selected area aperture centered over the amorphous and nanocrystalline region and shows a diffuse amorphous halo with diffraction spots arising from nanocrystalline grains with a size range of 5-30 nm. The HRTEM micrograph and fast Fourier transform (FFT), as shown in Fig. 30(b), confirm the presence of a mixture of nanocrystalline grains within an amorphous matrix. In this case the fuel/oxygen ratio is moderate (HVOF-G2) so this duplex microstructure can be explained by quenching of semi-molten particles when impinged to the cold substrate. Therefore, some unmelted particles crystallized inside the HVOF flame to yield nanocrystalline grains which embedded within the amorphous matrix. A nanocrystalline structure with equiaxed nanograins was obtained in case of HVOF-G3 coating (Fig. 30c). In this condition the fuel/oxygen ratio has a minimum value and the HVOF flame temperature is the lowest so the most of the individual powder particles were unmelted and crystallized inside the HVOF flame. Moreover, the cooling rate was sufficiently high to avoid grain coarsening and yielded nanocrystalline structure (Movahedi et al; 2010c, 2011).

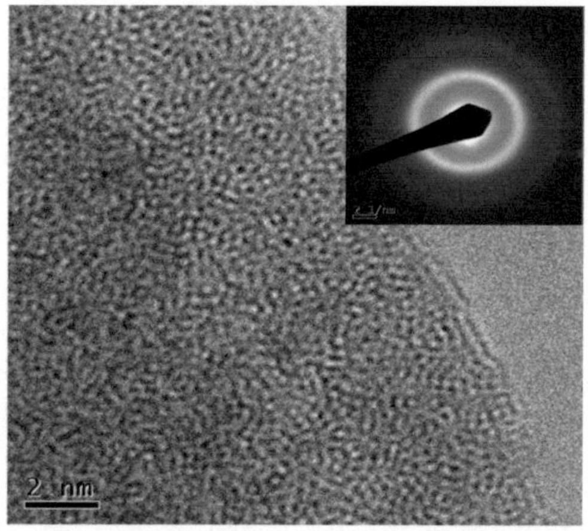

Fig. 29. HRTEM micrograph and SADP of amorphous Fe-Cr-Mo-P-B-C-Si HVOF coating (HVOF-G1) (Movahedi et al; 2010c, 2011).

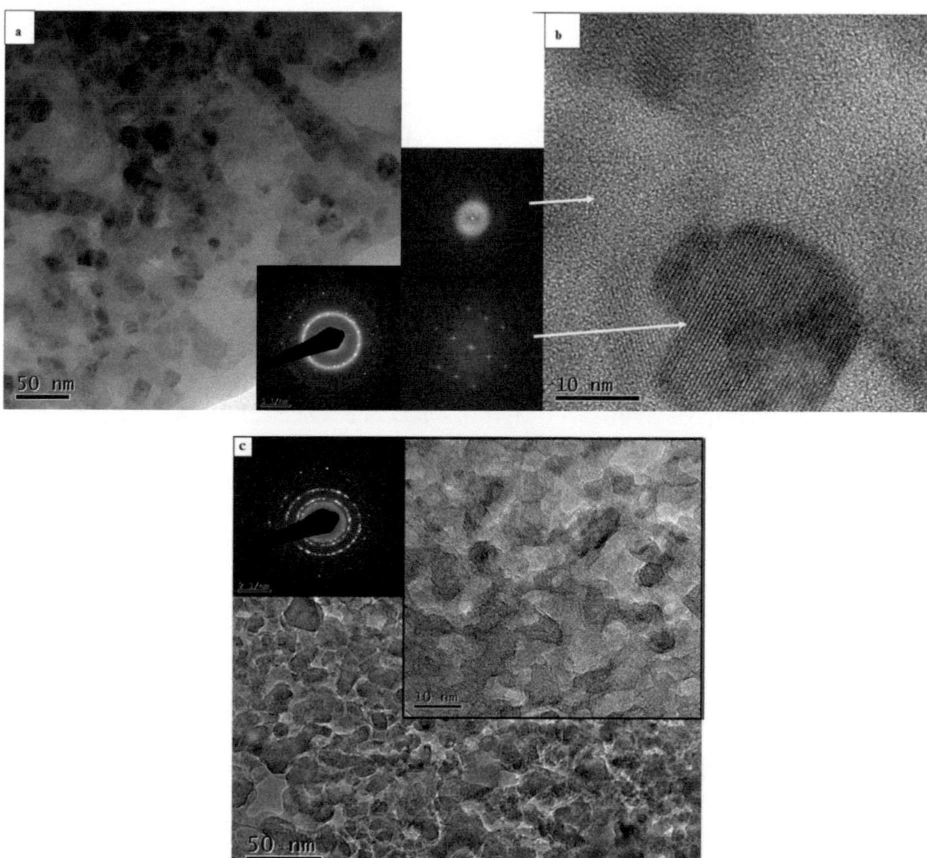

Fig. 30. (a) TEM and (b) HRTEM micrographs, SADP and FFT of amorphous-nanocrystalline Fe-Cr-Mo-P-B-C-Si HVOF coatings (HVOF-G2), and (c) TEM, HRTEM micrographs and SADP of nanocrystalline of HVOF-G3 (Movahedi et al; 2010c, 2011).

Movahedi, et al. reported that the microhardness value of HVOF coatings are all have a high hardness of around 800 to 1200 HV. The value of hardness was similar to that of the bulk Fe-based metallic glass (Kobayashi et al; 2008) but it was higher than that of electroplated chromium and the Ni-based amorphous coatings (Ni et al; 2009). The difference in hardness value of the three groups of HVOF coatings were developed by Movahedi, et al. is attributed mainly to the difference in volume fraction of amorphous and nanocrystalline phases. A fully amorphous coating has lower hardness (830 HV) compared to the duplex amorphous-nanocrystalline coating (950 HV). The fully nanocrystalline coating has the highest microhardness (1230 HV), probably due to precipitation of some carbides such as $Fe_{23}(C, B)_6$ and Fe_5C_2 during crystallization (Movahedi, 2010c). Some researchers suggested that the hardness of the amorphous Fe-base coating increases after crystallization (Branagan et al; 2005). In contrast Kishitake et al. reported that the duplex microstructure consisting of the both amorphous and nanocrystalline structure exhibits a higher hardness than fully amorphous or nanocrystalline structure (Kishitake et al; 1996). The difference is mainly attributable to the difference of decomposition of amorphous phase during crystallization.

4. Conclusion

This chapter reviews the solid state synthesis and characterization of advanced feedstock powders used in various thermal spray techniques, and processing and characterization of amorphous-nanostructured related coatings. The published results show that mechanical milling can be effectively used to synthesize advanced materials and nanocomposite powders. On the other hand, whether a composite or a single-phase starting powder is involved, mechanical milling leads to the formation of nanocrystalline structure under certain milling conditions.

5. References

Atzmon, M. (1990). In Situ Thermal Observation of Explosive Compound-Formation Reaction During Mechanical Alloying, *Physical Review Letter*, 64, pp. 487-490

Berndt, C.C. (1992). *Materials Production for Thermal Spray Processes*, Education Module on Thermal Spray, Pub. ASM International, OH, Materials Park, USA

Boldyrev, V.V. & Tkacova, K. (2000). Mechanochemistry of Solids: Past, Present and Prospects, *Journal of Materials Synthesis and Processing*, 8, pp. 121-132

Branagan, D.J; Swank, W.D; Haggard, D.C. & Fincke, J.R. (2001). Wear-Resistant Amorphous and Nanocomposite Steel Coatings, *Metallurgical and Materials Transcation A*, 32, pp. 2615-2621

Branagan, D.J; Breitsameter, M; Meacham, B.E. & Belashchenko, V. (2005). High-Performance Nanoscale Composite Coatings for Boiler Applications, *Journal of Thermal Spray Technology*, 14, 2, pp. 196-204

Chen, T; Hampikian, J.M. & Thadhani, N.N. (1999). Synthesis and Characterization of Mechanically Alloyed and Shock-Consolidated Nanocrystalline NiAl Intermetallic, *Acta Materialia*, 47, 8, pp. 2567–2579.

Chen, Q.J; Fan, H.B; Shen, J; Sun, J.F. & Lu, Z.P. (2006). Critical Cooling Rate and Thermal Stability of Fe–Co–Zr–Y–Cr–Mo–B Amorphous Alloy, *Journal of Alloys and Compounds*, 407, pp. 125-128

Chen, Q.J; Fan, H.B; Ye, L; Ringer, S; Sun, J.F; Shen, J. & McCartney, D.G. (2005). Enhanced Glass Forming Ability of Fe–Co–Zr–Mo–W–B Alloys with Ni Addition, *Materials Science and Engineering A*, 402, pp. 188-192

Chung, C.Y; Zhu, M. & Man, C.H. (2002). Effect of Mechanical Alloying on the Solid State Reaction Processing of Ni-36.5 at.%Al Alloy, *Intermetallics*, 10, pp. 865-871

Deevi, S.C; Sikka, V.K; Swindeman, C.J. & Seals, R.D. (1997). Reactive Spraying of Nickel-Aluminide Coatings, *Journal of Thermal Spray Technology*, 6, pp. 335-344

Eigen, N; Klassen, T. & Aust, E. (2003). Production of Nanocrystalline Cermet Thermal Spray Powders for Wear Resistant Coatings by High-Energy Milling, *Materials Science and Engineering A*, 356, pp. 114–121.

Eigen, N; Gartner, F; Klassen, T; Aust, E; Bormann, R. & Kreye, H. (2005). Microstructures and Properties of Nanostructured Thermal Sprayed Coatings Using High-Energy Milled Cermet Powders, *Surface and Coatings Technology*, 195, pp. 344– 357

El-Eskandarany, M.S. (2001). *Mechanical Alloying for Fabrication of Advanced Engineering Materials*, Noyes Publications, 0-8155-1462-X, New York

Enayati, M.H; Karimzadeh, F; Tavoosi, M; Movahedi, B. & Tahvilian, A. (2011). Microstructural Characterization of Nanostructured NiAl Coatings Prepared by Mechanical Alloying and HVOF Technique, *Journal of Thermal Spray Technology*, 20, 5, pp. 440-446

Enayati, M.H; Sadeghian, Z; Salehi, M. & Saidi, A. (2004). The Effect of Milling Parameters on the Synthesis of Ni_3Al Intermetallic Compound by Mechanical Alloying, *Materials Science and Engineering A*, 375-377, pp. 809-811

Enayati, M.H; Karimzadeh, F. & Anvari, S.Z. (2008). Synthesis of Nanocrystalline NiAl by Mechanical Alloying, *Journal of Materials Processing and Technology*, 200, pp. 312-315

Filho, A.F; Bolfarini, C; Xu, Y. & Kiminami, C.S. (2000). Amorphous Phase Formation in Fe-6.0wt%Si Alloy by Mechanical Alloying, *Scripta Materialia*, 42, pp. 213-217

Greer, A.L; Rutherford, K.L. & Hutchings, I.M. (2002). Wear Resistance of Amorphous Alloys and Related materials, *International Materials Review*, 47, pp. 87-112

Goldenstein, H; Silva, Y.N. & Yoshimura, H.N. (2004). Designing a New Family of High Temperature Wear Resistant Alloys Based on Ni_3Al-IC: Experimental Results and Thermodynamic Modeling, *Intermetallics*, 12, pp. 963-968

He, J; Ice, M. & Lavernia, E.J. (1998). Synthesis and characterization of nanostructured Cr_3C_2-NiCr, *Nanostructured Materials*, 10, pp. 1271-1283

Hearley, J.A; Little, J.A. & Sturgeon, A.J. (2000). The Effect of Spray Parameters on the Properties of High Velocity Oxy-Fuel NiAl Intermetallic Coatings, *Surface and Coating Technology*, 123, pp. 210-218

Hearley, J.A; Little, J.A. & Sturgeon, A.J. (1999). The Erosion Behaviour of NiAl Intermetallic Coatings Produced by High Velocity Oxy-Fuel Thermal Spraying, *Wear*, 233, pp. 328-333

He, J. & Schoenung, J.M. (2002). Nanostructured Coatings, *Materials Science and Engineering A*, 336, 274-319.

Horlock, A.J; Sadeghian, Z; McCartney, D.G. & Shipway, P.H. (2005). High-Velocity Oxyfuel Reactive Spraying of Mechanically Alloyed Ni-Ti-C Powders, *Journal of Thermal Spray Technology*, 14, 1, pp. 77-84

Houben, J.M. & Zaat, J.H. (1973). Investigations into the Mechanism of Exothermically Reacting Nickel-Aluminum Spraying Materials, *Proceedings of Seventh International Metal Spray Conference,* pp. 77-88, London, Welding Institute, Sept 1973

Inoue, M; Suganuma, K. & Nihara, K. (2000). Fracture Mechanism of Ni_3Al Alloys and their Composites with Ceramic Particle at Elevated Temperatures, *Intermetallics,* 8, pp. 365-370

Inoue, A. & Wang, X.M. (2000). Bulk Amorphous Fc20(Fe-C-Si) Alloys with Small Amounts of B and Their Crystallized Structure and Mechanical Properties, *Acta Materialia,* 48, pp. 1383-1395

Ji, G; Elkedim, O. & Grosdidier, T. (2005). Deposition and Corrosion Resistance of HVOF Sprayed Nanocrystalline Iron Aluminide Coatings, *Surface and Coating Technology,* 190, pp. 406-416

Kim, Y.S; Kim, K.T; Kim, B.T. & Bae, J.I. (2007). Microstructure and Wear Behavior of Thermally Sprayed Fe-based Amorphous Coating, *Key Engineering Materials,* 353-358, pp. 848-851

Kishitake, K; Era, H. & Otsubo, F. (1996). Thermal-Sprayed Fe-10Cr-13P-7C Amorphous Coatings Possessing Excellent Corrosion Resistance, *Journal of Thermal Spray Technology,* 5, 4, pp. 476-482

Koch, C.C. (1997). Synthesis of Nanostructured Materials by Mechanical Milling: Problems and Opportunities, *Nanostructured Materials,* 9, pp. 13-22

Kobayashi, A; Yano, S; Kimura, H. & Inoue, A. (2008). Mechanical Property of Fe-base Metallic Glass Coating formed by Gas Tunnel Type Plasma Spraying, *Surface and Coating Technology,* 202, pp. 2513-2518

Knotek, O; Lugscheider, E. & Eschnauer, H.R. (1973). Reactive Kinetic Observations for Spraying with Ni-Al Powder, *Proceedings of Seventh International Metal Spray Conference,* pp. 72-76, London, Welding Institute, Sept 1973

Lee, D.B. & Kim, D. (2001). The Oxidation of Ni_3Al Containing Decomposed SiC Particles, *Intermetallics,* 9, pp. 51-56

Maric, R; Ishihara, K.N. & Shingu, P.H. (1996). Structural changes during low energy ball milling in the Al-Ni system, *Journal of Materials Science Letter,* 15, pp. 1180–1183

Mashreghi, A. & Moshksar, M.M. (2009). Partial Martensitic Transformation of Nanocrystalline NiAl Intermetallic During Mechanical Alloying, *Journal of Alloys and Compounds,* 482, pp. 196-198

Minic, D.M; Maricic, A. & Adnadevic, B. (2009). Crystallization of α-Fe Phase in Amorphous Fe81B13Si4C2 Alloy, *Journal of Alloys and Compounds,* 473, pp. 363-367

Morris, D.G. (1998). Possibilities for High-Temperature Strengthening in Iron Aluminides, *Intermetallic,* 6, pp. 753-758

Morsi, K. (2001). Reaction Synthesis Processing of Ni-Al Intermetallic Materials, *Materials Science and Engineering A,* 299, pp. 1-15

Movahedi, B; Enayati, M.H. & Salehi, M. (2009). Thermal Spray Coatings of Ni-10wt.%Al Composite Powder Synthesis by Low Energy Mechanical Milling, *Surface Engineering,* 25, 4, pp. 276-283

Movahedi, B; Enayati, M.H. & Salehi, M. (2005). Synthesis of Ni-Al Composite Coating in Thermal Spray Applications by Utilizing Low Energy Ball Milling Powder, *Proceeding of Ninth Iranian Metallurgical Engineering Society Conference,* Shiraz University, Shiraz, Oct. 2005a

Movahedi, B; Enayati, M.H. & Salehi, M. (2005). Investigation of Adhesion Strength Thermal Spray Coating of Mechanical Alloying Ni-Al Powders, *Proceeding of Eighth Heat Treatment and Surface Engineering National Conference*, Isfahan University of Technology, Isfahan, May. 2005b

Movahedi, B; Enayati, M.H. & Wong, C.C. (2010). Study on Nanocrystallisation and Amorphisation in Fe-Cr-Mo-B-P-Si-C System during Mechanical Alloying, *Journal of Materials Science and Engineering B*, 172, pp. 50-54 a

Movahedi, B; Enayati, M.H. & Wong, C.C. (2010). On the Crystallization Behavior of Amorphous Fe-Cr-Mo-B-P-Si-C Powder Prepared by Mechanical Alloying, *Materials letter*, 24, 9, pp. 1055-1058 b

Movahedi, B; Enayati, M.H. & Wong, C.C. (2010). Structural and Thermal Behavior of Fe-Cr-Mo-P-B-C-Si Amorphous and Nanocrystalline HVOF Coatings, *Journal of Thermal Spray Technology*, 19, 5, pp. 1093-1099 c

Movahedi, B. & Enayati, M.H. (2011). Fe-Based Amorphous-Nanocrystalline Thermal Spray Coatings, *2011 TMS Annual Meeting and Exhibition*, pp. 17-24, San Diego, 27 Feb-3Mar 2011

Ni, H.S; Liu, X.H; Chang, X.C; Hou, W.L; Liu, W. & Wang, J.Q. (2009). High Performance Amorphous Steel Coating Prepared by HVOF Thermal Spraying, *Journal of Alloys and Compounds*, 467, pp. 163-167

Olofinjana, A.O. & Tan, K.S. (2007). Thermal Devitrification and Formation of Single Phase Nano-Crystalline Structure in Fe-Based Metallic Glass Alloys, *Journal of Materials Processing and Technology*, 191, pp. 377-380

Otsubo, F; Era, H. & Kishitake, K. (2000). Formation of Amorphous Fe-Cr-Mo-8P-2C Coatings by the High Velocity Oxy-Fuel Process, *Journal of Thermal Spray Technology*, 9, 4, pp. 494-498

Pawlowski, L. (2008). *The Science and Engineering of Thermal Spray Coatings*, Wiley, ISBN: 978-0-471-49049-4, England

Pang, S.J; Zhang, T; Asami, K. & Inoue, A. (2002). Synthesis of Fe-Cr-Mo-C-B-P Bulk Metallic Glasses with High Corrosion Resistance, *Acta Materialia*, 50, pp. 489-497

Rickerby, D.S. & Matthews, A. (1991). *Advanced Surface Coatings: A Handbook of Surface Engineering*, Chapman and Hall, New York, 1991

Sauthoff, G. (1995). *Intermetallics*, VCH, Germany

Sampath, S; Gudrnundsson, B; Tiwari, R. & Herman, H. (1990). Plasma Spray Consolidation of Ni-Al intermetallics, *Thermal Spray Research and Applications,Proceeding of third National Thermal Spray Conference*, pp. 357-361, Long Beach, ASM International, May 1990

Sampath, S; Jiang, X.Y. & Matejicek, J. (2004). Role of Thermal Spray Processing Method on the Microstructure, Residual Stress and Properties of Coatings: an Integrated Study for Ni-5wt.%Al Bond Coats, *Materials Science and Engineering A*, 236, pp. 216-232

Schmalzried, H. (1995). *Chemical Kinetics of Solids*, Wiley-VCH, Weinheim

Schuh, C.A; Hufnagel, T.C. & Ramamurty, U. (2007). Mechanical Behavior of Amorphous Alloys, *Acta Materialia*, 55, 12, pp. 4067-4109

Shin, D.I; Gitzhofer, F. & Moreau, C. (2007). Properties of Induction Plasma Sprayed Iron Based Nanostructured Alloy Coatings for Metal Based Thermal Barrier Coatings, *Journal of Thermal Spray Technology*, 16, 1, pp. 118-127

Stein, A.; Keller, S.W. & Mallouk, T.E. (1993). Turning Down the Heat: Design and Mechanism in Solid-State Synthesis. *Science*, 259, pp. 1558-1564

Stoloff, N.S; Liu, C.T. & Deevi, S.C. (2000). Emerging Applications of Intermetallics, *Intermetallics*, 8, pp. 1313-1320

Totemeier, T.C. (2005). Effect of High-Velocity Oxygen-Fuel Thermal Spraying on the Physical and Mechanical Properties of Type 316 Stainless Steel, *Journal of Thermal Spray Technology*, 14, 3, pp 369-372

Steenkiste, T.H; Smith, J.R. & Teets, R.E. (2002). Aluminium Coatings via Kinetic Spray with Relatively Large Powder Particle, *Surface and Coating Technology*, 154, pp. 237–252

Schwarz, R.B. & Johnson, W.L. (1983). Formation of an Amorphous Alloy by Solid-State Reaction of the Pure Polycrystalline Metals, *Physical Review Letter*, 51, pp. 415-422

Sunol, J.J; Clavaguera, N. & Cavaguera-Mora, M.T. (2001). Comparison of Fe–Ni–P–Si Alloys Prepared by Ball Milling, *Journal of Non-Crystalline Solids*, 287, pp. 114-119

Suryanarayana, C. (2001). Mechanical Alloying and Milling, *Progress in Materials Science*, 46, PP. 1-184

Suryanarayana, C. (2004). *Mechanical Alloying and Milling*, Marcel Dekker, 0-8247-4103-X, New York, USA

Wu, Y; Lin, P; Xie, G; Hu, J. & Cao, M. (2006). Formation of Amorphous and Nanocrystalline Phases in High Velocity Oxy-Fuel Thermally Sprayed a Fe-Cr-Si-B-Mn Alloy, *Materials Science and Engineering A*, 430, pp. 34-39

Zhang, Q; Li, C.J; Wang, X.R; Ren, Z.L; Li, C.X. & Yang, G.J. (2008). Formation of NiAl Intermetallic Compound by Cold Spraying of Ball-Milled Ni/Al Alloy Powder Through Post Annealing Treatment, *Journal of Thermal Spray Technology*, 17, 5-6, pp. 715-720

Zhang, D.L; Liang, J. & Wu, J. (2004). Processing Ti_3Al-SiC Nanocomposites High Energy Mechanical Milling, *Materials Science and Engineering A*, 375-377, pp 911-916

Zhang, D.L. (2004). Processing of Advanced Materials Using High-Energy Mechanical Milling, *Progress in Materials Science* 49, pp. 537-560

Part 4

Plasma Spray in Polymer Applications

9

Atmospheric Pressure Plasma Jet Induced Graft-Polymerization for Flame Retardant Silk

Dheerawan Boonyawan
Chiang Mai University,
Thailand

1. Introduction

Improving the flame retardant property of textiles become necessary to minimize the fire hazard under many circumstances (Wichman, 2003). Since fire accidents cause injuries and fatalities and also devastate property, considerable efforts have been made to develop flame-retardant textiles. Silk is one of the most commonly used textiles for interior decoration, such as upholsteries, curtains, and beddings, for its luxurious appearance. It is therefore of primary significance to improve the flame retardant property of silk fabrics in which the safety regulations are concerned. Flame retardant fabrics are typically prepared by treating the fabrics chemically with flame retardant agents. Halogen-based flame retardant agent is one of the most efficient reduction of the fire hazard for fabrics. However, because of their corrosivity, the presence of dioxin, a carcinogen, and suspected smoke toxity by products, there are legislated regulations to restrict halogen-based flame-retarded textile products. The non-halogen-based flame retardants have subsequently been replaced. Phosphorous-based compounds are the most extensively used (H. Yang & C. Yang, 2005) (Gaan & Sun, 2007), (Wu & C. Yang, 2007), (Horrocks & Price, 2001). For natural fiber textiles, a number of studies focus on flame retardant property of cotton fabrics (Reddy *et al.*, 2005), (Wu & C. Yang, 2006), (Tsafack & Levalois-Grützmacher, 2006) and silk fabrics (Achwal *et al.*, 1987), (Kako & Katayama, 1995), (Guan *et al.*, 2009). It was shown that a high level of flame retardancy could be achieved when silk fabric was treated by a reaction mixture of urea and phosphoric acid through pad/dry process (Achwal *et al.*, 1987). However, the treated silk had limited laundering durability. The flame retardant agent under the commercial name "Pyrovatex CP" which is N-hydroxymethyl (3-dimethylphosphono) propionamide (HDPP) was applied to induce flame retandancy on silk (Kako & Katayama, 1995) and (Guan & G. Chen, 2006). This compound needs formaldehyde, which is one of human carcinogens, as the bonding agent. Recently, the use of formaldehyde-free flame retardant finishing process was developed (Guan & G. Chen, 2006). The treated silk shown improved flame reatadancy with limited laundering durability. Although varying degrees of flame retardancy were obtained, the durability is difficult to solve due to the water solubility of the agent. It is even more problematic when the textiles are from natural origins. The development of satisfactory, durable flame retardant silk is indeed challenging and the alternative eco-friendly processes have to be considered.

Plasma treatment is a potential technique to impart flame retardant properties to textiles. The reactive species in the plasma interact with the surface atoms or molecules and modify the surface properties without affecting bulk properties. Recently, it was reported that microwave plasma had been employed in the flame retardant finishing process (Tsafack & Levalois-Grützmacher, 2006a) and (Tsafack & Levalois-Grützmacher, 2006b). However, low pressure plasma systems need to operate under vacuum which, in turns, add the cost and complexity to the process. Atmospheric pressure plasma source is an alternative system. A few different designs have been developed and employed to modify the surface of materials (Cheng *et al.*, 2006), (Schafer *et al.*, 2008), (Guimin *et al.*, 2009) and (Osaki *et al.*, 2003). The system is promising to industrial application since the vacuum system is eliminated.

In this work (Chaiwong *et al.*, 2010) we utilized an atmospheric pressure plasma jet to graft phosphorus-based flame retardant agent onto silk. The treated silk fabrics were submitted to 45° flammability test. The incorporation of phosphorus was studied via quantum simulations and Energy-Dispersive X-ray spectroscopy (EDS). The durability of the treatment was evaluated.

2. The setup

2.1 The silk

Silk, which is derived from the silk moth *Bombyx mori*, has a heavy chain that consists mainly of glycine (44%) and alanine (30%) (Dhavalikar, 1962). Silk yarn is scoured (degummed) to remove sericin, a gummy deposit on silk fibers. The crystal structure of silk fibroin has been examined by several research groups using the constrained least-squares refinement (Takahashi *et al.*, 1999). The simplest model consistent with the X-ray scattering pattern is Gly-Ala or Ala-Gly. Although these structures were solved earlier (Tranter, 1953, 1956) and (Naganathan & Venkatesan, 1972) the Gly-Ala structure is polymorphic, indicating the flexibility and the potential of possible alternate structures.

Silk fabric (Grazie™) of a density 52.9 g/m^2 used in this study, has a warp density and a weft density of 129 and 99 per inch respectively. The air penetration resistance was 98.4 $cm^3/cm^2 \cdot s$ indicating that the silk has high air resistance. The fabric was cut into 5 cm×17 cm samples which size fit to the flame spread test.

2.2 Plasma jet system

A self-made plasma jet system used is shown schematically in Fig. 2. The inner hollow electrode covered with a quartz tube was centred at the axis of the outer electrode. The inner electrode was connected to a 50 kHz, 0–10 kV voltage source whereas the outer electrode was grounded. High purity Ar was used as a plasma gas with adjustable flow rate from 2 to 10 standard litre per minute (slm). The gas flow was controlled by a gas flow controller. The operating voltage was set to 8 kV to keep constant input power to plasma.

The plasma jet was monitored by using a S2000 fibre optics spectrometer (Ocean Optics Inc, USA). The fiber optics probe was placed at right angle to the jet axis at a distance of 5 mm away. The emission spectrum of the plasma was collected at 0.3 nm resolution. The emission spectrum of the Ar plasma jet measured at the sample position which is 5 mm from the jet nozzle is shown in Fig. 3. It can be seen that the spectrum in the wavelength range of 250–

850 nm was dominated by excited argon (Ar I) peaks. In addition, reactive radical peaks including hydroxyl (OH) and atomic oxygen were found at 308.9 nm and 777.1 nm, respectively. Ambient species, such as N_2, were also observed. The presence of these radicals was undesirable since they might react with the surface of the samples. However, the emergence of these species could be controlled by the system parameters. For example, the OH band was drastically suppressed if the discharge voltage increased. The Ar flow rate was one of the parameters that affect the presence of radical species. It was found that excited N_2 peaks appeared more intense than OH radicals if the Ar flow rate was over 6 slm.

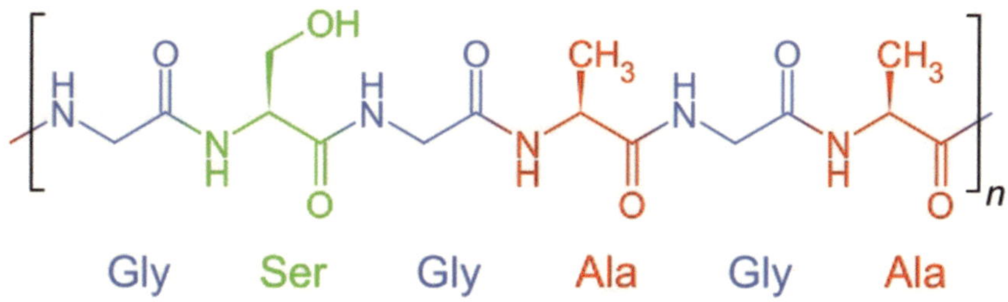

formula	code	Z	R factor	density /g.cm^{-3}	space group	cell parameter a/Å	b/Å	c/Å	α/°	β/°	γ/°
$C_6N_2O_3H_{14}$	AG	2	0.059	1.439	$P2_1$	5.28	11.81	5.51	90	101	90
$C_6N_2O_3H_{14}$	GA	4	0.045	1.549	$P2_12_12_1$	9.68	7.53	9.53	90	90	90

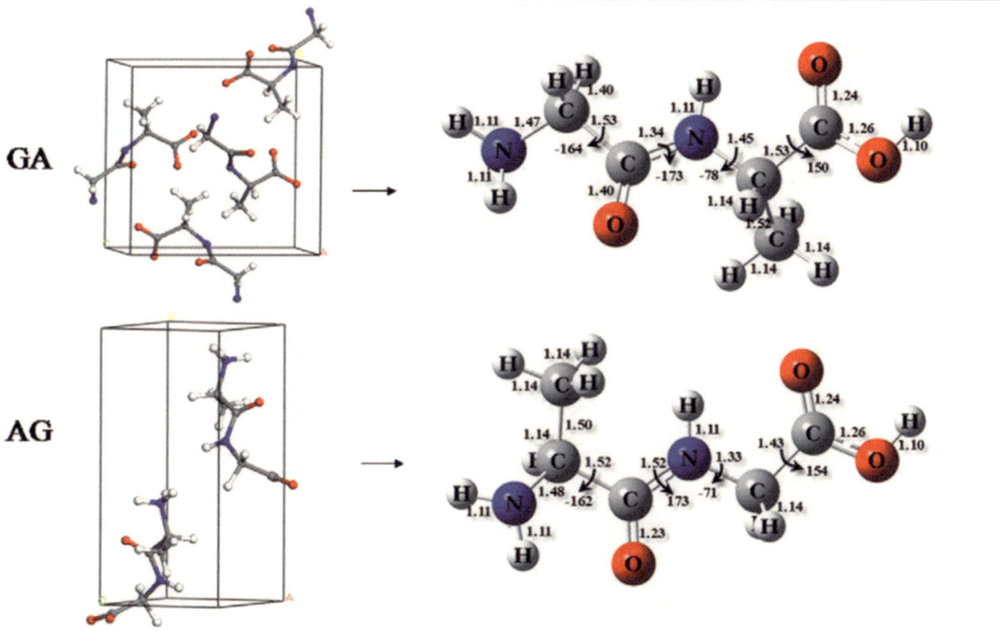

Fig. 1. The crystal structures of untreated GA and AG and corresponding cell parameters using single crystal X-ray diffraction technique (Sangprasert *et al.*, 2010).

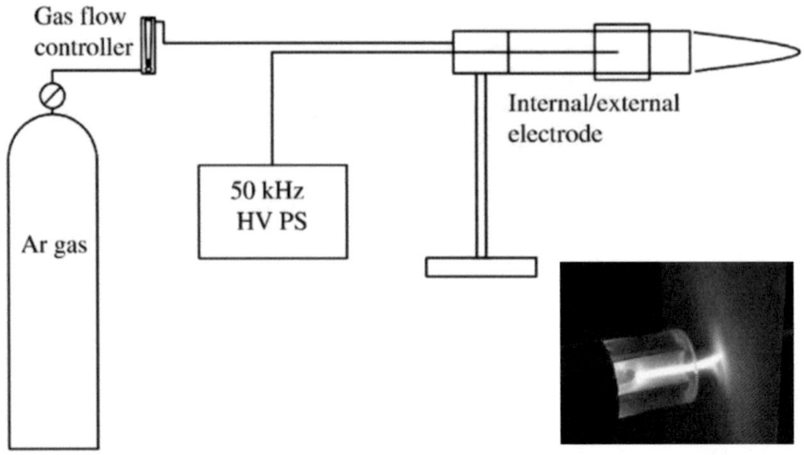

Fig. 2. Schematic view of self-made plasma jet system and the treatment (inset).

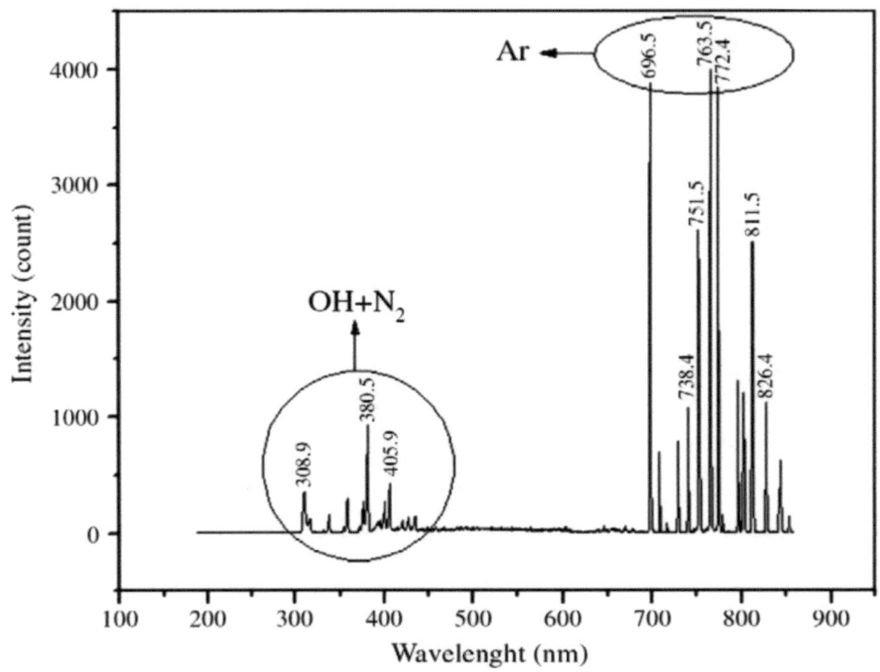

Fig. 3. Ar plasma emission spectrum at sample position from plasma jet, 6 slm flow.

2.3 The electron temperature

One important parameter to describe a plasma is its electron temperature, T_e. The plasma jet is one of a non-equilibrium plasma which electron temperature is a magnitude greater than the ion and the gas temperature. Determination of the electron temperature of our atmospheric plasma jet by optical emission spectroscopy (OES), the Boltzmann-plot method was applied. For this purpse a plot of ln $(I\lambda/gA)$ versus E_k should result a straight line with a slope of $-1/T_e$. I_k is the intensity of the emitted light, λ_k is the wavelength, g_k is the statistical weight, A_k is the transition probability, and E_k is the energy of the upper level. Table 1 shows the most intense Ar lines observed in the plasma and their characteristics.

Line	λ (nm)	Upper state (i)	Low state (j)	E_i (eV)	E_j (eV)	g_i	g_j	A_{ij} ($10^8 s^{-1}$)
Ar I	415.859	5p	4s	14.56	11.55	5	5	0.0140
Ar I	416.418	5p	4s	14.53	11.55	3	5	0.00288
Ar I	418.188	5p	4s	14.69	11.72	3	1	0.00561
Ar I	419.071	5p	4s	14.51	11.55	5	5	0.00280
Ar I	419.832	5p	4s	14.58	11.62	1	3	0.0257
Ar I	420.068	5p	4s	14.50	11.55	7	5	0.00967
Ar I	425.936	5p	4s	14.74	11.83	1	3	0.0398
Ar I	427.217	5p	4s	14.52	11.62	1	1	0.00797
Ar I	452.232	5p	4s	14.46	11.72	3	1	0.000898
Ar I	706.722	4p	4s	13.30	11.55	5	5	0.0380
Ar I	714.704	4p	4s	13.28	11.55	3	5	0.00625
Ar I	727.294	4p	4s	13.33	11.62	3	3	0.0183
Ar I	738.398	4p	4s	13.30	11.62	5	3	0.0847
Ar I	750.387	4p	4s	13.48	11.83	1	3	0.445
Ar I	751.465	4p	4s	13.27	11.62	1	3	0.402
Ar I	763.511	4p	4s	13.17	11.55	5	5	0.245
Ar I	794.818	4p	4s	13.28	11.72	3	1	0.186
Ar I	800.616	4p	4s	13.17	11.62	5	3	0.0490
Ar I	801.479	4p	4s	13.09	11.55	5	5	0.0928
Ar I	810.369	4p	4s	13.15	11.62	3	3	0.250
Ar I	811.531	4p	4s	13.08	11.55	7	5	0.331

Table 1. the most intense Ar lines observed in the plasma and their characteristics (NIST, 2009).

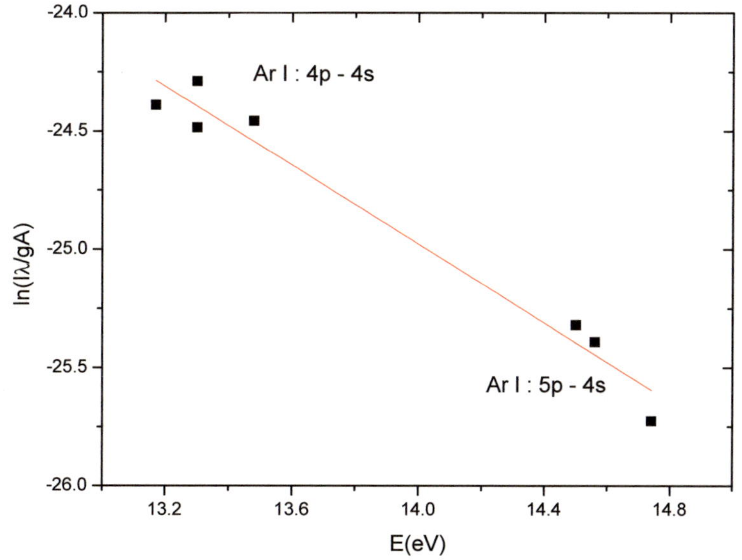

Fig. 4. Boltzman plot to determine the electron temperature of the plasma jet.

Atmospheric plasmas jet was generally in near-partial-local thermodynamic equilibrium (PLTE) (Griem, 1964). Thus, the distribution of atoms and ions in different excited states could be described by the Boltzmann distribution function, and the well-known method of Boltzmann-plot. From the OES in Fig.3, two groups of distribution points for Ar I transitions: 4p–4s and 5p–4s have been observed. The electron temperature has derived from fitting procedure for all of these points. The slope revealed the electron temperature of 1.3 eV. The electron temperature of this setup varied from 1.0-1.3 eV depends on the argon flow rate. The presence of Ar II lines has not shown since these lines are commonly observed in low-pressure plasmas but not in high-pressure discharges. In a DC microplasma plasma jet (Sismanoglu *et al.*, 2009), reported the presence of Ar II lines with a hollow anode configuration.

2.4 Treatment of the silk

Plasma-Surface Interactions

Like others plasma sources, many fundamental processes take place at the plasma-substrate interface as shown in Table 2. The surface is reached by fast electrons, ions, and free radicals, combined with the continued electromagnetic radiation emission in the UV-vis spectrum enhancing chemical-physical reactions. The minimum energy required to remove an electron from the highest filled level in the FERMI distribution of a solid into vacuum (to a point immediately outside the solid surface) is given by the work function eφ, with φ being the electron emission potential. The energy can be provided thermally (phonons, kBT), photons (ħω) or from the internal potential energy or kinetic energy of atoms and ions or metastable excited states.

Reactions	Description
AB + C(solid) → A + BC(gas)	Etching
AB(gas) + C(solid) → A(gas) + BC(solid)	Deposition
$e^- + A^+ \rightarrow A$	Recombination
A* → A	De-excitation
A* → A + e^- (from surface)	Secondary Emission
A* (fast) → A + e^- (from surface)	Secondary Emission

Table 2. Plasma-surface reactions (adapted from Braithwaite, 2000).

Presence and concentration of plasma active species is strongly dictated by the operational parameters of the plasma discharge used. Since electrons initialize ionization, changes of the electron gas (density, temperature, electron energy distribution function, (EEDF)) strongly influence the formation, the concentration and chemical reaction rate of reactive species and the intensities of the different wavelength emissions. The electron gas parameters in turn depend on the operational parameters of the plasma such as power, excitation frequency, gas flow and pressure.

Electrons

It is known that plasma electrons are not mono energetic. This is important as the rates of plasma-chemical reactions depend on the number of electrons with energy equal or higher to the reaction-specific threshold. The probability density for an electron having a specific energy ε can be described by means of the electron energy density function. The EEDF strongly depends on the electric field and the gas composition in a plasma and often is very far from being a real equilibrium distribution. Due to the various assumptions made in the quasi-equilibrium Maxwell-Boltzmann approximation, the EEDF of non- local thermodynamic equilibrium (LTE) plasmas is often better approximated by the Druyvesteyn distribution function.

$$f(\varepsilon)=1.04\langle\varepsilon\rangle^{-3/2}\varepsilon^{1/2}\exp\left(-\frac{0.55\varepsilon^{2}}{\langle\varepsilon^{2}\rangle}\right)$$

As can be seen in Fig. 5, the Druyvesteyn distribution function is characterized by a shift toward higher electron energies.

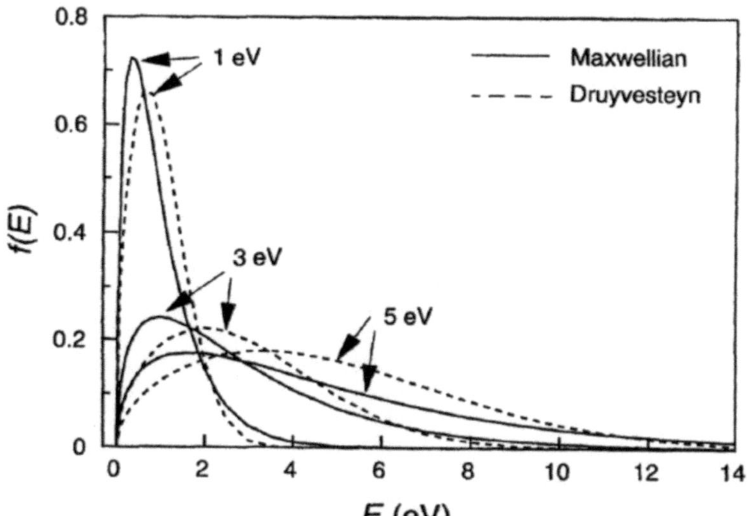

Fig. 5. Electron energy distributions according to Druyvesteyn and Maxwell. The numbers indicate the average electron energy for each distribution (Grill, 1994).

Both energy distributions however, regardless of the adopted approximation, show an important fact: While the majority of the electrons in non-LTE plasma have a low electron energy range (0.5-4 eV), there exist a very small but significant number of electrons characterized by a depleted high-energy tail region (8-15 eV). Though small in numbers, these electrons significantly influence the overall reaction rates in plasma, contributing to reactions, requiring a specific energy threshold value. Most of the electrons in this kind of plasma have energies high enough to dissociate almost all chemical bonds as shown in Table 3.

Ions

As ionization rates of ions in non thermal plasma are much lower than those for molecular dissociation, radical species density can be orders of magnitude higher than that of ions.

Therefore, plasma chemistry was inferred of being mainly governed by radical reactions, or by photochemical means. Due to the often high kinetic energy they gain in the plasma sheath, ions are considered to substantially contribute to plasma-chemical kinetics (Becker *et al.*, 2004). Ion formation reactions have been regularly illustrated by excitation, ionization, dissociation, and further electron impact reactions, like dissociative ionization or dissociative attachment by plasma electrons. Consequently, the various loss channels of positive and negative ions are as listed in Table 4.

Bond type	Bond energy ($kJmol^{-1}$)	Bond energy (eV)
C-H	411	4.25
C-C	346	3.56
C-N	276	2.86
C-O	358	3.70
C-S	272	2.80
C=C	602	6.23
C=O	724	7.50
C≡C	835	8.65
N-H	385	3.99
O-H	456	4.73

Table 3. Dissociation energies of organic compounds (Mathew *et al.*, 2008).

Reactions	Description
$e^- + AB^+ \rightarrow (AB)^* \rightarrow A + B^*$	Dissociative Electron-Ion Recombination
$e^- + A^+ \rightarrow A^* \rightarrow A + \hbar\omega$	Radiative Electron-Ion Recombination
$2\,e^- + A^+ \rightarrow A^* + e^-$	Trimolecular Electron-Ion Recombination
$A^- + B^+ \rightarrow A + B^*$	Bimolecular Ion-Ion Recombination
$A^- + B^+ + M \rightarrow A + B + M$	Trimolecular Ion-Ion Recombination
$e^- + M \rightarrow (M^-)^* \rightarrow M^- + \hbar\omega$	Radiative Electron Attachment
$e^- + A + B \rightarrow A^- + B$	Trimolecular Electron Attachment
$A^+ + B \rightarrow A + B^+$	Ion-Atom Charge Transfer
$A^+ + 2\,A \rightarrow A_2^+ + A$	Ion Conversion
$e^- + AB \rightarrow A^+ + B^- + e^-$	Polar Dissociation
$e^- + A^- \rightarrow A + 2\,e^-$	Electron Attachment

Table 4. Loss reactions of positive and negative ions in plasma (Fridman, 2008).

Ion chemistry of atmospheric plasmas is said to be rich. One example is the ion-induced formation of dangling bonds, acting as chemisorption sites for alkyl or any other free radicals (Von Keudell & Jacob, 2004). Their formation from impinging energetic ions has been recently demonstrated by particle beam experiments (Kylian *et al.*, 2009) and (Raballand *et al.*, 2008). Fig. 6 demonstrates surface active sites can be attacked by oxygen species (atomic or molecular oxygen) which leads either to fast passivation of the surface defect structure giving rise to various oxygen functional groups. Or a gradual volatilization occurs, namely of H_2O, ·OH, CO and CO_2, which diffuse from the bulk to the surface and desorb (Coburn & Winters, 1979).

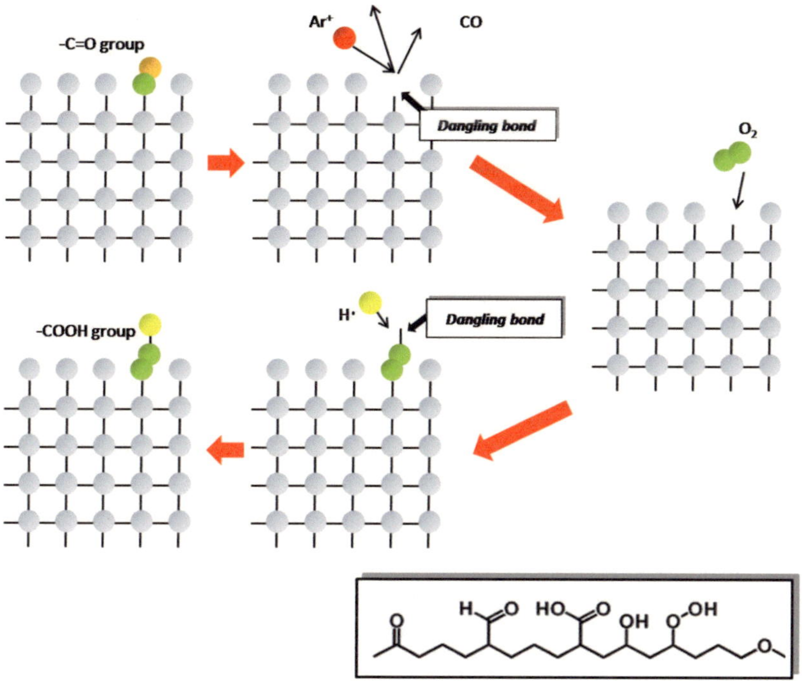

Fig. 6. Radicals (ions) in plasma interact with adjacent solid surface perform a functional surface. (Grzegorzewski, 2011)

Although the generation of dangling bonds is a strongly endothermic process, the high ion kinetic energy is usually sufficient for homolytical cleaving chemical bonds (Table 3). For atmospheric plasma jet, ion energy up to hundred eV is possibly obtained by a few hundred-negatively biased of the substrate.

Photons

At high pressures, single-collision conditions do not longer prevail. Beyond binary collisions, three-body interactions take place, leading to the formation of excimers. Rare gas excimer formation usually proceeds via electron-impact ionization or directly by metastable rare gas atom excitation. In either case, the initial step is a three-body collision process in which two ground state atoms interact with an excited state atom (metastable state or resonance, Table 5). Efficient excimer formation requires both a sufficiently large number of electrons with energies above the threshold for the metastable formation (or ionization), and a pressure that is high enough to have a sufficiently high rate of three-body collisions (Kurunczi *et al.*, 2001). In case of Ar, the minimum energy needed to form a metastable Ar atom by electron impact on ground-state Ar is about 12 eV.

It is due to this unique environment that non-LTE plasmas are not only able to increase the efficiency of traditional chemical processes. They offer as well alternative approaches to in conventional chemical synthesis otherwise inaccessible reaction pathways, often by changing the symmetry of the molecule's electronic configuration. The initiation of novel reaction channels at moderate bulk temperatures might lead to new transient and secondary products, which is an often highly desired and already exploited result of plasma treatment.

However, the generation of high chemically active species harbors as well the risk of not only uncontrollable but as well undesired plasma-chemical synthesis. A thorough knowledge of plasma reaction chemistry therefore is mandatory for any industrial application.

Reaction	Description
$e^- + X \rightarrow X^+ + 2e^-$	Electron-impact ionization
$X^+ + 2X \rightarrow X_2^+ + X$	
$X_2^+ + e^- \rightarrow X^* + X$	
$X^* + 2X \rightarrow X_2^* + X$	
$e^- + X \rightarrow e^- + X^*$	Electron-impact excitation
$X^* + 2X \rightarrow X_2^* + X$	
$X_2^* \rightarrow 2X + \hbar\omega$	Radiative decay

Table 5. Rare gas excimer formation

2.5 Flame retardant compound grafting

Plasma induced grafting is the two-step process. Prior grafting, the free radicals formation by using inert gas plasma is included. The active sites for further reaction are generated on the surface by the subsequent plasma species as mentioned above. In this case, Ar plasma jet was initiated at 8 kV with 4 slm. These parameters were kept constant for all of the Ar treatments. The sample surfaces was pre-activate for 5 min with Ar plasma. The distance between the nozzle and the sample was set at 5 mm. After Ar pre-treatment, the samples were immersed in the finishing solution of PBS for 10 s and air dried at 60 °C for 10 min. Graft polymerization was performed with Ar plasma for 5 min. These samples were designated as Ar-PBS-Ar silk. The samples were finally immersed in ethanol to remove the residual un-grafted molecules and dried in air at room temperature. For comparison, samples without Ar pre-treatment, directly immersed in the PBS solution were prepared and designated as PBS silk.

2.6 The test

2.6.1 Washing stability testing

To evaluate the laundering durability of the flame retardancy, the samples were washed according to TIS-121 (3–1975) in an 1 g/L solution of commercial non-ionic detergent and tap water and at 35 °C for 30 min. The samples were air dried and stored in a desiccator until required.

2.6.2 Surface and chemical composition analysis

Scanning electron microscopy (SEM) and energy dispersive x-ray spectroscopy (EDS) were used to examine the surface of the samples and well as the chemical composition before and after the washing process. The SEM used in this work was a JSM 633S (Jeol, Japan) equipped with EDS. Additionally, Fourier transform infrared spectroscopy (FTIR) was done to extract the chemical bonding on the surface of the samples. The IR spectra were obtained by using a Nicolet 6700 FTIR spectrophotometer (Bruker, Germany) operated in attenuated total reflectance (ATR) mode. The spectra were collected by averaging 64 scans at a resolution of 4 cm^{-1} from 400–4000 cm^{-1}.

By comparing the SEM micrographs of the PBS silk (Fig. 7(a)) and the washed Ar-PBS-Ar silk Fig. 7(b), the grafting of PBS can be observed. As shown in Fig. 7(a), PBS particles deposited locally on the knot of the silk yarn. The surface topography along the yarn was relatively smooth. In contrast, the yarn of the washed Ar-PBS-Ar silk was rough and uniformly covered with the PBS particles. It is evident that the durable flame retardant property of silk can be obtained via Ar plasma grafting.

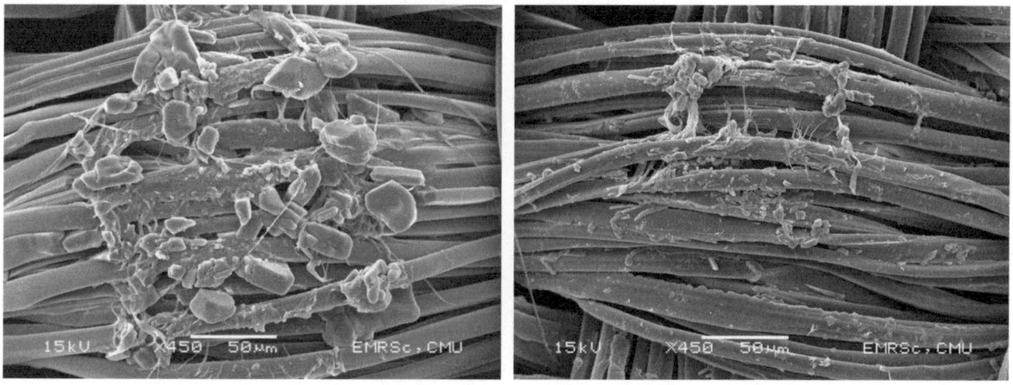

a b

Fig. 7. SEM micrograph of; a) PBS silk and; b) washed Ar-PBS-Ar silk

Fig. 8 shows the EDS spectrum obtained from the deposit on the yarn knot of the PBS silk. The spectrum showed evidence of phosphorus arising from PBS compound. Peaks of silk compositions, such as N, C, and O, were revealed. Calcium is one of the fingerprints of natural silk. Quantitative analysis of phosphorus content in the samples was done by means of EDS. The phosphorus content in the Ar-PBS-Ar was found to be 11% weight higher than that in the PBS silk, whose phosphorus content was 7% weight. This high level of phosphorus content in the Ar-PBS-Ar silk remained constant after the washing process. The results clearly indicate that in order to achieve durable flame retardant property, graft polymerization is necessary. The Ar plasma jet used in this work allowed us to bind covalently the flame retardant compound to the silk fabric. One can say that after the washing process, the Ar-PBS-Ar sample was similar to the ordinary silk with addition flame retardant property.

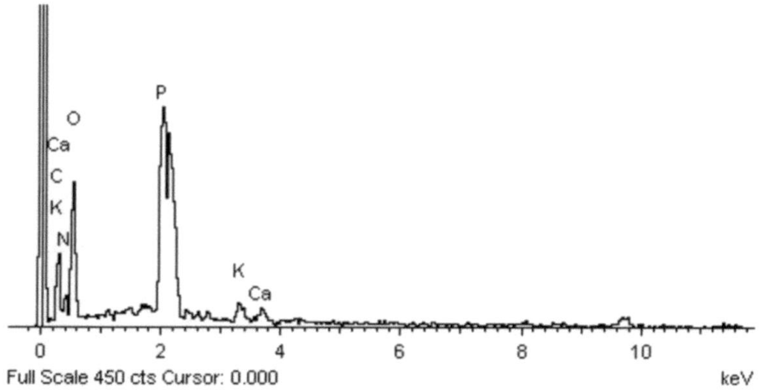

Fig. 8. The EDS spectrum of washed Ar-PBS-Ar silk (Au peak was not subtracted.).

The washed Ar-PBS-Ar silk sample has been characterized by ATR-FTIR in comparison with the untreated as shown in Fig. 9. Graft-polymerization via Ar plasma was indicated by the presence of bands at 1196 cm^{-1} (C–O stretching vibration), 1078 cm^{-1} and 919 cm^{-1} (P–O–C stretching vibration). The P═O stretching vibration that indicates the PBS compound overlapped within the C–O band. The IR peak intensity changes seem relatively low indicated the very thin layer of graft-PBS on silk surface from plasma treatment.

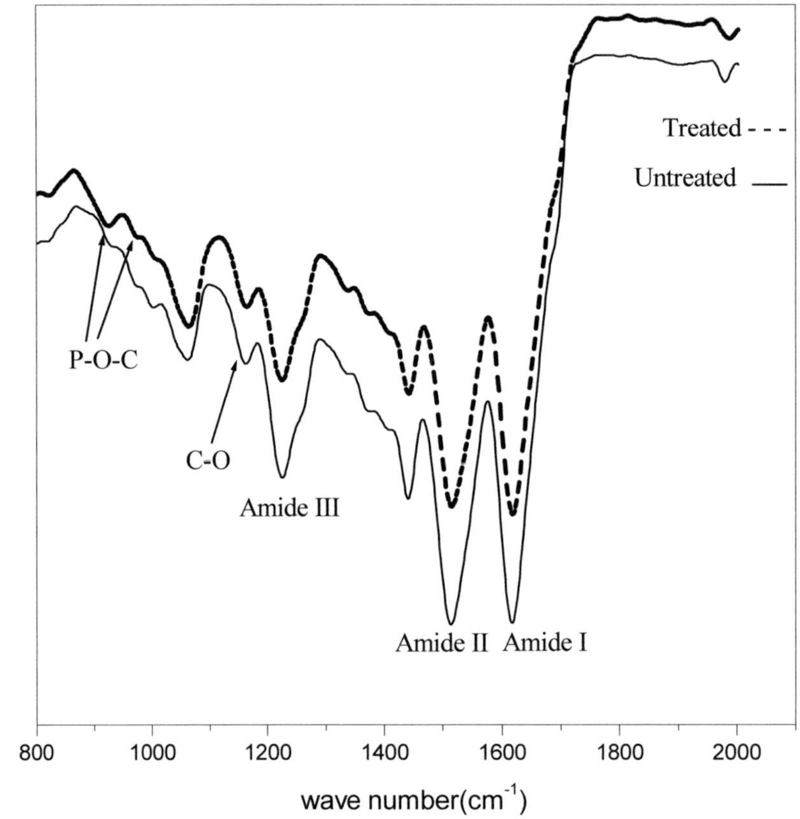

Fig. 9. ATR-FTIR spectra of untreated and washed Ar-PBS-Ar silk.

2.6.3 Flame retardancy testing

Burning behavior and 45-degree flame spread rate of untreated and treated cotton fabrics before and after washing were examined using 45° Flammability Tester according to ASTM D1230 with the impingement time of 5 sec at 30°C and 62 +/- 3% RH. The burning behavior and flame spread was recorded by a digital video camera. The flame spread time is the time taken for any flaming to proceed a distance of 12.7 cm (5″) up the fabric, and is automatically recorded by the burning of a stop cord. Fig. 10 shows the burning behavior of the silk samples. The samples prepared with different procedures were tested. In the case of untreated silk, the sample ignited instantly with a rapid flame spread of 1.43 cm/s. The flame extended to the entire sample without burning smoke. For the sample directly

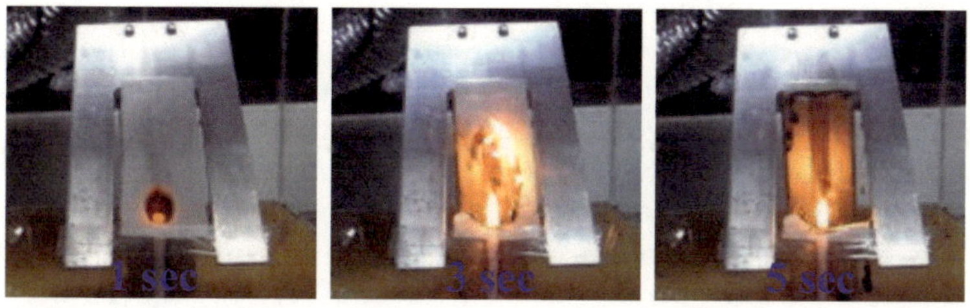

(a)

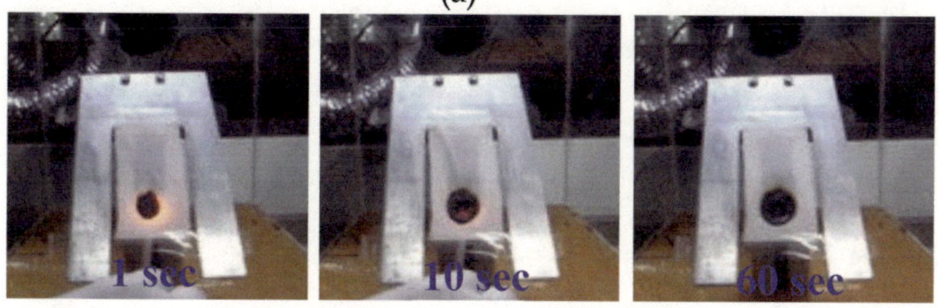

(b)

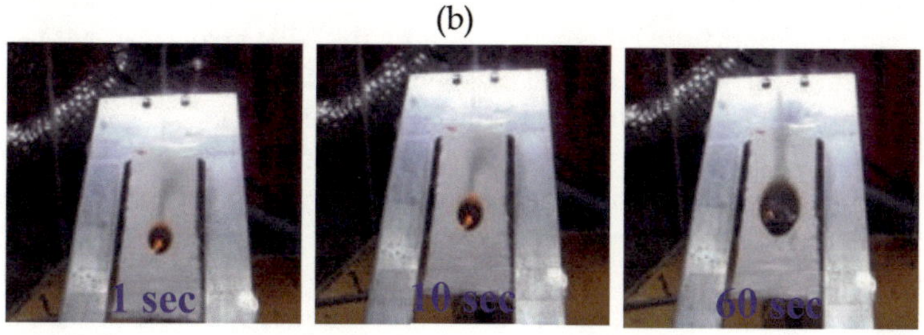

(c)

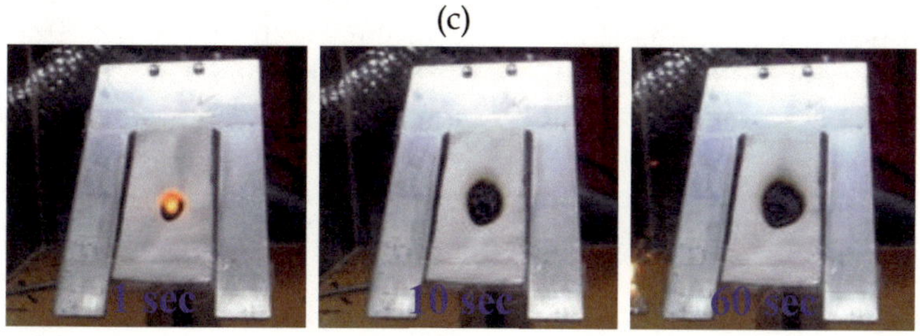

(d)

Fig. 10. Burning behavior of; a) untreated silk, completely burned within 7 s; b) PBS silk; c) washed Ar-PBS-Ar silk and; d) ethanol wash only.

immersed in PBS solution (PBS silk), the ignition character was identical to that of the untreated sample but the flame spread terminated immediately. The sample did not exhibit the afterglow. Burning smoke, as a consequence of char formation, was observed. The char formation is an indication of phosphorus containing residue on the surface of the sample (Tsafack & Levalois-Grützmacher, 2006a). The compound decomposed to polyphosphoric acid when heated and formed a viscous surface layer. This layer prevents oxygen to reach the silk fiber. As a consequence the fiber decomposition is inhibited.

After the washing process the burning behavior of the PBS silk was similar to that of the untreated sample. Some burning smoke was observed. This is due to the fact that PBS is water soluble, thus it can be removed from the silk during washing process. The smoke indicated that some PBS remained in the silk. In contrast, the Ar-PBS-Ar silk behaved differently. Its flame spread was higher than the PBS sample. However, the flame vanished immediately without the afterglow. The char formation was observed. This small amount of phosphorus catalyzes the oxidation of the carbon char to carbon monoxide instead of carbon dioxide during pyrolysis. The burning smoke was dramatically reduced to the amount that is close to the untreated sample. Since burning smoke mainly comes from the residual PBS on the surface of the sample, it can be said that most of the PBS molecules were grafted homogeneously into the silk molecular chains by the Ar plasma. Washing process might take away the un-grafted PBS molecules from the silk structure but the majority remained intact in the silk structure. Hence, with adequate level of grafted PBS molecules, silk samples can generate char to prevent flame spread without excess burning smoke.

2.7 The simulation

Molecular dynamic (MD) simulation of silk structure

To study the chemical bonding between PBS and the silk structure, MD simulation was performed. The simulation to predict the IR spectrum of silk after the incorporation of PBS was carried out to envisage the interactions. Silk model was generated using repeating glycine–alanine unit as discussed in the previous study (Khomhoi *et al.* 2010). Material Studio 4.3 software was used to build the model and perform energy minimization and MD simulations of the macroscopic structure of silk polymer containing 5 chains of 10-unit glycine–alanine in a periodic box of 30 × 30 × 30 Å using COMPASS forced field. Energy minimization was carried out to eliminate the potential energy which might arise as a result of the interaction with the neighboring chains with conjugate gradient method. After the minimized cell was obtained, the simulated annealing with Metropolis Monte Carlo (MC) method of Sorption module was designed to simulate the interaction between PBS and the silk model. The cut off distance was set at 12.5 Å for micro canonical ensemble. Trajectories from the MC simulation were collected for radial distribution analysis. To predict the IR spectrum of silk after plasma treatment process, quantum calculation of silk model compound modified by PBS predicted product from MD simulation was performed using GAUSSIAN 03 (Frisch *et al.* 2004). B3LYP/6-31G (d) level of density functional theory (DFT) was used to calculate optimized structure and IR frequencies.

The interactions between PBS and silk was investigated through MC simulation using model shown in Fig. 11. The most probable structure from MC simulation indicated that the reactive oxygen atom in P=O and P-O-N part of PBS molecule tend to react with silk polymer surface at methyl group of alanine unit.

The radial distribution function (RDF) plot (Fig. 12) of H_{meth}–$O_{O=P}$ and H_{meth}–$O_{P\text{-}O\text{-}N}$ represent P=O and P-O-N in PBS surrounding methyl group in silk. RDF calculated from collected trajectories suggest that the distribution of PBS around silk was contributed from strong interaction of P=O and P-O-N in PBS with methyl group in silk. The graph infers that H_{meth}–$O_{O=P}$ dominate intermolecular interaction in term of hydrogen bonding from strongest electrostatic interaction of partial negative oxygen and partial positive hydrogen with shell of interaction at 3.25 Å. On the other hand, H_{meth}–$O_{P\text{-}O\text{-}N}$interaction is mostly diffuse with radius around 4-9 Å. Therefore P=O group of PBS should react with methyl group of alanine residue in silk.

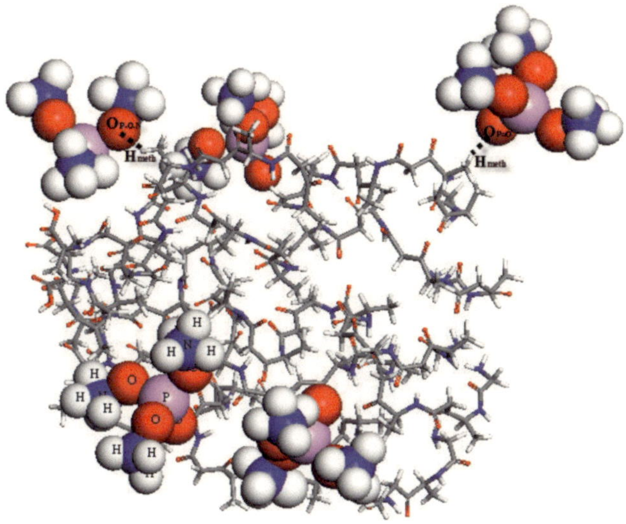

Fig. 11. Complex structure of silk model and PBS molecules system from Monte Carlo Simulated Annealing, dash line indicate strong interaction of P=O and P-O-N in PBS with methyl group in silk at distance 2.50-2.60 Å.

Product of PBS reacting with silk was deduced using above mentioned evident as shown in Fig. 13 in comparison with silk model. The use of calculations level at B3LYP/6-31G(d) show C=O stretching at 1777 and 1844 cm^{-1}, the stretching of C-O bond presents at 1196 cm^{-1} while group of N-H bending and C-N stretching was found in range of 1200-1700 cm^{-1} for both untreated and PBS silk. The P-O-C stretching vibration at 1078 cm^{-1} and medium peak of P-O-C stretching vibration at 919 cm^{-1} were found correlated well with previous studies (Zanini *et al.* 2008) and (Zou *et al.* 2002).

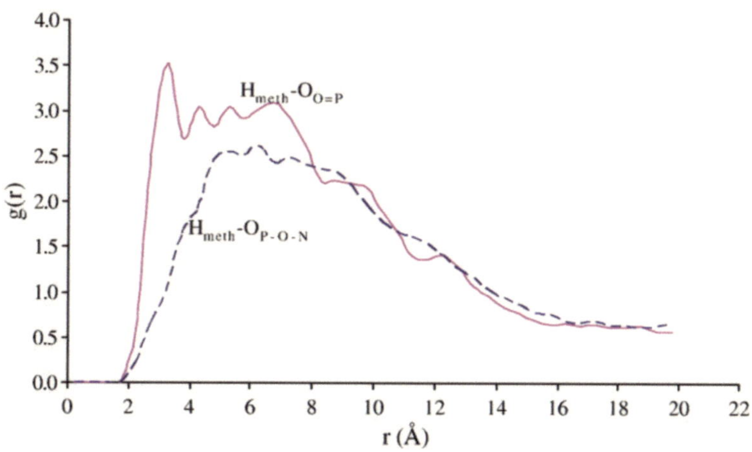

Fig. 12. Radial distribution function of H_{meth}-$O_{O=P}$ and H_{meth}-O_{P-O-N} represent P=O and P-O-N in PBS surrounding methyl group in silk.

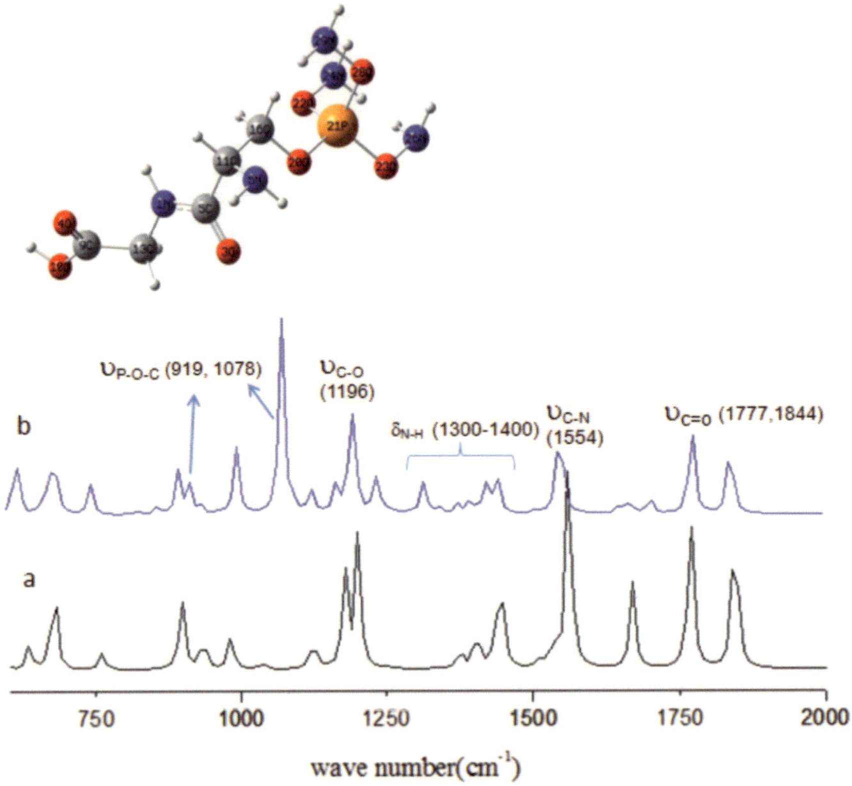

Fig. 13. Vibration spectrum of silk and propose product after treated structure calculated by using B3LYP/6-31G(d); a) untreated silk and; b) PBS silk.

3. Conclusions

Non-equilibrium atmospheric plasma jet has been proved for its benefit on high-density active species generation, low gas-surface temperature, direct treatment and low cost system. The reactive species in the plasma interact with the surface atoms or molecules and modify the surface without affecting bulk properties. Also penetration of jet plasma into woven materials such as textile fabrics can be achieved. This near-room temperature modification makes plasma jet of industrial interests. Phosphorus-based flame retardants becomes a major source of interest to replace halogen compounds because of their environmentally friendly by-products and their low toxicity.

The graft polymerization was needed because of phosphorus-based flame retardant agent is non-durable to washing. Plasma jet with selective ions can be used to replace chemical agent(s) with time-consuming pad-dry processes. The flame retardant property of silk fabrics induced by grafting of flame retardant compound using Ar plasma jet has been investigated. It has been shown that Ar plasma grafting is a necessary procedure to achieve the durable flame retardant property. Ar plasma grafting conferred endurable flame retardant property to silk fabric. The good washing stability could be attributed to the presence of phosphorus that was covalently bound to the silk structure.

The Ar-grafted PBS silk showed a higher level of flame retardancy as investigated by burning behavior and 45° flammability test. Carbonaceous char was formed and after glow was suppressed by PBS grafting. It was found that Ar plasma conferred durable flame retardancy to the silk yarn since the flame retardant character retained to washing. In the study, the Ar-grafted PBS silk improved the durability of the flame retardancy by decreasing 45-degree flame spread rate about 9.5 times. Also their low production of smoke in fire furthers their appeal. These compounds promote dehydration and char formation on their substrate preventing flame spread. Scanning electron microscopy results illustrated that the yarn of the Ar-grafted silk was uniformly covered with the PBS particles whereas PBS deposited locally on the non-graft silk. Energy dispersive X-ray spectroscopy showed the presence of phosphorus up to 11 wt.% in the Ar-grafted silk. Fourier transform infrared spectroscopy exhibited the bondings between phosphorus and the silk molecular chains. The molecular dynamics simulations affirmed the incorporation of phosphorus in the structure of silk at methyl group of alanine unit since the predicted IR spectrum agree well with the measured one.

4. Acknowledgment

DB would like to thank Chiang Mai University for research funding and laboratory supports. Thanks and appreciations also go to colleagues and students in the project and people who have willingly helped me out with their abilities.

5. References

Achwal, W.B. Mahapatrao, C.R. and Kaduska, P.S. (1987) *Colourage* 6, p. 16.

Becker, K.H. Kogelschatz, U. Schoenbach, K. H. Barker, R.J.(2004) *Non-equilibrium air plasmas at atmospheric pressure; Series in Plasma Physics*, Taylor & Francis: New York

Chaiwong, C. Tunma, S. Sangprasert, W. Nimmanpipug, P. Boonyawan, D. (2010) *Surf. Coat. Technol.* 204, p. 2991.

Cheng, C. Liye, Z. and Zhan, R.-J.(2006) *Surf. Coat. Technol.* 200, p. 6659

Coburn, J.W. and Winters, H.F. (1979) *J. Appl. Phys.*, *50*(5) p. 3189

Dhavalaikar, R.S. (1962) *J. Scientific and Industrial Res.* 21(C), p. 261

Fridman, A. (2008) *Plasma Chemistry*; Cambridge University Press: New York

Frisch, M. J. Trucks, G. W. Schlegel, H. B. Scuseria, G. E. Robb, M. A. Cheeseman, J. R. Montgomery, Jr., J. A. Vreven, T. Kudin, K. N. Burant, J. C. Millam, J. M. Iyengar, S. S. Tomasi, J. Barone, V. Mennucci, B. Cossi, M. Scalmani, G. Rega, N. Petersson, G. A. Nakatsuji, H. Hada, M. Ehara, M. Toyota, K. Fukuda, R. Hasegawa, J. Ishida, M. Nakajima, T. Honda, Y. Kitao, O. Nakai, H. Klene, M. Li, X. Knox, J. E. Hratchian, H. P. Cross, J. B. Bakken, V. Adamo, C. Jaramillo, J. Gomperts, R. Stratmann, R. E. Yazyev, O. Austin, A. J. Cammi, R. Pomelli, C. Ochterski, J. W. Ayala, P. Y. Morokuma, K. Voth, G. A. Salvador, P. Dannenberg, J. J. Zakrzewski, V. G. Dapprich, S. Daniels, A. D. Strain, M. C. Farkas, O. Malick, D. K. Rabuck, A. D. Raghavachari, K. Foresman, J. B. Ortiz, J. V. Cui, Q. Baboul, A. G. Clifford, S. Cioslowski, J. Stefanov, B. B. Liu, G. Liashenko, A. Piskorz, P. Komaromi, I. Martin, R. L. Fox, D. J. Keith, T. Al-Laham, M. A. Peng, C. Y. Nanayakkara, A.; Challacombe, M. Gill, P. M. W. Johnson, B. Chen, W. Wong, M. W. Gonzalez, C. and Pople, J. A. (2004) *Gaussian 03, Revision C.02,* Gaussian, Inc., Wallingford, CT.

Gaan, S. and Sun, G. *J.* (2007) *Anal. Appl. Pyrol.* 78 , p. 371.
Griem, H.R. (1964) *Plasma Spectroscopy.* McGraw-Hill, New York
Grill, A. (1994) *Cold Plasma Materials Fabrication: From Fundamentals to Applications;* Wiley-IEEE Press: New York
Grzegorzewski, F. (2011) *PhD. Dissertation,* University of Berlin, p. 29
Guan, J. and Chen, G.Q. (2006) *Fire Mater.* 30, p. 415.
Guan, J. Yang, C.Q. and Chen, G. (2009) *Polym. Degrad. Stabil.* 94, p. 450.
Guimin, X. Guanjun, Z. Xingmin, S. Yue, M. Ning, W. and Yuan, L. (2009) *Plasma Sci. Technol.* 83.
Horrocks, A.R. and Price, D. (2001) *Camb. Woodh. Publ. Limit.* ISBN:1855734192.
Kako, T. and Katayama, A. (1995) *Nippon Sanshigakul Zasshi.* 64, p. 124.
Khomhoi, P. Sangprasert, W. Lee, V.S. Nimmanpipug, P. (2010) *Chiang Mai J. Sci.* 37 (1) 106.
Kurunczi, P. Lopez, J. Shah, H. Becker, K. (2001) *Int. J. Mass Spec., 205*(1-3), 277
Kylian, O. Benedikt, J. Sirghi, L. Reuter, R. Rauscher, H. Von Keudell, A. Rossi, F. (2009) *Plasma Process. Polym. 6,* p.255.
Mathew, T. Datta, R.N. Dierkes, W.K. Noordermeer, J.W.M. Van Ooij, Mechanistic, W.J. (2008) *Plasma Chem. Plasma Proc. 28,* 273.
Naganathan, P.S. and Venkatesan, K. (1972) *Acta Crystallographica Section B* 28, p. 552
Osaki, K. Fujimoto, S. and Fukumasa, O. (2003) *Thin Solid Films Proceedings of the Joint International Plasma Symposium of the 6th APCPST, the 15th SPSM and the 11th Kapra Symposia* vol. 435, p. 56.
Raballand, V. Benedikt, J. Wunderlich, J. Von Keudell, A. (2008) *J. Phys. D: Appl. Phys.* 41, p.115207.
Reddy, P.R.S. Agathian, G. and Kumar, A. (2005) *Radiat. Phys. Chem.* 72, p. 511.
Sangprasert W., Boonyawan D. and Nimmanpipug P. (2010). *Journal of Molecular Structure,* 963 (2-3) 130-136
Schafer, J. Foest, R. Quade, A. Ohl, A. and Weltmann, K.-D. (2008) *J. Appl. Phys.* 194010.
Sismanoglu, B. N. Amorim, J. A Souza-Corrêa, A. Oliveira, C. Gomes, M.P. (2009) *Spectrochimica Acta Part B* 64, 1287-1293
Takahashi, Y. Gehoh, M. And Yuzuriha, K. (1999) *Int. J. Biological Molecules* 24, p.127
Tanter, T.C. (1953) *Acta Crystallographica Section B* 26, p. 805
Tanter, T.C. (1956) *Nature* 177, p. 37
Tsafack, M.J. and Levalois-Grützmacher, (2006) J. *Surf. Coat. Technol.* 201, p. 2599.
Tsafack, M.J. and Levalois-Grützmacher, J. (2006a) *Surf. Coat. Technol.* 200, p. 3503.
Von Keudell, A. and Jacob, W. (2004) *Prog. Surf. Sci.* 76, p.21
Wichman, I.S. (2003) *Prog. Energ. Combust.* 29, p. 247.
Wu, W. and Yang, C.Q. (2006) *Polym. Degrad. Stabil.* 91 , p. 2541
Wu, W. and Yang, C.Q. (2007) *Polym. Degrad. Stabil.* 92, p. 363.
www.NIST.gov (2009) accessed via internet
Yang, H. and Yang, C.Q., (2005) *Polym. Degrad. Stabil.* 88, p. 363.
Zanini, S. Riccardi, C. Orlandi, M. Colombo, C. and Croccolo, F. (2008) *Polym. Degrad Stabil.* 93, p. 1158.
Zou, X.P. Kang, E.T. and Neoh, K.G. (2002) *Surf. Coat. Technol.* 149, p.119.